THE
UNIVERSE
A Challenge to the Mind

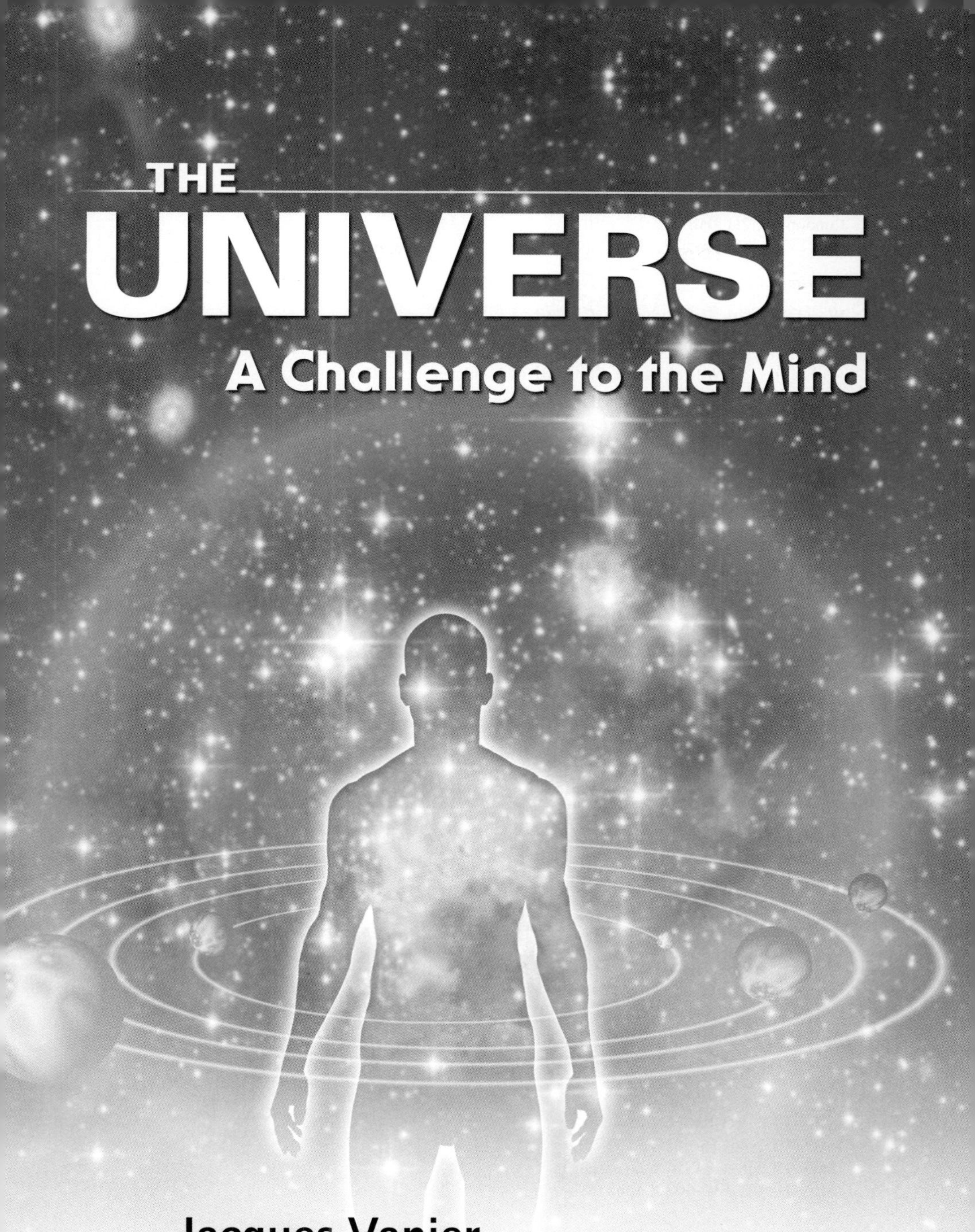

THE UNIVERSE

A Challenge to the Mind

Jacques Vanier

University of Montreal, Canada

ICP

Imperial College Press

Published by

Imperial College Press
57 Shelton Street
Covent Garden
London WC2H 9HE

Distributed by

World Scientific Publishing Co. Pte. Ltd.
5 Toh Tuck Link, Singapore 596224
USA office: 27 Warren Street, Suite 401-402, Hackensack, NJ 07601
UK office: 57 Shelton Street, Covent Garden, London WC2H 9HE

British Library Cataloguing-in-Publication Data
A catalogue record for this book is available from the British Library.

THE UNIVERSE
A Challenge to the Mind

ISBN-13 978-1-84816-601-1
ISBN-10 1-84816-601-X
ISBN-13 978-1-84816-602-8 (pbk)
ISBN-10 1-84816-602-8 (pbk)

Typeset by Stallion Press
Email: enquiries@stallionpress.com

Printed in Singapore by Mainland Press Pte Ltd.

Table of Contents

Preface

When I was in college, like many other kids, I was fascinated by the Universe we live in. There were those large-scale phenomena taking place around us, such as the beautiful aurora borealis in the northern sky. There were those frightening lightning flashes during electric storms, and there were, among many other things, the small-scale phenomena and effects such as the spinning top and the selective attraction of materials by a magnet. There was that mysterious world of the atom we were reading about in some popular books, with its nucleus and orbiting electrons that had been discovered relatively recently and that seemed to be fundamental to most of the visual observations in our world; there was the discovery of nuclear fission that promised enormous amount of cheap energy but was used first for the creation of the atomic bomb. There was that other world above our heads, much larger, with all those bright stars, which we were told were like our Sun and which were so far away that we could not really comprehend their distances. And this led to the deep question troubling my young mind: if that space was so vast, what was its dimension? Was it finite or infinite?

I remember that in those days I used to record some of these questions in a book. One of the questions was about the size of that Universe and I had drawn a cartoon, which from memory represented a human being walking away from an agglomeration of stars and falling off at the border. My question was: if I keep going in a straight line among the stars and the world is finite, will I get to the end and fall off the same as in my cartoon? Of course, I did not have much understanding of the working of that world, of the interplay of the various forces present, in particular, gravity. What was at the end? How could I fall off? The teachers I had did not dare venture into such matters. It is true that I was educated in a college run by catholic priests and that religion, with its absolute focus on faith, was not so concerned with providing 'rational' answers regarding the nature of the Universe. I remember a philosophical explanation that I had read somewhere. It said that space was the relation between material objects; no objects, no space. Like many other explanations of a philosophical nature that I was exposed to in those days, I did not really comprehend the full meaning of it. However, I was surprised to learn later that Leibniz, a great mathematician contemporary of Newton, had essentially relied upon that concept when facing the difficulty of defining space. I learned later that Aristotle had also reflected on the question. For him space was an extension of tangible objects. If questions of that nature were troubling the mind of such giants I concluded that they were probably deeper than I thought. Questions about the nature and the dimension of space have troubled the human mind at all stages of its evolution towards a better understanding of our world.

Even today we do not have an answer to that question. Actually, as yet we have no answers to many basic questions regarding the functioning of the Universe. We have made models and have found that in many instances we can write rules or laws that appear to explain *how* the mechanics work when we create compartments in the ensemble. We often think that we understand *why* it works the way it does, only to find out that there is another level in our understanding that raises yet another question. The true *why* seems to be of a nature that is not accessible to us.

Our physical laws are based on experimental facts: they must agree with observation even if it does not appear to make sense, as in atomic world, where we have to accept that an electron, for example, can be both a wave and a particle. We observe various phenomena and we find that we can explain why nature behaves in that particular manner if we assume that the electron is either a wave or a particle depending on the circumstances. But then, we don't know *why* the electron may be a wave or a particle. We do not have a theory that encompasses the whole ensemble. We simply observe that nature operates according to specific rules and we have developed sophisticated models that are based on those rules and relatively simple principles that we have to accept. We have developed mathematical tools that provide an elegant framework for the logical description of those models. Those models are most attractive and as the fruit of our intelligence we are extremely proud of them. However, all those models are made in response to experimental observations. They are the creation of our mind and they appear to fulfil that limited goal: selected measurements are made in particular circumstances within well defined borders and conditions, and a mathematical model is made to describe the observations. The measurements can be repeated over our planet, the Earth, and its close surroundings at different locations and different times. The same results are obtained and we simply extrapolate to the whole world, leading us to assume that the same rules apply over the whole Universe.

Are these models a full and complete realistic description of the world? The answer, of course, is: if a model works it must be right or there must be at least something right in it. However, this is true only in the compartments and borders that we have defined. Since 'the totality of experiments is never accessible to us, we must renounce all hope of finding anything like the correct theory.'[1] Many models were developed in the past that described observations made in particular circumstances well. Upon more advanced and more accurate measurements, they appeared to be incomplete and sometimes totally wrong. In such circumstances, a new model is then developed that appears to better describe the observation, at least on a broader scale. This is the case for the theory of relativity that brought a point of view totally different from the Newtonian approach to gravity and motion, as described by classical physics until that time. This is the case in quantum mechanics, which replaced the first theory of atomic structure that could provide only a partial description of the atomic world. This is the case for a very recently developed theory, given the strange name 'quantum chromodynamics', which totally changed our description of what we use to call the field of elementary particles. The list could go on and it is still a fact today that the Universe is hiding many of the basic laws of its functioning. At the end of the 19th century it was widely believed that very soon everything would be known and that there would be little to be done in fundamental physical research. There have never been as many questions about the functioning of our world as at the beginning of the present millennium. We have approached our quest for a description of the Universe by compartmentalising the whole into small parts that we can handle, sectors that appear to behave as a whole. But we have found in many cases that those sectors are simply parts of a broader context that can be described by a more sophisticated model. The best example of such a development is the joining of electric and magnetic phenomena under a single model called electromagnetic theory as developed by Maxwell at the end of the 19th century. The search for a final theory that would cover the whole field of observations is at present the main challenge for the scientific community. It is called the theory of everything, or TOE. It is possible that a model could be found that would describe by a single mathematical set of equations the behaviour of the whole Universe. Such a theory could very well provide a logical framework for a description of the components and physical laws governing their behaviour without answering the *why*. One last step would be a theory that answers the basic question as to the origin of elementary concepts such as the nature of mass, the nature of electric charge, the nature of elementary particles called quarks, and all the forces that originate through the presence of those elementary entities; and the most fundamental of all, the nature and origin of time and space and so-called fundamental constants that appear to be a property of that space. Are we basically limited to the description of the world by means of compartmentalised models, restricted to separate sets of mathematical equations — beautiful as they may be — and including so-called fundamental constants that we accept as such? Will we be able one day to base our models on something of a different nature that helps us understand *why* it is the way it is?

There is sometimes a division in the physics community between those working on abstract ideas or models, which may offer a framework describing the operation of our Universe, and those working in laboratories with the real world of physics or still making observation on the larger scale of the Universe. These two sectors are identified as theoretical physics and applied physics and sometimes may appear artificially partitioned. Fortunately, the division is more apparent than real. The two groups rely on progress made by each other. There is also a large grey area between these sectors. We see many scientists doing the experimental work, even implementing useful devices and at the same time developing by themselves the theory at the base of their findings. Many subjects provide examples of this: magnetic resonance that led to Magnetic Resonance Imaging (MRI), a most useful tool in medicine; the Maser that led to the development of the atomic clock, a most precise device that was used to verify a prediction of relativity to the greatest accuracy and is now used in one of the most accurate navigation systems; the Laser, the applications of which are sufficient to fill several books from surveying to medicine; the transistor, the basis of operation of solid state devices, including those integrated circuits that have made possible instruments such as computers that have transformed our lives. The list could go on. Scientists involved in research leading to those creations knew very well the physics connected to their work and in most cases developed themselves the theory necessary for giving appropriate research direction and for explaining and understanding the results obtained. Some may see a great difference between working on string theory or quantum gravity theories and a theory related to a particular, more mundane phenomenon in solid state physics. It is the accumulation of all the data collected within a given compartment that makes a theory within that compartment interesting. And it is from those experiments and observations that the scientific community is awoken to the fundamental physics problems. It is most interesting to observe that the majority of Nobel prizes in physics during the last 25 years have been given to scientists working in those grey sectors. These are not secondary sectors and are the ones leading to questions regarding the validity of advanced theories. That grey area has been very productive and the success of the whole field is certainly due to the interaction amongst its members and their openness.

I have been teaching physics a great part of my life, either at universities or on invitation to give lectures at summer schools, symposia or conferences. That teaching, covering classical physics, thermodynamics, electrodynamics, quantum physics and solid state physics, reflected closely the research I was doing on atomic physics, a compartment that has been my passion since I was in college. In teaching, I always tried to explain things as simply as possible with elementary concepts supported by a lot of figures, drawings, models, and even gadgets. The research I have been involved in was primarily concerned with applications of quantum physics, particularly in the field of quantum electronics, which deals, for example, with lasers and masers and is part of applied physics. This will certainly become evident in reading this text. The theories I worked on were essentially extensions and applications of existing theories and they were used to 'explain' the experimental results I had obtained with my colleagues. This is the grey area I was talking about. Consequently, this book sometimes puts emphasis on experiments with devices, either practical or imagined, in order to provide a better understanding of the world we live in.

There has been so much written on the general field of physics, and particularly the Universe, that it feels like sometimes the most an author can do is to say things in a different way. However, I wrote this text with the intention of providing some answers to questions that young students may ask themselves when exposed to scientific courses in colleges and universities and provide them with a global view of the whole field. I assume that they are not different from the person I was when I was their age. The teaching method used very often in schools of various levels is based on a compartmentalising of the field of physics without connection between the compartments. In changing grades or moving from one school to another the teaching approach may be drastically different and the student may be lost in the details. I will try to connect and interlace, when possible, these compartments and provide an overall view of the field of physics in order that it does not appear an endless litany of separate subfields. I will also describe sectors of physics that some may consider more mundane. Those sectors, for example, are connected to applications and provide a framework, or tools, for venturing into the larger picture. They are also sectors that have provided us with the tools that have become part of our daily life. I just hope that such a description may create in many a feeling of admiration for the beauty of this Universe as well as an admiration for those geniuses who have imagined the models we now have to describe its functioning and who have developed the instruments that have improved our lives. If I accomplish that task,

I will consider that my goal has been reached. I believe, however, that by its very nature, the book may be of interest to a broader spectrum of readers with some general education who are curious about our present understanding of the functioning of the Universe. In this sense, it should interest engineers and scientists who have specialized in a specific field or who have been exposed to the subjects treated in the present text only through various unconnected articles in popular magazines. This book should offer them an integrated view of the physics involved in the scenarios proposed for explaining the evolution of the Universe. If the book should interest members of the general public with some curiosity about the present state of our understanding of physical science, I will consider it a bonus beyond my first goal.

The time when an educated person could be considered an expert in all fields of physics has long gone. Physics covers subjects from the very small, the atomic world, to the very large, the world of galaxies. The experimental data accumulated every year is, to say the least, astronomical, and the theories developed to explain those observations are becoming more complex year on year. Just to stay current in the most important developments requires a great effort. So no one can claim to be an expert in all fields. As for me, the experience I have accumulated has been through teaching general physics courses and doing research in atomic physics and quantum electronics. In the description of subjects outside of those sectors I have attempted to utilise the techniques I have used in my teaching experience, that is to say, to simplify matters as much as possible. This is sometimes dangerous since the process of simplification can introduce undesired distortions in the explanation of the concepts presented. I only hope that there are not too many of these.

This book is essentially a non-mathematical description of the laws that describe the functioning of our world, laws that the human being has so brilliantly uncovered. It contains models supported by graphics and figures that describe our present understanding without complex mathematics and equations. This will most probably make a large group of readers happy. A short Annex is provided at the end supporting with equations some of the models developed in the main text. The reader may very well concentrate on the main text without reading the Annex. It is a modest summary of the most important laws of physics, at least those referred to in the main text. This is done in general for cases where it is possible to do so in an elementary way. Consequently, for some part of that Annex, the reader requires only an elementary knowledge of mathematics. However, in some other parts the mathematics is more involved. This is a consequence of our limitation: we require a technique to make sure that what we describe with words agrees quantitatively with observations. We have found only one way of doing this: with advanced mathematics. It is unfortunate, and those readers that are not familiar with calculus or vector analysis, for example, may be baffled by the complexity of the language and notation used. We cannot do anything about it. In such a case, the reader has to accept the conclusions exposed in common language in the main text. Those conclusions are reached through extensive work done by numerous scientists over the years and verified by many others. Research groups are generally found to be in civilized competition between each other and such an approach guarantees the objectivity of the conclusions reached. I think that scientists form one of the most open communities, always ready to accept the conclusions of others when the proofs are solid. It is that attitude that leads to the credibility of the conclusions reached and bears witness to the integrity of the whole community. I just hope that the present text will also lead the reader to a re-enforcement of that sentiment.

I would like to thank my colleague Robert Vessot who has volunteered to read the text in one of its first drafts and has made comments on it. I am indebted to him in many ways. Bob has introduced me to the wonderful world of atomic frequency standards, which have revolutionized the world of measurement, transformed the International System of Units and provided insight into the physics regulating the atomic world as well as proofs of fundamental conclusions reached by some of the theories presented here. Bob is the physicist who made the most precise measurement of the gravitational frequency shift of a clock as predicted by Einstein's theory of general relativity, providing a proof of the theory at an unsurpassed level.

I wish to thank R. Boily for his deep philosophical comments on the nature of the questions addressed in such a text, which reinforce some of the expressed points of view. I wish to thank M. Wertheimer for his positive and encouraging words. Thanks are due to C. Cohen-Tannoudji who gave his opinion on some of the sections particularly in connection with the field of quantum physics. Thanks are also due to G. Michaud for his general comments on matters related to cosmology. I would like to thank Mr. Shaun Barrington for editing my work. His efforts have considerably improved the style and language. I am very much indebted to him. My son Pierre and daughter Lyne have many times challenged me with deep questions on the dynamics of our Universe; the writing of this book is partly in answer to their questions. I thank them also for their comments on the manuscript. Finally, I wish to thank my wife Lucie who has tolerated those long periods of isolation when I was totally coccooned in the abstract sphere that surrounded my studies and writing.

Jacques Vanier
December 2009

I

Introduction

It is important to understand clearly what we mean by such terms as science and scientific theory. Many articles have been written on that subject. We could write a whole chapter on the definition of science and consequent ramifications of a philosophical nature, such as the meaning of our existence; but we will examine it in a very practical way. We may say that science is the human activity that attempts to understand by means of natural explanations how the world we live in functions and that uses that knowledge to improve our existence in that world. That is sufficient for the purposes of this book. Science as it is understood here is based on the observation of phenomena taking place in the Universe we know and the development of theoretical models to describe their behaviour. And the Universe includes a piece of chalk as well as a faraway galaxy. There is no such thing as a science that is based on beliefs or on imagination alone. They may play a role in the development of a scientific theory but experimental facts dictate finally if the imagined concepts and beliefs are correct: a theory must absolutely be in agreement with experimental observations to have scientific value. One might have a dream and develop a theory about the occurrence of a phenomenon but if it does not agree with experimental measurements made of the characteristics of the phenomenon, the theory is not useful to science. It may be logical and attractive, but it remains an exercise in the abstract. To be right a scientific theory must be in agreement with experimental observations either made in the laboratory or with observations made in the Universe at large. In the first case, the scientist has general control on the phenomenon he is observing and measurements can be repeated at will, at different times and in different locations. In the second case, when the scientist has no control over the object observed — such as with celestial phenomena — present behaviour and prediction of the characteristics of future observations and their confirmation by means of measurements, play a vital role in establishing the validity of the theory developed.

A scientific theory may also be based on observation of past events by means of the present state of residual artefacts that can be studied at will. This looks very much like history. This is often a more delicate subject since we know well how history can be distorted. But it is the task of the scientist to use judgment and discernment and accept only the most convincing arguments and artefacts of known origins, with the support of all aspects of other scientific facts coming into play. The theory that illustrates this best is evolutionism as developed by Darwin. That theory is as scientific as the one about the orbits of planets around the Sun, since it is based on observations made in the field that can be verified by anyone and is furthermore supported by present-day laboratory observations in molecular biology and results obtained from the examination of the genes of the great mammals.

A theory that is based on a single fact that no one can reproduce remains a curiosity and is not very useful to science. An example of this was the laboratory observation of cold fusion several years ago. No one could reproduce in the laboratory the results reported by the researchers who had claimed such an accomplishment. There was most probably an uncontrolled parameter that was not taken into account, or errors were made in the measurements. The observation of a given phenomenon that happened once and left no residual traces cannot play an important role in science either, unless we can reproduce it. This applies for example to the observation of UFOs reported since time immemorial. It is not impossible that such objects exist but there is no credible proof of their existence and the assertion that they come from outer space, from a possible alien civilization is pure speculation and of no value to science. The observations are not credible and this is not science.[1] Another type of so-called science that has no scientific value is astrology. Sometimes it 'works' for a very specific event by coincidence. We do not hear about the times it does not work. All those ventures are not science. They are observations of peculiar events that cannot be reproduced and from which predictions cannot be made. We can only say that they are interesting. They belong to fiction.

How is a scientific theory developed? As mentioned above, physics is based on observations and it is on those observations that a theory is developed. One person may have an idea about an observation and develop a theory around that idea. Is the theory right? The reply from a scientist is: test its prediction in the laboratory or compare those predictions to results of other measurements made on the same phenomenon. This is the only way to verify its validity. If it agrees with the tests within certain limits, the theory is most probably correct and is accepted. If someone ever comes in and observes by means of careful measurements that a particular observation does not agree with the theory already developed for the phenomenon studied, then a new approach is needed and either a correction must be made to the theory or a new theory must be developed. It usually turns out that the previous theory was an approximation of the new theory and applied in special circumstances. Einstein at the beginning of the 20th century developed a new conceptual theory about gravity. It did not invalidate Newton's law of interaction between masses, which became an approximation of Einstein's theory of gravitation. However, it predicted new effects. Measurements agreed with the predictions and the theory was accepted as valid. Is the theory unique or could a different theory be developed that would explain the same observations? This is a more delicate problem and the question is usually resolved by means of the predictions made by one of the competing theories and verified at a later time.

Questions are constantly raised by scientists concerning the validity of the assumptions used to establish a theory. There is always some scepticism about any theory. It is the way scientific development takes place. In the 20th century and into the 21st considerable efforts have been made to attempt to develop a theory that would explain the behaviour of the Universe as a whole, to synthesise gravitational attraction with electromagnetism, quantum theory, and nuclear phenomena. The aim is to unify all the compartmentalised theories that we now have into a single one, providing a logical framework for explaining all observations made in the Universe — a theory of everything (TOE). Recently such an attempt has taken form under the name of 'String theory' in which all forms of matter in the Universe are assumed to originate from microscopic strings in various modes of vibration. The theory has demanded a large-scale effort and polarised the community of theoreticians for several years. However, it is not supported yet by experimental data and, at this time, it is not accepted at large.[2]

Scientific theories are not formulated out of nothing. Earlier observations act as a trigger for a new physics concept and theory. As is often said, scientists stand on the shoulders of their predecessors. Sometimes, these theories appear to be hugely complex, with mathematics that only a few can fully understand and for the person not familiar with the mathematics involved, they appear to be totally disconnected from the real world. However, when examined carefully the development of such theories is justified by deep questions raised by observations. Scientists involved in those theories are driven by a passion to reach a deeper understanding of the functioning of the Universe. They deserve our respect. The number of possible approaches, however, is extremely large and we face at present a difficulty in achieving a framework that would unify the theories to explain the physical world now defined in separate compartments, such as the world of macroscopic objects, the stars, and the world of the small objects, the atoms. This is extremely complex. To approach the subject requires so much knowledge that I would suggest that if one has a dream and wakes up with a luminous idea on how the Universe works, there is an overwhelming probability that the idea is wrong. The number of possible avenues in that approach is essentially infinite. Such ventures are reported rather often and they generally have value only in the mind of their inventor.

Another fundamental point about science is that its development is independent of belief or 'faith' in extraterrestrial or spiritual intervention. There is presently a belief held by several groups that the complexity of certain objects or beings in the Universe cannot be explained by Darwin's theory of evolution. Living beings, for example, are considered too complex to make acceptable the idea that they have appeared by simple random mutations and natural selection. It is claimed that an external agent was necessary, leading to the existence of a supernatural being responsible for the implementation of the steps required for their complex organization. This is sometimes called 'Intelligent design'. The believers want their 'hypothesis' to be accepted as a scientific theory and taught in schools alongside Darwin's evolution theory. No scientific basis, as described above, has been found in the description made in such theories or hypotheses. The concepts developed in the description of the Universe by those adepts are based on faith and have no scientific value. We may say that faith is part of another 'faculty' of the mind. Science is based on scepticism and uncertainty. Its development is generally achieved by questioning previously accepted theories. Dogmas are irreconcilable with science.

In this book, I will try to provide a concise outline of our present understanding of the functioning of the Universe through physical models and how that view is supported by experiments and measurements. It will become clear that our knowledge is essentially limited to *How* it works and cannot encompass the essential fact of *Why* it is that way. It seems that the *how* sometimes feels like a *why*, but every time a deeper investigation is made, one finds that a new concept at a higher level or a new parameter not thought of previously needs to be introduced in the theory and has to be accepted. Nevertheless, if we continue in developing our understanding of how it works, either at the small-scale laboratory or the large-scale cosmic levels, it seems that we may one day approach the basic existentialist concept of why the world is the way it is and why we are here. It may be a question of our level of understanding, but this is not a certainty: the Universe may very well always hide its existentialist inner side. Human beings, living on a small planet, satellite of a modest star, negligible companion of billions of them, inside a standard copy of billions of galaxies composing the Universe, may have to live with their limitations as accidental passive observers that do not play any role in the whole scenario. Some may find that this last remark is irreconcilable with their faith. There is no particular intention in making it. It is just a conclusion that seems to follow from the observation that the Universe has been evolving very well for most of its existence without us living on that speck of dust that we call Earth and that our influence is restricted to its immediate environment.

II

The Universe Landscape

Our Universe

When we look at the sky at night, we are amazed at the vastness of what is in front of us. Even without being aware of the size and nature of the objects seen, the first humans for sure, even with their limited capacity for enquiry, must have been amazed by the vision of stars. We have come a long way in our observation of what constitutes that large-scale visible world and have invented tools to study its constituents and understand the physical laws ruling their behaviour. But our Universe is not limited to that night vision. The mysterious beauty of all the objects that surround us here, on planet Earth, is also the subject of our questioning on the small scale. We like to know why objects have different colours or have different texture. We will examine those points of view by making an imaginary journey examining the large and the small scale of our Universe.

The Very Large Scale

What is the essence of this Universe? It appears that in the world we know and have direct access to, we are the only living organism whose brain is large enough to be conscious of its existence and who can reflect on such a paradigm. Let us use our imagination, go into space and examine the landscape. Let us use a very special spaceship and go on a trip to the confines of the space surrounding us. We will be able to observe objects we see here on Earth with complex instruments like large telescopes. Our spaceship has multiple gears and can travel at tremendous speeds. We will travel great distances to such places that we see from Earth as they were hundreds, thousands and even millions of years ago.

There is a basic observation, that we will detail later, that says that light travels at a certain speed, which is of the order of 300,000 km per second and in physics is called 'c'. If we turn on a flashlight in the night, the beam of light seems to appear instantaneously. However, it is limited to that speed c and it is only because our sense of vision is limited to a change of things at a slower pace that we see the light transmission as instantaneous. This has an awkward consequence for our vision of space. If the object is so far that the light it emitted took a million years to reach us, we thus see the object as it was a million years ago. The object, at the moment of emission, was at a distance from us that corresponds to the distance travelled by light in one million years: we will often express distances in light years, or 'ly'. A ly corresponds to a distance of about 10,000,000,000,000 kilometres. We use a very simple notation to express that — 10^{13} km — the exponent 13 to the number 10 meaning a 1 followed by 13 zeros. Notation is one of the most useful tools. In physics, when things get a little messy, one invents a new notation that simplifies the writing. Great progress has been accomplished by means of simple notations and it should not be disdained. The ly is a distance that is out of proportion with our familiar dimensions. We have to accept it and just think that it is large. But how could we go to such a place, a million ly away? We will see later that the speed of light is an absolute limit: we cannot exceed that speed whatever the energy we spend to accelerate our spaceship. Consequently, it appears ridiculous to talk about travelling to such distant objects: we would not live long enough. However, our Universe is subtle and has a bag of tricks. We will see later that time and space are intertwined creating situations and paradoxes that we can disentangle to our advantage; space travel across huge distances may not be totally ruled out. For the moment, however, we will assume that we make a virtual trip, in a virtual ship. We will visit celestial objects and observe them as we would see them on Earth with a super telescope, that is to say, we will see the objects at the time they emitted their light. We will not see the world as it is now, because the object that we see may have emitted its light millions of years ago and may very well not be in existence any longer. However, such a trip will provide us with a better grasp of the dimensions involved.

Let's go! Our rocket has sufficient speed, about 11 km per second, to make us escape from the Earth's attraction. At a few hundred kilometres above the Earth's surface, at the altitude of the International Space Station, we start losing track of the details of cities, streets, lakes, forests. The horizon already shows a slight curvature. The sky is black because the vacuum of space does not contain sufficient objects or elements to reflect solar light. We keep going and arrive at the Moon, and looking back: we see a blue ball floating in space partly covered with clouds, our own planet Earth, our home. Figure 2.1 shows that beautiful view as seen by astronauts in the Apollo program.

We head towards the Sun that will give us a gravitational boost to fly through the whole planetary system. We are escaping the gravitational attraction of Earth. We first pass by our inner neighbour Venus, covered with bright white clouds, and see dull

Figure 2.1. Our home planet, Earth, as seen from the distance of our satellite, the Moon. The foreground is the surface of the Moon. (Credit, Apollo 10, NASA.)

Mercury covered with craters and closer to the Sun. We don't stay in those surroundings long. We see immense jets of hot gas blown away from the surface of the Sun. We conclude that they are probably larger than the Earth. They bend and turn in back to the surface of the Sun. It looks as if a strong force were acting on those jets. In fact the jets are made of electrically charged particles forming a strange kind of substance called a plasma. The strong magnetic field that exists all around, caused by electric currents within the Sun itself pointing in various directions, bends the jets back and forces them to return to the surface. Some of the constituents or particles of the jets escape and heads towards Earth. It is those particles that, when they arrive at Earth, get deflected by Earth's magnetic field and interact with our atmosphere to cause the beautiful aurora borealis. Our star produces immense quantities of energy, something like 3.85×10^{20} megawatts, emitted as light, other radiation at various wavelengths, and atomic particles. Our most powerful hydroelectric plants on Earth produce energy at the rate of a few thousand megawatts. The Sun energy output corresponds to 10^{15} such hydroelectric plants. At the Earth's distance from the Sun, 146 million km, the intensity of the radiation corresponds to an incidence of about 1.37 kW per square metre. Some of it is reflected by clouds, some is absorbed in the atmosphere. About half of it reaches Earth's surface. That energy keeps our planet at an average temperature of about 14 degrees Celsius. The Sun surface temperature is about 5,500 Celsius. This is called its effective temperature, deduced from the characteristics of the radiation that it emits and that reaches us. The emitted radiation corresponds to energy loss. As we will see, that energy originates from nuclear fusion processes that take place in the Sun's core. Part of the core mass is transformed into energy in the form of radiation. It corresponds to a loss of mass of the order of 1.25×10^{17} kg per year. It radiates the equivalent of one Earth mass every 50 million years. However, it is big, with a diameter of 1.4 million km and a mass of 2×10^{30} kg it could contain 330,000 Earth masses (6×10^{24} kg). Since its birth, some 5 billion years ago, it has radiated about 100 times a mass equivalent to that of the Earth. It is expected that it will still exist as a star for another 10 billion years. Our generation does not have to worry about a temperature drop on Earth's surface.

Our ship is getting hot. We are too close to it. However, due to the Sun's gravitational attraction we are going faster and faster. We circle it, get a gravitational boost and travel out, like being shot by a sling. We are travelling like a comet, but without the tail. We go in the direction of planet Mars. We see the Earth as a small blue dot and the Sun already looks smaller than when it is seen from Earth. We are now at 100 millions km from our base planet and we see clearly the red planet Mars. If we could land, we would see that Mars really is made of red stuff, as we see it from Earth, and rather dry. In fact, earlier robot explorations have shown that the surface contains large quantities of iron oxide. There is no water except probably at the poles where water could be frozen, intermixed with solid carbon dioxide, and maybe under the ground as permafrost as we have on Earth in polar regions. We could stay for a sunset but the Sun would be much smaller than seen on Earth, as shown in Figure 2.2, a photograph taken from one of NASA's robots sent several years ago. From our ship we see Olympus Mons, a giant volcano, much larger than anything we have on Earth. We see also Mariner Valley, a canyon the size of a country like the US. Since Mars has a diameter only two-thirds that of Earth, it looks as if everything is out of proportion. We do not see straight structures or canals as imagined

Figure 2.2. Sunset on Mars as seen from Rover Spirit, May 19 2005. (Credit: Mars Exploration Rover Mission, Texas A&M, Cornell, JPL, NASA.)

in those early telescope observations by astronomers Schiaparelli and Lowell. And it looks like a cold dry place, not very hospitable with its low pressure atmosphere made mostly of carbon dioxide.

We keep going towards the outer planets, and look back to see where we came from. Earth, Mars, Venus and Mercury are tiny dots close to the Sun. We profit from a rare conjunction of the planets almost all on the same side of the Sun. We pass giant Jupiter with its colourful cloud cover, beautiful Saturn with its wonderful rings. We feel the presence of these giants. Our trajectory is deflected by their gravitational pull and we are sent towards blue Uranus and Neptune, looking pretty but still very far away. It takes a long time to reach them. We do not see Pluto. It is somewhere on the other side of the Sun and is too small. As an insult, it was recently demoted and now, according to our classification, it belongs to the group called 'dwarf planets'. We get to a distance of over 5 billion km from Earth and we see the whole of the Sun's planetary system. The planets are dots, just barely visible with our limited instruments on board, with the Sun very small in the distance as shown in Figure 2.3.

We went through the solar system and during the trip we had a chance to observe and compare the size of those planets relative to the Sun and to each other. The winner for size is Jupiter, a celestial body that almost became a star. As shown in Figure 2.4, our blue home looks a negligible object compared to the other large planets.

We are now at over 10 billion km from Earth and the stars are still appearing as points in the sky. We keep going and find ourselves in a dark location where there is nothing to see but stars as point sources. For us, being used to the presence of the Sun and the Moon, it looks like total emptiness. This makes us realize the immensity of the surrounding space. Our Sun and its planets is a minuscule system no longer visible in the vastness of that empty space.

If we want to see something else we better hurry up because those stars in front of us seem rather distant. We have been travelling at an average speed of about 40,000 km/h, which was our escape velocity from Earth's attraction. It is now reduced quite a lot due to the attraction of the total solar system on us. In a real trip we would

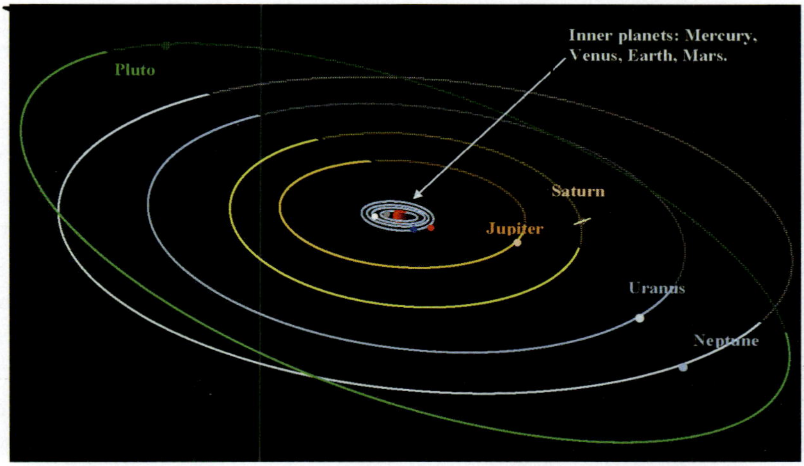

Figure 2.3. Our solar system as seen from outer space, billions of kilometres away. From that distance the inner planets, Mercury, Venus, Earth and Mars appear to orbit very close to the Sun.

Figure 2.4. Relative size of the solar system's planets. The Sun is so large that it does not fit in the picture: relative sizes are close to reality but distances between the objects have not been respected. (Credit: NASA.)

have been away from home for over twenty years to reach that point. At that speed before we get to another planetary system around one of those visible stars it would be a long, boring trip. We better put the space ship in 'light-year gears' rather than travelling in terms of km. We thus set our cruise control in the light-year mode. That way we will gain in two respects: first simply through speed but secondly, if we go fast enough, through reducing our aging rate relative to friends we have left on our home planet.[1] We need that effect because we would not live long enough to get to one of those stars in our lifetime.

We now reach a distance of 1 light year from Earth and we still see stars as points, the Sun included. At a distance of five light years we head towards the constellation Canis Major. We approach the nearby star, Sirius. It is about 8.6 ly from the Sun. It is not much bigger than the Sun (1.7 times larger) but looks much more luminous (25 times). Does it have planets circling it like our own? We do not know. With the largest telescopes on Earth, even with space telescopes, it is difficult to see such small objects as planets circling close to a star. In our spaceship we could certainly go and check that, but our travel plan is fixed and does not allow us to do so. We leave it as a space secret.[2]

We use the pull of Sirius to swing our ship in another direction. We could have gone towards constellation Orion to see Betelgeuse and Rigel but they are far away in another direction and there are some other stars close to the Sun that seem to be quite interesting. We swing towards a small constellation, Bootes. We travel a long time and pass many stars in the distance. We see also that the constellations we are familiar with when seen from Earth change shape and some are now barely recognisable. It is because those constellations as observed from Earth are seen as if the stars forming them were on a sphere. But in space these stars may be very far from each other, distributed in three dimensions at various distances from the Sun. We finally see a star growing in front of us. It is Arcturus, much bigger than the Sun, about 25 times its diameter and much brighter. It is 36.7 ly from the Sun. Again, we use the pull of Arcturus to swing the spaceship towards constellation Scorpius. It looks like a long trip, but we notice at a distance another rather bright star growing up on us. It is Antares at 600 ly from the Sun. We better stay a bit distant from that one: what a beauty, it is about 800 times larger than the Sun and 65,000 times more luminous. If our Sun were that big it is evident that life could not exist on planet Earth. Our planet would volatize, its orbit being within the star! The relative size of those monsters astonishes us. Our Sun appears ridiculous in size and the sketch in Figure 2.5 makes this explicit.

If we want a pull for swinging us around, Antares is the chosen candidate. We get the pull and heads towards one of the most beautiful shows in the cosmos: we head towards the 'Pillars of Creation' in the Eagle Nebula. It is a long trip, 7,000 ly away, but it is worth it. What a sight! It is shown in Figure 2.6. It is a very large cloud with dimensions of the order of 50 to 70 ly constituted of hydrogen gas, charged particles and dust. The whole thing is held together by gravity. The concentration of mass is much larger at certain places within the nebula and gravity is much stronger in those areas. The Pillars are like some natural structures, sometimes called monuments, existing in some deserts on planet Earth. On Earth, these monuments are protected by the nature of the material they are made of. They are the remnants of hills, of which the surrounding material has been eroded by wind and water. In the case of the celestial Pillars the effect of erosion comes from ultraviolet light originating from a star outside the picture on the top right and illuminating the whole structure. The gas cloud formed mostly of molecular hydrogen evaporates slowly on the surface exposing the inner parts of the Pillars. The inside is very dense and through gravitational attraction forms dense globules: the effect leads to bright spots. They are stars in formation called protostars. These globules may also be exposed by the evaporation process. They may dissipate slowly into space before they have a chance to grow bigger and become a star.

And now we let ourselves go and we travel like this for thousands of light years, and see stars just passing by all around us, grouped in unfamiliar new arrangements, constellations to which we could give

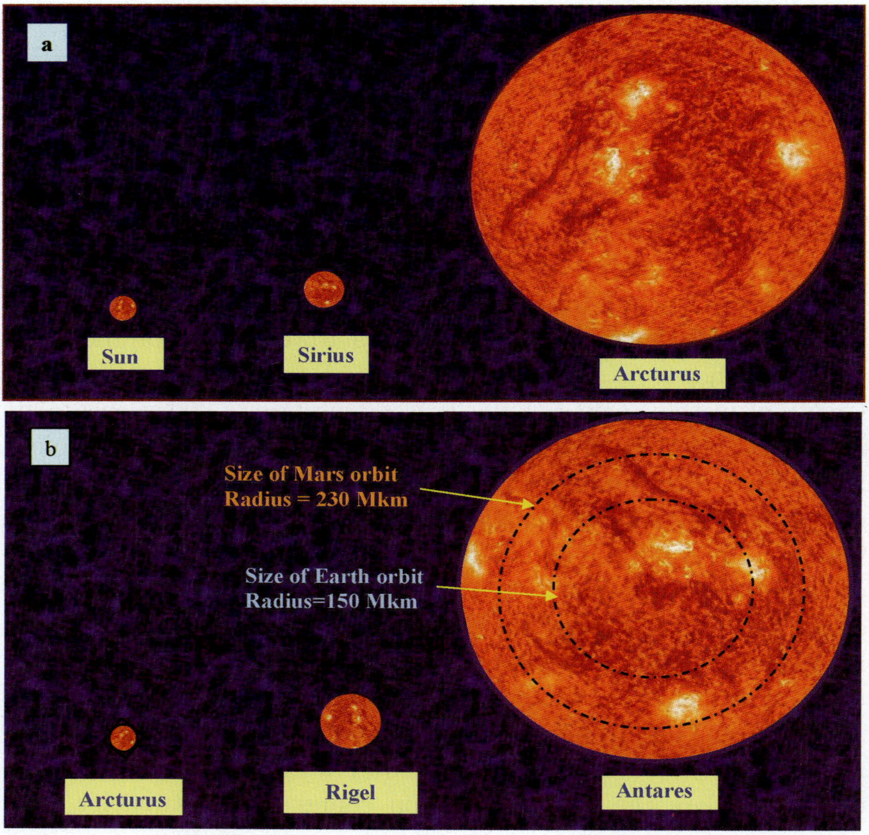

Figure 2.5. Relative size of close-by stars making explicit the small size of our Sun. The relative size of the orbits of Mars and Earth around the Sun are superimposed on Antares to make explicit the size of that star. The Sun would be an almost invisible dot in Figure 2.4(b). The distance from the Sun to each of these stars is: Sirius: 8.6 ly; Arcturus: 36.7 ly; Rigel: 800 ly; Antares: 600 ly.

names providing the basis for a new astrological pseudo-science. It seems without end and it becomes boring. But, gradually, they all seem to get away from us. The further we go the less brilliant they look and they appear to organize themselves in fuzzy arrangements, in long arms spiralling towards a central large core. We are now at several hundred thousand light-years from home and we see the whole galaxy, our Milky Way. It looks like a flat disk about 100,000 ly

Figure 2.6. The 'Pillars of creation' in the Eagle nebula within constellation Serpens. We are looking at objects mostly of gas and dust, in the range of several light years in size, 7,000 light years from the Sun. (Credit: J. Hester, P. Scowen, (ASU), HST, NASA.)

in diameter and may be 1,000 ly thick. Our solar system is there, invisible and lost at the end of one arm. That galaxy contains several hundreds of billions of stars, probably with planets around them forming systems like our own solar system. Maybe some are inhabited! At the centre, there is a kind of large, brilliant yellow bulge. The arms appear to form a spiral motion towards that bulge. But in fact we know that the stars are travelling around that central bulge somewhat like in orbit. What is inside that yellow bulge? We think there is a very heavy structure called a black hole. From Earth we have an idea of the whole Milky Way by means of pictures of the night sky. It looks as in Figure 2.7. This picture being taken from Earth within the galaxy, we do not have a clear view of the ensemble.

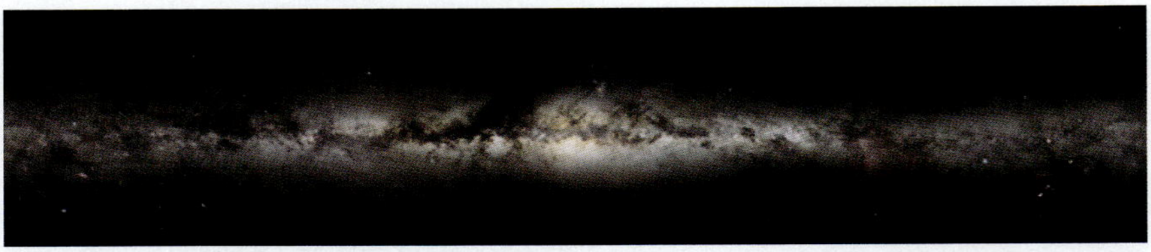

Figure 2.7. Milky Way as seen from planet Earth. We are part of the galaxy and it is thus seen on edge. The dark veil in front of the galaxy is assumed to be formed of regions containing dust masking our view of the details. The central bulge is partly visible. (Credit: Digital Sky.)

Figure 2.8. Andromeda galaxy as seen from a terrestrial telescope. The view from Earth, partly on edge, provides an interesting perspective of the galaxy. (Credit & Copyright: Davide De Martin, 'Digitized Sky Survey', with permission.)

We see it along its plane, there is a lot of dust in the way and we do not see the arms individually.

When we turn around towards space in front of us we see very little, only fuzzy spots here and there. Now this is real space. How long do we need to travel to start to see something interesting? We keep going and arrive at one of those fuzzy spots. It is a galaxy, called Andromeda, very much looking like our Milky Way. It is shown in Figure 2.8. It is 4 million light-years (Mly) away from ours and is of the order of 100,000 ly across. It contains, also like the Milky Way, tens or hundreds of billions of stars and a central core.

It is unimaginable. In space it goes on like that: dimensions are not at the scale of the human being. Fortunately, our intelligence makes a bit of sense out of it. Those closest galaxies form a so-called proximity group separated from other similar groups by hundreds of millions of light-years. And it keeps going like that for hundreds and hundreds of millions of light years, adding up to thousands of millions of light years. In the very far distance we see a very brilliant object: a Quasar. From measurements made on Earth we understand that being so far and so bright, it must emit the equivalent energy of a trillion Suns. What can cause that? What can be the source? We do not know. If we kept going, would we reach an end? No. When we look at the heavens from Earth with our powerful telescopes, make some observations and analyse them, we come to the conclusion that there is a line, a 'horizon' at a large distance of billions of light-years that appears to be a limit to visible light coming towards us. Up to that point everything in all directions is the same with galaxies distributed seemingly at random but forming a very homogeneous medium. But like on Earth, the horizon is not the end of land or sea. It is a question of geometry. We can suspect that the space horizon has a similar property. If we stand on the space horizon, what do we see? It should not be more of a privileged spot than our position on Earth and we should still see the same landscape in all directions. Is the Universe infinite? We cannot answer now. We will address that question later because it is a real puzzle.

We are very far and anxious to get back home, on our planet Earth. We better stop our trip. We zip back to Earth on our virtual spaceship under full light-year cruise control.

The Small and Ultra Small Landscape

We went in the heavens in the direction of large objects and found splendour. How about examining now, right here, the objects around us? What do we see? There is evidence that a grand divide exists: living beings against inanimate objects. Are they as different as they appear? The Greeks said that everything was made of Air,

Earth, Fire and Water. That was not a bad observation and at the time it was very practical. One could pursue most of his or her normal activities, such as eating, drinking, breathing, keeping warm, lodging and so on, using those concepts. When a living being dies, its constituents decompose and the residues go back to the soil as 'earth', water and air. Leaving aside Duncan Macdougall's experiment attempting to prove that a dead human weighed less than when he was alive, providing a measurement of the weight of his soul, we will assume that living and inanimate beings are made of the same stuff. How inanimate matter became alive at one point of the Universe's evolution is a hot question. It appears that at certain stage of complexity and organisation of the basic constituents that we are going to describe, life is an inevitable natural process.[3] We will not delve into this subject at this time and we will only examine the basic constituents of matter that we can see. The Greeks were looking at the bulk of matter. We are going to look at the details. We now need a means to examine that matter, but in the very small. We construct an instrument, an ideal microscope, whose magnifying power can be adjusted at will and assume that we will have light appropriate for observing even the ultra small objects. We will see later that this is possible only down to a certain size of object, but for the purpose of simplifying things we will assume that such an instrument is feasible. Our instrument is again a virtual one replacing the multitude of instruments used for the purpose of studying the very small, such as visible light microscopes, X ray interferometers, electron microscopes, tunnel microscopes and the ultimate, powerful particle accelerators. To examine our Universe in the small scale region, let us use a snow flake, an object that we know rather well as made of crystallized frozen water. Without magnification, it looks like a small, highly symmetrical structure that exists in various forms. We look at one of its extruding points with our virtual microscope and we see a very regular structure inside. It is a little complicated but we clearly see a repetition of an assembly of small fuzzy balls organized in what is called a crystal structure. In fact we see that a group of three balls distribute themselves in a very symmetrical manner throughout the crystal. They form small hexagonal packages that we call cells. They are very small, about one thousandth of one millionth of a metre on their sides. The number we just wrote is 0.000,000,001 metre. To make things simpler, we need to introduce a new notation for that type of number. In our so-called scientific notation we write this as a power of 10 as before for large numbers, but with a minus sign in the exponent. Thus, we write our number as 1×10^{-9} metre, meaning that the '1' in front appears as the ninth digit after the point. It is just an arbitrary notation and once we have accepted it, it reduces the amount of writing needed for small numbers. What are these three little balls? They are atoms: two are hydrogen and one is oxygen, so-called elements that cannot be divided into parts without changing their personality. These atoms are the same entities that the Greek Democritus had imagined more than two millennia ago. They are held together in that marvellous arrangement by forces of electrical and quantum origins at an angle of about 104 degrees as shown in Figure 2.9.

We increase the magnifying power of our instrument and look at the hydrogen atom: we see first a fuzzy little ball, which we call an electron, dancing around within a limited stage. It is quite a dance! It wanders around a very small object, which it circles. This small

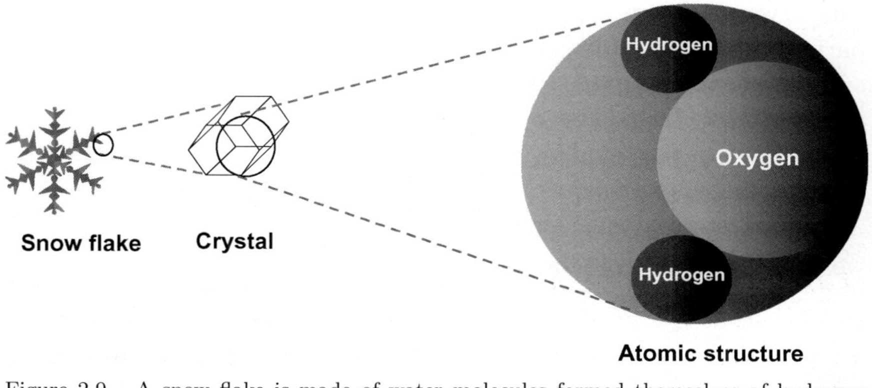

Figure 2.9. A snow flake is made of water molecules formed themselves of hydrogen and oxygen atoms organized in hexagonal symmetrical cells that reproduce themselves to give the varying and beautiful symmetry of the flake bulk structure.

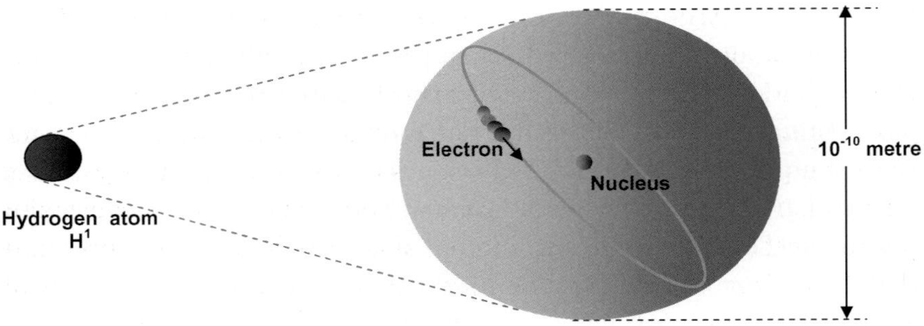

Figure 2.10. Naive representation of the hydrogen atom. In reality it is believed that the electron is not circling the nucleus as in an orbit but envelops it, continuously exchanging information with it by means of virtual photons.

object is called the nucleus. This is illustrated in Figure 2.10. In reality, the electron circling the nucleus is not very well localised. We cannot tell very well how it looks. It is simply impossible to see it. To our eye it looks very much like a small cloud because it jiggles around very rapidly and our eyes are too slow to follow its motion. It carries a *negative electric charge* in its dance. It also has a very tiny mass. If we try to examine that electron more closely, it appears as a ghost. Sure, it has that electric charge because physicists have measured it very accurately. It is one of those fundamental constants of nature, or natural constant, called 'e'. That charge is quantized, which means in simple words that it is a unique quantity that cannot be cut in smaller parts. Millikan and Fletcher were the first to make the demonstration of its properties in the 1910s in their famous experiment in which they studied the behaviour of electrons attached to oil drops falling in the Earth's field while submitted to an opposing force created by an applied electric field.[4]

Sometimes the electron looks like a wavy structure. Actually we will see that a whole theory based on the assumption that it is a wave has been developed. The theory is found to be in full agreement with observations. Sometimes it looks like a very tiny particle. Some experiments show without doubt that it can also have that property. If we try to measure its distance from the nucleus with a ruler it speeds away as if it were very shy and unwilling to expose its true personality. Since we cannot locate it well we also cannot determine exactly its speed. This phenomenon results from a basic principle of quantum mechanics that we will examine later. It simply says that it is not possible to measure at the same time the position and the speed of a particle to an accuracy better than a certain value, which although very small, sets a constraint on any measurement process. It is called the uncertainty principle and is an intrinsic property of nature, not of the measurement apparatus or process. The electron does not seem to have real size. Worst of all, it seems to emit extremely short bursts of radiation all around itself, so-called *photons*, and it reabsorbs them. It looks as if it was inspecting its surroundings to adjust its behaviour. What a strange object, moving randomly, looking like a particle and avoiding observation. It is living in a world of its own: *the quantum world*. What is the size of the sphere in which it is wandering? It is about 10^{-10} metre. If we try to remove the electron from that environment, we can do it but with some special tweezers. What would be such tweezers? A particle of light, a photon, would do: a photon with sufficient energy would grab it, knock it out of its orbit, and send it through space. The same could be done with another electron shot at it with sufficient energy. The force holding the electron to the nucleus is electromagnetic and moderately strong, at least much stronger than gravity. It is the force exerted by the nucleus carrying a positive electrical charge attracting the electron, which carries a negative charge. Consequently the tweezers must have sufficient holding or knocking power. Is the electron an elementary particle? But, what is an elementary particle? One definition easy to understand is that a particle is elementary if it cannot be divided into smaller parts. Another definition could be that an elementary particle is an object that cannot be excited internally into some kind of vibration mode, since there are no internal components or gears to set in motion. An elementary particle would then always be internally identical to itself. On the other hand, if a particle has size it means that it is made of some stuff and that stuff being distributed could be set in some kind of relative motion. Thus an elementary particle seems to be condemned to be a point in space without dimension.

As you can see, the identification of the true character of a particle depends really on the ability that we have to study its inner parts. There was a time when more than 100 elementary particles were identified; they could not be split into parts with the tools available at the time. They appeared, however, to behave strangely. Such a number of elementary particles did not make much sense and it became evident that with some newly available tools, theoretical and experimental, we could prove that they were far from being elementary. However, since its discovery by Thomson at the end of the 19th century, the electron has remained identical to itself under all circumstances, is not divisible and does not seem to be made of other stuff. It looks like a point particle, if that concept means anything.

On the other hand the nucleus of the hydrogen atom has a dimension of the order of 10^{-15} m and is thus 100,000 times smaller than the whole atomic cloud enveloping it. It is very small indeed. If we could enlarge the volume of activity of the electron to the size of a large stadium, the nucleus would look like a mosquito standing in mid air. The electron would be on the roof, looking like a drop of rain. The atomic space thus appears even more empty than the celestial space we have visited with our spaceship. If we look at that nucleus with low magnifying power it appears to be made of only one particle, but as we will see, this is an illusion. That particle has been called a *proton.* The proton is loaded with one positive electric charge 'e' and has a mass about 1800 times larger than that of the electron. The electric charge on the proton is exactly the same as the charge on the electron but of opposite sign. It is a very puzzling coincidence with dramatic effects. In the hydrogen atom they cancel each other exactly, totally, absolutely, without any measurable residue. The atom is totally neutral. This is quite an achievement from Mother Nature since, as we will see, the charge on both particles is of different origin, at least in appearance. Nature has accomplished quite a feat.

If we return to our snow flake and examine carefully its elementary cells, we find that for about every hundred hydrogen atoms one of them has a bigger nucleus than the others. If we examine that nucleus carefully with our virtual microscope we find it is made of two particles, but the atom as a whole looks the same as the other atoms around in the snowflake. One of those two particles is a proton and the other looks the same, but does not carry an electric charge. It does not seem to do anything except adding mass to the atom. That atom is a so-called isotope of hydrogen; it is called heavy hydrogen or deuterium and this new particle in the nucleus has been given the name *neutron* for the obvious reason of not being electrically charged. The arrangement is shown in Figure 2.11. The neutron has a mass a little larger than that of the proton. It sticks like a parasite to the proton and in fact is held there by a new force. It is the nuclear force and is much stronger than the electrical forces mentioned above attaching the electron to the proton. If we try to remove it with the tweezers we described above, it does not work. The retaining force is too strong and requires a real shock like throwing particles of the same nature at the nucleus. It is like when we want to dislodge a piece of rock from a structure; we need to throw at it a rock of a similar size with sufficient energy. Thus, in that case, the particles to be used are either protons or neutrons that are travelling at a very high velocity. If we succeed at dislodging a neutron, we find

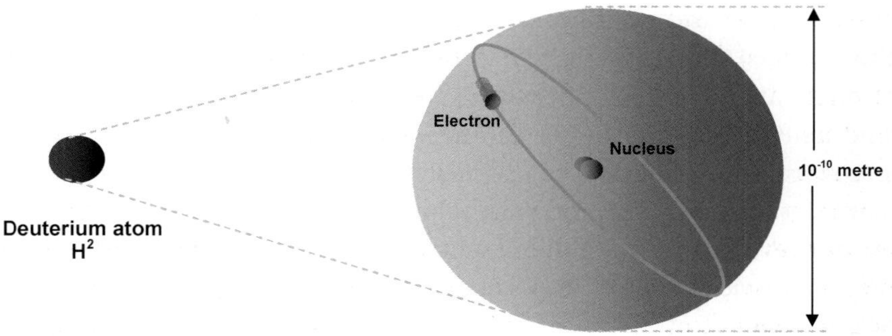

Figure 2.11. A deuterium atom is similar to a hydrogen atom, but its nucleus contains a neutron in addition to the proton. Its mass is thus twice as large as that of the hydrogen atom.

that it disappears in about 15 minutes. A proton and an electron appear in the process. Furthermore, for reasons that will be explained later, conservation properties regarding quantities such as energy and momentum require that another object appear in the process: that object looks like a phantom particle and has been given the name *neutrino*. It has no electric charge and has a very small mass if any. It is extremely difficult to detect. It was predicted in the 1930s by Pauli and detected only in the 1950s. The decay of the neutron has important effects on the evolution of the Universe as we will see later. In nature, more complex atoms are made of more electrons and more protons in the nucleus. If we look at our molecule of water we find a more complex atom made of eight electrons wandering around a nucleus containing eight protons and eight neutrons. It is the oxygen atom attached to the two hydrogen atoms as shown in Figure 2.9. We also find some of those atoms distributed at random, dissolved in the water from which the snowflake was grown. It is free oxygen, dissolved in water, that keeps the fish alive providing them with the element necessary for sustaining life through a kind of breathing particular to them.

In their normal state, all atoms contain the same number of electrons as there are protons in the particular nucleus. Due to the exact same charge on the proton and the electron all atoms are neutral. But what is the neutron role? What is it doing in the nucleus, aside from adding mass? It brings stability to the nucleus, which otherwise would be made entirely of positive charges. In the absence of neutrons, the protons would repel each other very strongly being in close proximity and all being positively charged. The neutrons and protons arrangement, due to the strong nuclear forces, counterbalances that repulsion. We will examine this effect in more details later and find that relativistic and quantum effects play an important role in holding nuclear particles together. However, it is quite clear that protons, neutrons, and electrons are the basic constituents of atoms.

But what are these neutrons and protons made of? Are they elementary particles? From what we saw just above the free neutron transforms itself into other particles after some time. It looks very much that it has internal parts. We need to look at it closer. In some experiments it is observed that the proton looks a little rough on its surface. After examination with more magnifying power we find three particles in there. They have been called *quarks*, a name without intrinsic meaning, which was chosen with a bit of humour from a published fiction story. Quarks are very amazing particles. They look somewhat like electrons enveloped in some kind of cloud, changing appearance and jumping around continuously almost free within the small volume less than 10^{-15} m in diameter occupied by the proton or the neutron. They feed that cloud with bursts of some substance or field. This is their way of staying together with some kind of glue that softens when they get very close to each other. The associated sticking particle that is exchanged is properly called a *gluon.* It is like the photon in the case of the electrons; the quarks are thus continuously interacting showing their presence and examining their surroundings by means of those gluons. The quarks do not have a full electrical charge. In the proton there are two quarks with a charge of $+2/3$ e and another one with a charge of $-1/3$ e. Thus the total charge is $+1$ e as it should be. The two quarks with the electric charge $2/3$ e have been given the strange name 'up' while the one with charge $-1/3$ e is called 'down'. The structure is illustrated in Figure 2.12.

It seems that in the discovery of new concepts, the community is getting short of names and symbols, and once a particle has been baptized it is extremely difficult to alter its name. We have to live with those names. The neutron is constructed very similarly: it consists of two 'down' quarks and one 'up' quark thus forming a neutral entity. They are the particles that give mass to the nucleus and to the atom since the electron is very light and does not count much in the total mass of any atom. Let us try to remove one quark with our tweezers to examine it. No way! Although they look free within a given small volume they are very strongly attached to their friends by strong elastic strings: these are called the strong forces of nature that are mediated by the gluons. The force acts on a very small range. When inside a certain volume, close to their partners, they move around freely but as soon as you pull on one of them,

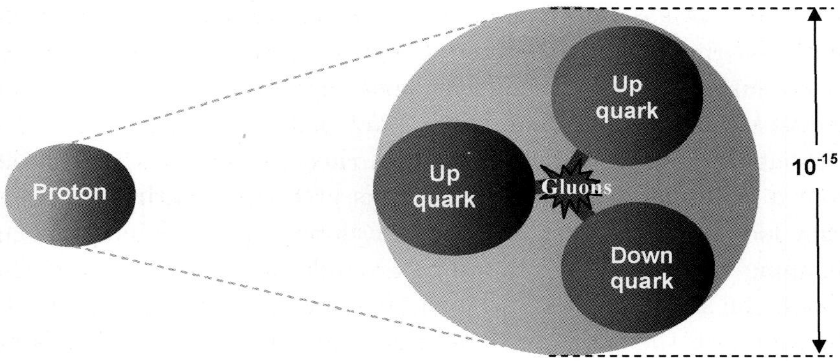

Figure 2.12. The proton is made of three quarks. Two of the type called "*up*", each with electric charge +2/3 e and one quark of the type called "*down*" with charge −1/3 e. In addition, quarks have another internal property called colour charge, responsible for the strong force of interaction, which has nothing to do with the meaning of that word in our macroscopic world.

the other two oppose. They then look like as if they were attached by means of elastic bands that are loose when they are close to each other. If we try to knock one off, say by bombarding the nucleus with another energetic proton, the elastic band gets stretched to the point that it breaks and new particles appear. The process will be explained in more detail later. The quark obviously does not like to be alone. There is no such thing as an isolated quark. It thus seems that we have seen the ultimate particle. Is it elementary? Although we cannot have one in isolation, it appears from experiments of various kinds done with very energetic collisions that they behave like elementary particles.

We have thus gone around our Universe. The objects that we have seen in the heavens are extremely large and seem to behave according to one rule: they aggregate or move around under the influence of gravitational attraction. The gravitational force, although weak, seems to be a terribly efficient dictator on the large scale. There is no way of escaping from it whatever the distance. More than that, it is not possible to create a shield against it. If one puts a planet between two objects, the effect of gravitational attraction between the two objects is not diminished: the gravitational attraction of the planet in between just adds some more gravitational effect of its own. On the large scale it is thus a world of gravity. However, the density of matter seems to be low and space looks very empty. In our travel we did not seem to be inconvenienced in our motion through space. This is because the density of matter is extremely low in free space, of the order of a particle, a hydrogen atom, per 10 cubic centimetres. Free space is thus a very good vacuum. But there are a lot of questions like the size of that space, its nature, its connection to time, its relation to the speed of light. At the other end of the landscape, in the small scale, nature seems to have problems in organising itself with simple rules. It seems to like to rely on mysteries, not deciding if an object is a wave or a particle, keeping a certain distance vis-à-vis measurements, preferring imprecision, and relying on probability rather than absolute dictatorship, as in the case of gravity. It is the world of quantum physics. Let us represent that whole landscape by means of one single drawing as in Figure 2.13.

The scale is frightening. That landscape is awfully large. There is a change in dimension by a factor of the order 10^{40} between the objects we have to deal with. This is the change in size from distances within nuclei to distances in the cosmos. We have outlined observations to provide a picture of the ensemble without venturing into detailed explanations of the *how* this whole system operates. Are there rules that makes the various components engage the way they do? What are these rules? Where does the Universe come from? Why does this landscape exist anyway? Where is it going? What is our

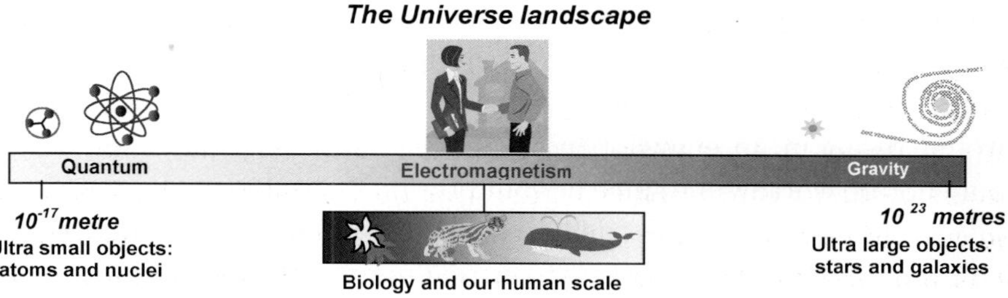

Figure 2.13. The Universe landscape that we observe, which extends from the very small, 10^{-17} metre, to the very large, 10^{23} metres.

role in that Universe? Some of these questions are of an existentialist nature. We may never be able to answer them. Are we fulfilling a purpose or are we a negligible speck of a greater design? Some other questions are of a nature that excites our curiosity relative to the overall picture and to the parts that make it what it is. This is the role of physics: to discover the mechanisms of the functioning of that world and the basic principles of operation of that large, complex machine. Can we handle that with the tools we have at our disposal? And what are those tools? They are our senses, our brain, our memory, our reasoning, our logic, and the mathematics we have developed.

The Framework of Our Understanding

When we walk in a scientific library, we are baffled by the diversity of book titles. Just going through my own bookshelves, for example, which is a miniscule example of collections that exist in a university or a national library, I find titles such as *Chemical Physics, Solid State Physics, Quantum Field Theory, Quantum Electrodynamics, Photons and Atoms, Ligand Field Theory, Quantum Physics, Quantum Mechanics, The Quantum Physics of Atomic Frequency Standards, Quantum Electronics, Magnetic Resonance, Principles of Physical Cosmology, Relativity, Space-time Physics*, and it goes on like that, book after book. Those books are written by specialists in their own field and fulfil precise goals, providing an understanding of a given subfield or outlining a recent progress in a given subject. If one had to know all details that are written in those books to understand the functioning of the Universe, it would be frightening and impossible. A single volume, describing in detail and in depth our present knowledge would require a huge effort of compilation. It would result in an encyclopaedia of an almost inconceivable size. It may not be worthwhile since it would become obsolete in a short time, due to the rate of progress achieved in our present day understanding and our realisations.[5] However, in reading those books it is readily observed that they very often overlap in the subjects they cover, the difference being the approach used to a subject. Furthermore, they use some of the same basic physical principles to analyse and explain the various phenomena and experimental observations that are the subject of the authors' interest. It is one of the goals of physics to find those common points in the compartments introduced by specialisation in the various fields of research.

In practice, it is possible to define large compartments that include many sectors that appear to have things in common, or that are even specialization of a broader field. For example electricity, magnetism, and radiation have many concepts in common and may be grouped under the title 'Electromagnetism'. There are also physical phenomena that seem to be of a more fundamental nature than others and seem to lead to basic principles that are common to many sectors. For example, quantum mechanical phenomena play a fundamental role in the study of light, atoms and nuclei while relativistic effects are observed at the atomic and nuclear levels as well as in the description of the dynamics of larger scale objects. On the other hand, gravity does not seem to play a role at the atomic level, that type of interaction being much too weak compared to the other interactions entering into play: it works effectively only on the large scale. In our travel through the Universe landscape we have been mentioning those various interactions. In that context, the practice has been to differentiate sectors by the strength of the interaction, or force acting between objects. We can identify at present four types of interaction that provide a rather practical framework for regrouping subfields. These interactions lead to forces that are well differentiated. In order to provide a proper background for the chapters to follow and to have an overall view of what those interactions are, we will introduce them immediately. Some of them may look rather strange at this stage, but they will take shape later through a more formal introduction.

Force of gravity

This is a force that is believed to take place between all matter in the Universe and leads to mutual attraction of objects with mass. It is a weak force described by the General theory of relativity of which

Newton's gravitational theory is an approximation. The gravitational effect of an object with mass is felt up to infinite distances.

Electromagnetic interaction

This type of interaction gives rise to a force that is studied in a theory that includes both the classical description of electric and magnetic fields and their quantized form. In the last case the model leads to the *photon* concept, a representation of light (or radiation) by means of particles. It deals with the photon-atom interaction and the supporting theory is given the name 'Quantum electrodynamics'. Quantum electrodynamics is part of a broader theory called 'Quantum field theory' that provides a framework in which all particles are understood as materialization of a supporting specific field. The strength of electromagnetic forces that result from this type of interaction is much larger than the strength of gravity and the force exerted by an electrically charged particle on another one extends also to infinity. The electrical repulsion between two electrons is larger than their gravitational attraction by a factor of the order of 10^{40}.

Weak interaction

This type of interaction takes place within the nucleus. We have mentioned above that a free neutron is unstable and changes into a proton by emitting an electron. Furthermore, a new neutral particle is created. It is called *neutrino.* It is called beta radioactivity. That phenomenon has been identified as originating from a type of interaction that is different from the electromagnetic interaction just mentioned. The phenomenon is also observed to take place in certain nuclei. The interaction taking place is very weak compared to the electromagnetic interaction. This can be visualised by realising that the neutrino, the central actor, has so little interaction with matter that it can traverse a block of lead kilometres thick without interacting with any of the nuclei within the block. Since the neutron radioactive decay involves a neutrino, an electron and a quark, an electromagnetic interaction is also present and a unified theory has been developed under the name 'Electro-weak theory'. The weak force is effective only at short ranges less than 10^{-15} metre.

Strong interaction

It gives rise to the force that keeps quarks together in the protons and neutrons and leads to the strong nuclear interaction forces that hold them together. The interaction is described by a theory that has been given the awful name 'Quantum chromodynamics'. We will see later why it was given that name. It seems to be related to a study of colour, but it has nothing to do with colour the way we understand it. The resulting force is much stronger than the three other forces just mentioned and is effective only at small distances of the order of 10^{-15} metre.

The weak and strong force cannot be compared as directly and as easily as in the case of the previous ones, electrical and gravitational. They are localised within a very small region. The strong force decreases exponentially with distance and becomes negligible at a distance of the order of a few femtometres ($\sim 10^{-15}$ m). It grows rapidly at shorter distances; it is several hundred times larger than, for example, the electromagnetic force. The weak interaction, causing the weak force between the neutrino and a quark is evaluated to be less than the strong force by a factor of the order of 10^{14}. If the gravitational force is normalized to unity we can thus say very roughly that the four fundamental forces that we know have relative strengths as:

Gravitational force:	1
Weak force:	10^{28}
Electromagnetic force:	10^{40}
Strong force:	10^{42}

Those numbers should not be taken as absolute since the laws involved in the various interactions are of a different nature. Comparison is not direct since they depend on actual physical situations, like distances. However, the comparison shows clearly that gravitation does not play a role at atomic and nuclear dimensions. Its role is limited to the macroscopic objects on the large scale.

At present, those are the only forces that we know to exist in the Universe. All other interactions or forces that we have given names to, such as friction, chemical, molecular and others, originate from an action of those four forces or a combination of them. We do not know any other type of interaction or force and theories about the Universe's evolution are based on an understanding of the rules governing the dynamics of those forces. It thus appears that in a sense we are privileged, the number of interaction types to be studied being limited to four. However, as will become clear in reading the present text, the understanding of the role and the effect of those forces on the evolution of the Universe is rather complex and is a real challenge to the mind attempting to compose with them.

On the other hand, there exist some basic underlying physical theories that are required to deal with those forces and to obtain a complete picture. For example, all the models mentioned use common physical principles to describe motion of objects in space and time. This is a subject of general physics given the name Classical Mechanics, as originally developed by Galileo and Newton and further elaborated by scientists such as Laplace, Hamilton, Lagrange and many others to a level that draws admiration. In the classical world of Newton and Galileo time and space are two independent entities. Another underlying theory is Special Relativity developed by Lorentz, Einstein, Poincaré and Minkowski in a most elegant framework. In their world, time and space are entangled and form a single entity called space-time. The conclusions reached by that theory infiltrates all the interactions mentioned above. It is concerned with motion with constant velocity and has fundamental effects in the description of atomic particles as well as celestial objects travelling at very high velocities.

It should also be mentioned that the theories supporting the description of the three interactions, electromagnetic, weak and strong, are intimately interlaced. Electromagnetic and quantum concepts are at the basis of all three theories. Those theories use also a model based on symmetry groups, which provides a very elegant approach. This has led to a magisterial grouping of the three theories describing three aspects of our Universe in a common language: it has been called the *Standard Model of particle physics*. On the other hand the description and evolution of the large scale Universe by means of general relativity is called the *Standard Model of Cosmology*.

Interestingly enough, the phenomenon of gravity stands by itself without connection with the other three interactions. Gravity does not play an important role in particle physics. The force is so weak compared to electromagnetic and nuclear forces that, although it is assumed to be present, it appears to be completely negligible. Furthermore, there is no connection between quantum physics that seems to regulate the dynamics of the atomic world and the concepts developed by Einstein regarding gravity: there is no quantum gravity theory that seems to be satisfying yet. When scientists talk about a grand unification of the laws of physics, this is what they mean: a single theory that would describe the dynamics of the whole Universe from the ultra small to the ultra large joining electromagnetism, quantum physics, and gravity under a single general concept. This is presently a hot topic of theoretical physics.[6]

In our travel in the small scale we have encountered particles such as protons, neutrons, quarks and electrons. We have also introduced the neutrino, a particle emitted as a by-product of the catastrophic decay of the neutron. It should be mentioned immediately that these are not the only particles that exist. We will see that the Universe is extremely versatile in creating new particles and assembling some of them in new entities similar to neutrons and protons. Most of those particles have very short lifetimes. We will introduce them in later chapters and we will see that it is possible to group them in classes and families, making explicit symmetries that in turn provide appropriate frameworks for developing the theories that we have mentioned in relation to the four forces enumerated above. We may mention immediately that neutrons and protons are called *nucleons* and are part of a group called *baryons* while electron and neutrinos are part of the group called *leptons*. We will see later that other particles are also part of those groups.

We have included biology in the Landscape diagram shown in Figure 2.13. But what is biology? Biology is atomic physics organising on a grand scale atomic structures of enormous complexity; it is a description of chemical reactions in living beings, reactions that are nothing else than complex electrical interactions between atoms. The question of the appearance of life on this planet is a subject by itself and raises existentialist question of the most basic nature. Although it appears that, at a certain level of complexity and organization, life is a natural destiny of the Universe, it is not a question we can address easily in the present state of our knowledge. Life is associated with the evolution of the complex molecular assembly DNA,[7] providing through its two enlacing helix the genetic instruction or code for the development of species and individuals. Experiments have been made in the laboratory in which a soup of various kinds of gases, such as those existing at the early stage of evolution of our planet, has been exposed to an electric arc similar to lightning during an electrical storm. Resulting chemicals, sticking to the walls of the container, appeared to be of a nature similar to those that exist in living organism.[8] But life was not created. Life requires an extra 'element' for arising. We do not know what that 'element' is.[9] The following chapters will not explore this subject. We will limit ourselves to inanimate objects. In the present exercise it is sufficient to understand that the proteins and other substances that are present in living beings are formed of the same elements and chemicals as those in inanimate objects.

III

Space, Time, Motion

Worlds and Dimensions

How, as human beings conscious of our surroundings and with reasoning, did we arrive at constructing this marvellous structure that we call physics, based primarily on observations? How did we arrive at understanding the mechanics of the Universe to the present level? What do we have at our disposal to observe the world? It is our senses. It is really the sight of things that wakes us up to the Universe, the things around us and the stars in the sky, our perception of objects relative to each other and their motion. But we also have this marvellous instrument at our disposal, our brain, the largest in the living organisms and the most powerful, which can filter these perceptions, organise them, relate them to each other and through our fabulous memory relate them to previous perceptions. Our brain analyses these observations, makes relations between them, makes judgments about them and sometimes reaches a conclusion. Our brain develops logical relations, invents new concepts, implements them in instruments, leading us into unknown territory, which we would not be able to observe solely with the senses that nature has given us: we have extended these senses to such an extent that the Universe is no longer able to hide all its secrets. Our brain, in some aspects, is much more 'intelligent' than any computer that has been constructed yet.

In perceiving objects around us we have invented the concept of 'dimension', that is, distances between objects that we see or touch. We have given ourselves a standard to measure these distances, a ruler, a bar of metal of a very specific length and we have called it the metre, a little longer than half the height of an average adult. I will come back later to that concept and define it exactly. What we have also found is that those objects are situated at certain distances from us and in different directions. We have called this surround, space. In observing that space, we have also observed that we can go forward, sideways, and up or down. In physics, in an attempt to simplify things we have formed a so-called system of representation by means of *axes* representing these directions at right angles to each other. It makes a lot of sense and that way we can locate objects uniquely in that space, like an apartment on a given floor in a house, on a street, within a city. Since in that world there are three axes we call it a *three-dimensional* world. In that context, a *zero-dimensional* world is a point, since there are no axes associated with that world. A world in which only two such points exist can support only one axis, the line joining the two points like a fine thread spread tight between two posts: it is obviously a *one-dimensional world.* It is standard practice to identify that axis as the 'x' axis, as shown in Figure 3.1. In that world, a being can only travel between the two points, forward or backward, no other movement can be made. For example, we may identify its position by x = 0.35 m when the being

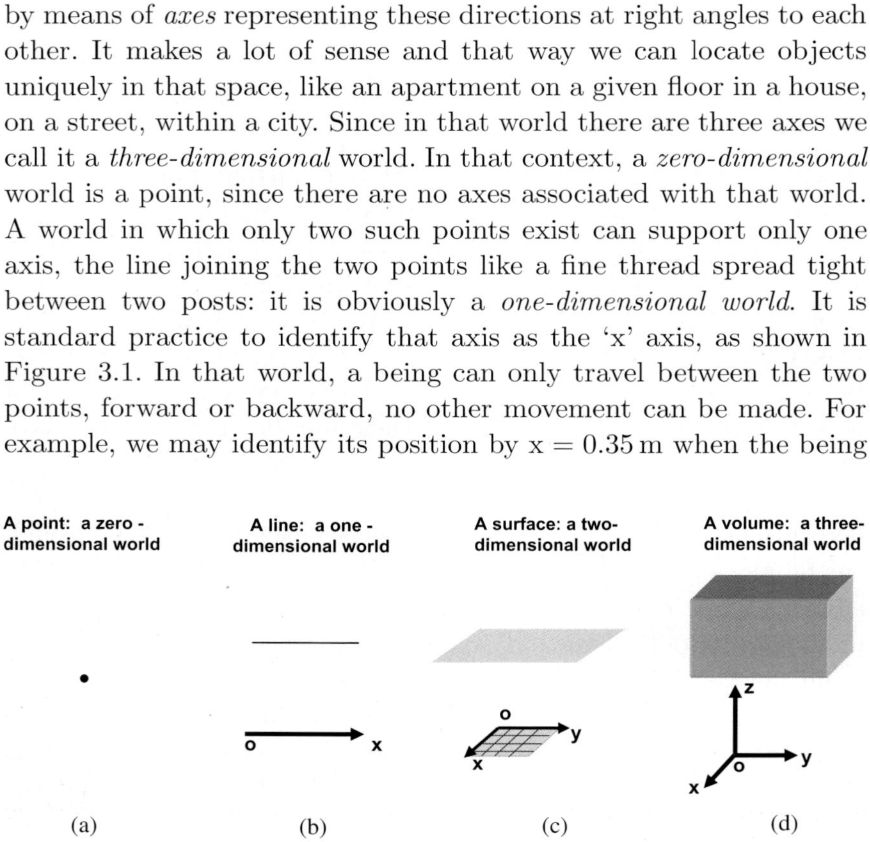

Figure 3.1. Examples of worlds with different dimensions. (a) A zero-dimensional world consists of a point. (b) In a one-dimensional world only one axis is possible: it is identified as 'x'. (c) A two-dimensional world consists of a surface like the top of a table; two axes are possible identified as 'x' and 'y'. Objects can be located by means of a grid like streets in a city. (d) Our visible world has three dimensions, with three axes represented as 'x', 'y', and 'z'. Position of objects can be located by means of grids in each plane as in the two dimensional case.

is situated at a distance of 0.35 m from the origin, called 0, which we may situate at one of the end of the axis. Just imagine two living beings in that world: there are bound to be collision problems! We note that the x axis could be infinite. In that case we would set arbitrarily the origin at a given point and measure everything from that position on the x axis. In another world in which it is possible to slide the line sideways, it is obvious that we have a situation where we can now move along a different direction: we have a surface and we have created a *two-dimensional* world, a much more interesting world. We may call it Flatland and flat beings can wander around without colliding. Finally, if we slide this surface in a perpendicular way we have our *three-dimensional* world, presumably the one we live in. We can move around in the three directions. It appears that this world is made for us or we have developed to fit in it. We note that in principle it can be infinite, expanding in all directions. We can also set the origin, 0, of the three axes arbitrarily and measure everything from that point. For example in a room we can think of one corner as the origin, 0, and measure position of an object in the volume of the room from that point, along the three axes formed by the three perpendicular lines originating from that corner and forming the limit between two of the walls adjacent to that corner and those walls and the floor. We do not question our existence in that world and it does not seem that we can comprehend easily an added dimension. But could we at least imagine such an added dimension? To illustrate it, let us look at the case of the being living in two dimensions, crawling in Flatland like an ant on a sheet of paper. Our fictitious ant is of course flat and we call her 'Flatant'! Flatant moves around and, herself a two dimension being, can perceive objects lying in her own plane, the sheet of paper. From our point of view she is very limited because she does not know there is another dimension possible. How would a three-dimensional object be perceived in that two-dimensional space? The space around Flatland could be very large, maybe infinite. Flatland essentially occupies no space at all and there could be an infinite volume around, which could contain an infinite number of objects. These objects would be 'felt' only when they cross the two dimensional plane as shown in Figure 3.2.

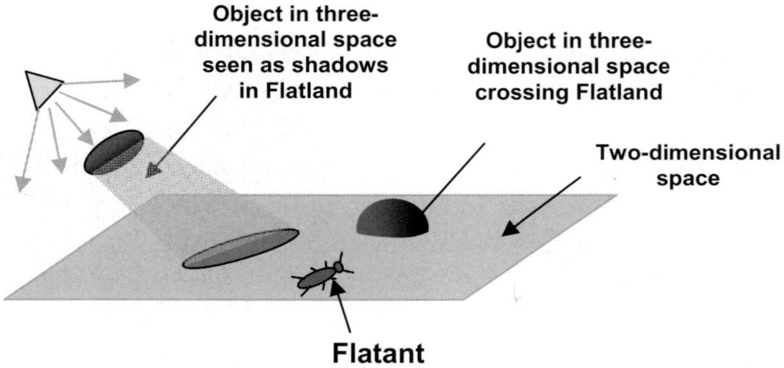

Figure 3.2. Objects in a three-dimensional world are seen as shadows or barriers in a two-dimensional world.

The section of the object crossing the plane would be perceived by the inhabitants of Flatland as a barrier they could or could not cross depending on the nature of the object. This barrier would change size and shape with the position of the object relative to the plane. An intelligent being could probably find some symmetry in the shape of the border that he observes as the object crosses the plane and deduce some laws that reproduce from time to time depending on the behaviour of the three dimensional object itself. For example, when a globe crosses the plane, Flatant would observe a barrier of circular shape. If the globe moves, the circle in Flatland would change diameter. With some imagination, Flatant, if she suspects a third dimension could conclude that a globe is crossing her plane. Another way that objects in the three dimensional space could be perceived in Flatland would be through their 'shadows', if light is coming from a certain distance. Again the information is very limited and it would take a great amount of imagination to determine the characteristics of the object in the other three-dimensional space to which the two-dimensional beings do not have access. If the object is moving, as in the case of the globe crossing the plane, more information could be obtained. Again, special symmetries of the object could be discovered and special laws could be written for the motion of those shadows. It would be very difficult for the two-dimensional beings to explain the origin of the laws: they would be assumed as laws of nature in the world in which they exist. There are many situations

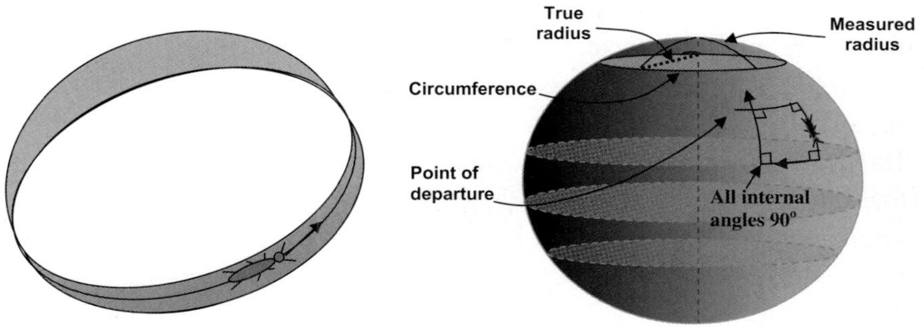

Figure 3.3. (a) Closed two-dimensional Universe strip. (b) A two-dimensional Universe as a sphere, the surface of a globe.

we could imagine that would complicate the life of Flatant in her two-dimensional world. For example, if the two-dimensional world consists of a ribbon closed back on itself as shown in Figure 3.3(a), Flatant, crawling in the same forward direction, would be rather puzzled that after a while she traces back the same path she started on. Crawling on the surface she has no way of knowing that her two-dimensional space is part of a three-dimensional environment she does not have access to. But coming back on her path may give her a clue that her world is part of something she does not have access to directly.

Similarly, the Universe of Flatant may be the surface of a sphere in a three-dimensional world, as in the case shown in Figure 3.3(b). She may attempt to make a square on her surface by travelling, say 100 metres in straight line, turning 90 degrees, travelling another 100 metres and so on. On a flat surface the path would close on itself and Flatant would find herself at the point of origin of her trip. But if the surface is that of a sphere, as shown, she will find that the path does not close because of the curvature. Similarly, if Flatant makes a circle on the surface and measures its radius she will find that the law for a circle, circumference $= 2\pi \times$ radius, is not obeyed. The true radius of the circle is inside the sphere, a place she does not have access to. The radius she measures on the surface is longer than that radius. These of course are crude examples of the conditions existing in the two-dimensional world. It is not our world.

We may go back in our three-dimensional space and wonder about the possibility of a fourth dimension and its visibility in our world. Could we expect effects such as 'pseudo-shadows' of 'objects' in that space? These pseudo shadows would appear three-dimensional. We would simply consider them as objects of our space. However, their behaviour would certainly be strange. By means of experimentation we could probably find laws that those objects obey and we would make theories about them as the two-dimensional Flatant would make theories about the shadows of three-dimensional objects in her two-dimensional world. Are the laws of nature, that we have discovered over the centuries and that we simply assume to be true without understanding the reason of their existence, shadows of the existence of a higher dimensional world? Are the fundamental constants of nature, that we observe and measure, traces of a world that is not accessible directly to us? A promising theory, although far from being accepted because of the lack of experimental confirmation, string theory, leads to a world of ten or even eleven dimensions and appears to describe some aspects of our three-dimensional world. On the other hand, could our three-dimensional space be curved in a higher dimensional space in a fashion reminding us of the case shown in Figure 3.3 for two dimensions? In that case, if we keep going in the same direction will we not come back to our origin like Flatant in its two-dimensional space of Figure 3.3(a). We will come back later to this space curvature concept when we discuss the theory of general relativity.

It appears that because of our nature we are in a three-dimensional world and we have to live with it. However, we are sensitive to something else. We see that there are changes that take place and the changes take place in a certain direction. Basins filled with water empty through holes created in their bottom and water flows down rivers. We hear our heart beat: we count these beats. We sense a direction of flow of some immaterial substance that the material objects are subjected to: we call it *time*. Time appears to be something that exists by itself because, if nothing happens, it appears to be still flowing. Or is it really? If there were no objects around us, if there were no material objects in the Universe, would there be such a thing as time flow? We do not know its origin any more than the origin of space. When we look above at the Moon we observe that it drifts slowly to the east every day and that before the whole process repeats or accomplishes a full cycle, the Sun re-appears about 29 times. And that process repeats with a regular period of 29 days. We have introduced the word cycle, something that repeats itself, and period, the length of that cycle. If we suspend a ball at the end of a thread just above the surface of the Earth (we call that a pendulum) and if we set it in motion, we observe that it oscillates back and forth at a very definite rate determined by the length of the thread used for suspending it. If the length is about 25 cm, we have a full oscillation back and forth, a cycle, 86,400 times between two sunrises and we say that the period is '1 second'. We thus have what is called a chronometer: we measure time lapses or periods and we have defined the *second*. It appears that Galileo was the first known to have used that phenomenon scientifically to make measurements. He observed that the suspended light in the church in Piazza Miraculi in Pisa was oscillating back and forth with the same period, whatever its amplitude. He presumably used his heart beat to make the first observations. If we accumulate the seconds from a given origin, count them and keep track of the addition by means of an electronic counter or needles on a dial we have a clock. Clocks play a most important role in our life and in our technology. Time is important for our daily activities and the stability of clocks, that is to say, their ability to keep the same second duration period after period, is most important in the field of communication and navigation. In fact the development of clocks is a fascinating field and has led to the development of the most beautiful pieces of art such as monumental grandpa clocks as well as the most sophisticated piece of equipment such as the atomic clock. We will come back to this subject in a future chapter.

We thus have two concepts: time and space and we are sensitive to both. We see space in front of us and we see changes taking place in the objects filling that space. The concepts of space and time were certainly a first perception of human beings in their early life in this hostile world and this perception may have stayed at a basic level of observation without much understanding for a long period. We have made great progress since those early times. We have found logical relations between phenomena related to space and time. Physics is precisely the science that exploits those relations to the maximum. We observe a phenomenon and we observe that it repeats itself always in the same circumstances: we make a principle out of it. We then use logic to relate phenomena to each other and make predictions about possible events in totally different situations. Unfortunately we cannot express everything in words: our common language is not sufficient. We need another tool. As was mentioned above the language most adapted to describe the physical world is mathematics. With that language we end up with numbers that we can compare to measurements. I will try in this book to explain phenomena without mathematics, but the picture will be qualitative and thus incomplete. The curious reader is referred to the Annex if he wishes a quantitative description of some of the more elementary concepts developed.

How do We Alter the Speed of an Object?

We observe that an object in motion continues in motion if we do not stop it by external means. A bowling ball rolling on a highly polished bowling alley or the granite disc used on ice in the curling game are good examples. They slow down because of the minute obstacles to motion created by air and small particles on the plane they roll or slide on. These are called friction forces. They are not new forces added to the four forces mentioned earlier; they are essentially of an electromagnetic nature. They oppose motion. Objects in free space travel and never stop unless they hit obstacles. Planets are evident examples. The density of particles in space is so small that their effect on the trajectory of observable objects is not visible except close to Earth where the atmosphere is dense enough to oppose motion. Furthermore, objects travel in the same direction unless a force is applied to them. These observations are so basic that a principle was made out of them. It is the *principle of inertia* first introduced by Galileo. If an object is left alone, undisturbed, it simply keeps going in the same direction. Newton made it the *First Principle.* We have never found it wrong and it seems very logical. We accept it as evidence and we may say that the whole of physics is based on it. We also define a new concept that characterises that motion. We have introduced above the concept of length, the metre, and of time, the second. We use them to characterise the motion and if we find that the object moving in a straight line has travelled 1 metre in a period of 1 second we say that it moves at a *speed* of 1 metre per second and we call it **v**. We have written in words a first equation. Anybody can repeat the experiments on the same object in motion and verify that the results obtained agree with the previous measurement. If there are no forces applied to the object, the same speed is measured at later times along the path of the object.

This concept of free motion was not always accepted as such. In fact, common sense seems to tell us that when we apply a force to an object it gets in motion: one is induced to think that force creates speed as was thought before Galileo's time since Aristotle. But that is only partly true. When the force is applied, the object goes, in a given frame of reference, from a rest situation to one in which it has acquired speed v. Consequently, it has changed its speed from zero to a value 'v' after the force was applied. If one removes the force the object keeps going in the same direction at the same speed. It does not stop moving. Thus, force either increases or decreases the speed of an object. A push on the bowling ball accelerates it while friction in the alley reduces its speed or decelerates it. We now have the concept of acceleration and its opposite or negative, deceleration.[1]

Thus, if we want to increase the speed of the ball first at rest in our hand we simply have to apply a force to it. This is what we do when we throw the ball on the alley. It was at rest in our hand relative to the floor and we set it in motion. The bowl is accelerated until it is thrown and then keeps going almost at the same speed to the end of the alley where it hits the pins. What is the rule that is obeyed in such a case? It is found that a larger ball made of the same material requires a greater push from the hand, a greater force, to be accelerated to the same speed. It is observed that if the ball is twice as large, its initial speed will be half as much, if the same push or force is applied to it. A concept was attached to that phenomenon, that of *mass*. What is mass? At this point of the discussion we can say that it is that 'substance' that pervades material objects and opposes their acceleration. It is measured in *kilograms*. What is a kilogram?

Figure 3.4. Copy No 50 of the International standard kilogram, maintained under bell jars at the National Research Council, Ottawa, Canada. It is a cylinder of an alloy of platinum (90%) and iridium (10%).

It is the mass of a piece of platinum kept in a vault in Sèvres at the International Bureau of Weights and Measures (BIPM). It is the International standard kilogram by definition. A copy of that artefact is shown in Figure 3.4. It seems that it took a lot of imagination to go from such a concept to the definition of the kilogram as a particular block of metal. At first the definition of the kilogram was the mass of a litre of pure water whose density, which is mass divided by volume, was thus 1. The mass of the standard kilogram was made as close as possible to the mass of a litre of pure water. There are many other concepts that were studied in parallel and those studies created a context appropriate for the adoption of such a definition. We will come back to it. Mass is thus simply the parameter that we need, to connect the force we apply to the resulting acceleration. We thus define mass as the substance that opposes acceleration. When in motion, mass does not affect in any way the speed of the object. Knowing that by definition the kilogram is that object, we can define

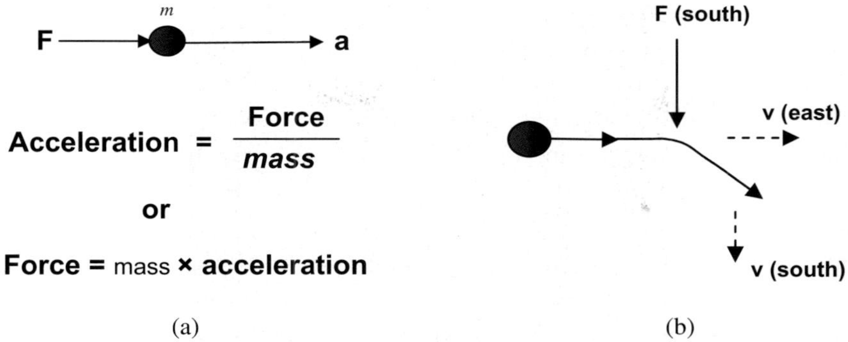

Figure 3.5. A principle of the functioning of the Universe. (a) A particle of mass m is accelerated from speed zero to speed **v** by the applied force **F** causing acceleration **a**. (b) A particle moving to the east sees its direction of motion changed by a force applied to the south. It conserves its speed to the east but acquires speed to the south.

a unit of force as that which gives a mass of 1 kilogram acceleration of 1 metre per squared second. We call that force a Newton, obviously a very wise choice of word. We have written in words one of the most important law of physics. It is the law that says how speed is altered by an applied force and it introduces mass as the substance that opposes changes in motion as illustrated in Figure 3.5.

This principle or law applies to forces in all directions. For example, an object moving to the east at a certain speed **v** may have its direction of motion altered to a motion to the south by an applied force as shown in Figure 3.5(b). The object speed to the east is not altered by the force since it is at a right angle to that motion but it acquires speed to the south by the action from the applied force. This is acceleration to the south. In a car, the engine is used to accelerate the vehicle. It is kept moving against the forces of friction of the tires on the road, against the friction of the bearings within the car, against the friction of air on the structure, by means of the continuous force applied by the engine. In a curving road the direction of the car is changed and the car is accelerated. We feel that acceleration by being pushed towards the door opposite to the new direction taken by the car as if our body wanted to keep going in straight line. The force required to change the car direction is applied through the steering wheel by the driver with the help of various compressors that reduce the demand on the driver's strength. Consequently, the law is direction sensitive: when we have a parameter that depends on direction or point in a given direction we call it a *vector*. Speed is a vector. Acceleration is a vector. Force is a vector. In drawings, that property may be made explicit by adding an arrow on top of the character representing those parameters or by setting them as bold characters. It is interesting to note that the military sometimes call their missiles vectors, since they point in a given direction. In the naming they were probably feeling nostalgic about earlier physics courses. Mass is called a *scalar* quantity since there is no direction associated with it and, whatever the direction of the force applied to it, the result is the same, the object is accelerated in the direction of the force. The equation of Figure 3.5 written in words, may be written as $\mathbf{F} = m\mathbf{a}$, where acceleration $\mathbf{a}$ and force $\mathbf{F}$ are written in bold characters, and the scalar 'm' as a normal character. This is one of the most basic laws and it is called the *Second Principle.* It is deduced from observations and is thus experimental. It is not derived from an abstract reasoning or logical arguments sometimes carrying evidence in themselves. It is directly connected to the law of inertia and it says that if no force is applied, the motion is constant with time.[2]

But what is mass really? In the present context, it is called inertial mass and it appears to be simply the substance that opposes changes in motion. We will see later on, that electrons, carriers of electric charge, have a mass, although very small; nucleons, protons and neutrons forming the nucleus of atoms have a much larger mass, about 1,800 times larger than that of the electrons surrounding them and, since matter around us is made of atoms, it is those nucleons that give mass to objects of our three-dimensional world. Light can be interpreted either as a wave phenomenon or a corpuscular phenomenon, the photon. The photon has no mass of the nature we have introduced above. The equation we have written in Figure 3.5(a) does not apply to it. Interestingly enough the photon speed cannot be altered in free space as will be discussed below. There is no way we can grab the photon. It does not have handles. The photon in a vacuum travels at the speed of light and the speed of light is a constant of nature.

In introducing the second principle, we have elevated a day-to-day question to a higher level, one where we introduced a concept that appears to answer the original question but has introduced another concept, mass, whose origin we cannot determine for the moment.

So again, what is mass? Is it a pseudo-shadow of a higher dimensional world, a world we do not have access to? Is mass the effect of the whole Universe inhibiting the object to alter its equilibrium position, through a force such as gravity? We will see that the theory of relativity developed by Einstein provides a partial answer to that question but again displaces the problem to a higher level.

The concept of force introduced previously appears to be somewhat abstract. Where does it originate from? It originates from other objects or particles. In the bowling alley example, the force is transmitted from our hand and other objects on the alley; all the action takes place at the atomic level, obeying some basic principle relative of atomic physics that we will examine in a later chapter. In the example chosen, the force is of electromagnetic origin. This, however, takes us to another principle regarding reaction of objects submitted to forces. The concept is easier to understand by examining the motion of billiard balls. Two balls of the same mass going at the same speed and colliding head on take off in precise directions, backward on their respective path. Each applies a force on the other. We may take ball 1 as acting and ball 2 as reacting. But we may take the opposite, ball 2 acting on ball 1, which reacts to the force. It is readily concluded that the forces acting on each ball by the other are equal. Upon collision they accelerate in reverse direction and go off at the same speed, but opposite to their original direction. There is an equal action and reaction. In a static case we may think for example of a person pushing on a wall. The person applies a force with his hand on a wall, but the wall applies an equal force on the hand of the person. These two simple examples lead to Newton's *Third Principle*: to any force applied by an object A on another object B there is an equal force of reaction on object A by object B. This is why there is recoil in firing a rifle. This is how rockets and jet airplanes operate. They send high speed gases through the exhaust and the airplane or rocket reacts by accelerating in the other direction. In the case of a rifle, the bullet acts on the rifle and pushes it backward on the shoulder of the sniper: it is only because of the difference in mass that the bullet goes at thousands of km per hour while the rifle recoils at a much lower speed. It is that third principle that makes possible the functioning of these machines.

What is so Special about the Speed of Light?

At the end of the 19th century a theory was developed describing how electricity and magnetism, thought to have different origins, were in fact phenomena essentially originating from the same fundamental properties of particles containing electric charges. That theory, called 'Electromagnetism', developed by the Scottish physicist Maxwell on the basis of several experimental phenomena and laws known at that time, led to a description in which all electrical and magnetic phenomena were connected together in an elegant formalism.[3] That process is called *unification* and we will see that the approach plays a central role in modern physical theories. The theory led to the concept of so-called electromagnetic waves propagating through space, waves formed of perpendicular oscillating electric and magnetic fields. The theory showed clearly that the waves were propagating at a well determined speed, called 'c', given exactly by a combination of two fundamental constants, one related to electrical phenomena, the other to magnetic phenomena. In that context, light was thus interpreted as electromagnetic radiation travelling through space at speed c. At the time Maxwell developed his theory, the speed of light had already been measured with some accuracy and the predictions of the theory were found to be in agreement with measurements. This is another aspect of the validity of a theory: it has to predict new things and be in agreement with observations. The theory was in agreement with so many observations that it was soon accepted. The speed c is just fantastic, about 300,000 km per second. You turn on a flashlight for a second and you create a light pulse of one second, consisting of a stack of electromagnetic radiation that propagates through space at speed c. If it does not meet obstacles that either absorb it or reflect it in another direction, that pulse of light will go on to the confines of space for billions of years! Stars emit radiation in all directions. Radio stations, television transmitters, our cellular phones, all emit radiation in space in all directions. It is only their wavelength or frequency that is different. Space is thus filled with electromagnetic waves and we are exposed to them. It is just amazing that in all that electromagnetic turbulence we can still detect clearly the voice of our partner on the cheap but sophisticated cellular phone that the telecommunication company gave us.[4]

At the time Maxwell's theory was developed, the concept that light had a wave nature was accepted as a fact because of all the interference experiments that were done with light and that were explained completely on the basis that light was a wave. However, it appeared difficult to understand that such electromagnetic waves could travel in space without a support. After all, waves on a lake or on a pond need water to support them. Sound waves propagate as perturbations of the pressure or density of air formed of nitrogen and oxygen molecules. In a vacuum no sound propagates. A mysterious medium, the so-called 'aether', was thus imagined as the support of electromagnetic waves. No one knew what it was. It was a medium occupying all of space like a canvas, but three-dimensional! It had no effect on the motion of the celestial bodies, and was totally transparent to the motion of the Earth, the Moon, the Sun, and the stars. As will be made explicit below, that aether was fulfilling the requirement of the presence of an absolute frame to which the motion of all objects could be referred to. This question of a stationary reference frame had been haunting scientists since Newton and Galileo. The question was about the free motion we introduced previously. The speed v of the ball on the bowling alley is relative to what? The alley and the player of course. But for an object moving in space, how do we determine its speed when we are ourselves moving at great speed with planet Earth orbiting the Sun, for example. This aether would provide a stationary reference frame filling all space in which light would propagate at speed c. Consequently, you could compare your own speed relative to the speed of light and determine in an absolute way if you are moving or not. The problem remained of doing such a comparison and thus proving the aether existence by experiments and observations. All the effects observed for the propagation of sound in air were well known and should have their equivalence in the case of propagation of electromagnetic radiation in the aether. But no one could measure such effects due to the speed involved with light as compared to the speed of sound that propagates in air at about 340 metres per second. In practice, in the case of sound, it was possible to implement experiments with objects that achieved speeds that were a good fraction of its speed, but not in the case of light. What was needed was the possibility of having a space vehicle moving at a velocity that was a good fraction of the speed of light to produce effects observable with the techniques available at the time. The Earth that travels at great velocity through space due to its revolution around the Sun, 30 km per second, appeared to be the ideal vehicle. It was expected that the Earth, even if its motion were complicated because it is attached to the Sun moving itself at great speed, would move at least at 30 km per second some time during the day or still some time during the year. Many experiments were done at the end of the 19th century and due to their complexity and sometime ambiguous results, a lot of confusion resulted about the existence of the aether. It is only with the advent of a clean experiment, thought of by Michelson, that the issue was finally resolved.

Michelson, and Morley, designed an instrument, a so-called interferometer, that now bears their name. It could measure exactly the effect of the motion of the Earth relative to that aether medium.[5] An interferometer is an instrument that measures the property of light through interferences of two beams that are forced to propagate along different paths within the device and recombine at a given point. The Michelson-Morley interferometer was made of two arms, 11 m long, perpendicular to each other with light originating from the same source and forced to travel back and forth in the two arms by means of mirrors. Since the arms were at a right angle, one could be oriented along the direction of motion of the Earth in space and the other one at right angle was essentially a reference. Since light would travel in that arm at right angle to the motion of the Earth it should not be affected in first order by the motion. The theory of operation of that interferometer in the context of drift within the aether and the effect of its velocity on the shape of the observed interference pattern is based on purely geometrical arguments, like Pythagoras' theorem, and involves simple algebra. It did not seem that basic mistakes could be done in the interpretation of the results. After tedious work and tremendous efforts to construct a device that would avoid any systematic errors, precise measurements were done with the interferometer oriented in various directions over several years. The results of the experiments were null: there was no visible effects on the observed interference fringes that could be associated with the motion of the Earth in the aether whatever its direction of motion. Actually the result expected was a fractional fringe shift of 0.4 while the observed shift was less than 0.01. The experiment was repeated with increased precision numerous times afterwards. The result was always negative.

That result was not compatible with the presence of an aether medium supporting the wave. There were several attempts to explain the observation, such as that advanced by H.A. Lorentz that it was possible that the arm of the interferometer in the direction of motion was shortened. The effect could be due to forces on the molecules of the arm presumably interacting with the aether wind. Other explanations were attempted, but as pointed out by Feynman,[6] the hypothesis used was created specifically to explain the phenomenon observed and seemed artificial. Something based on principles of a more fundamental nature was required to convince the community. In fact, in all attempts there seemed to be a conspiracy in nature that prevented the experimenter from measuring the speed of the Earth in the aether and, as pointed out by Poincaré, it seemed better to accept the 'conspiracy' as a law of nature. The aether idea was abandoned. As Joos said, 'Experiment has thus decided against our acoustic analogy and against the existence of a stationary medium carrying light; i.e. the existence of a cosmic aether or absolute space is disproved'.[7] As will be seen below this finding had great consequences for the understanding of the basic laws of physics.

On the other hand, development of physics in the first part of the 20th century took another crucial step. Observations that had been made in the laboratory during the previous century led to the conclusion that matter was made of small discrete particles, atoms, obeying laws that were radically different from those obeyed by the macroscopic world. Furthermore, electromagnetic radiation was not behaving as expected from the theoretical understanding of the time regarding radiation emitted by incandescent bodies, for example. This radiation is known as black body radiation. It was not possible to explain the radiation spectrum emitted by such objects at a given temperature with the classical laws known at that time. The problem was solved by Planck who made the assumption that the radiation field could take up energies only in small quantities, hν at a time. Up to that time, the evidence carried by experimental results was leading to the interpretation of the nature of radiation as a wave phenomenon whose energy could take any continuous value. The success of Maxwell's theory was re-enforcing that conviction. It is in that context that Einstein, extending Planck's assumption, made a radical step in assuming that the electromagnetic field could also be interpreted as a collection of particles. He used that new hypothesis to explain totally a phenomenon, called photoelectric effect, that is observed when light is incident on a piece of material such as a metal. In such experiments it is found that electrons are knocked off the piece of metal only if the incident radiation has a wavelength shorter than a critical value. Furthermore, if the radiation wavelength is longer than that value, no electron can be knocked off whatever the power of the radiation incident on the metal. The effect could only be explained if light was made of particles having a precise energy inversely proportional to its wavelength. The concept of light existing as a particle, later to be called a photon, was thus resurrected from the early ideas developed by Newton. It was now reborn on solid grounds and the dilemma of the wave-corpuscle nature of light was going to haunt the physics community up to this day and probably forever. Observers may use either of the approaches, wave or particle, to analyse observations. It depends on the particular experimental arrangement. For example a photodetector appears to detect light as photons. A single photon falling on a photodetector produces a clear 'click' in the amplification system to which the photodetector is connected. However, if light is made to pass through two slits separated by a short distance comparable to the light wavelength, interferences are observed on a screen behind the slits. The whole pattern observed can be explained with a simple calculation assuming that the light is a wave passing equally well through the two slits. It is similar to interferences observed with surface waves on water facing obstacles and small openings. Thus, light appears to have two facets, depending on the way we detect its presence: it may react either as a wave or a corpuscle, that is to say a photon.

But what are the main characteristics of the photon? One of them is its energy, which is just a measure of the amount of work it can do when in interaction with matter. This energy is given by the product of the frequency nu ('ν') of the radiation it represents by the so-called Planck constant 'h'.[8] In the International System of Units and in scientific notation, h is equal to 0.6626×10^{-33} joule second, a small number indeed. On the other hand, when we deal with visible light the frequency of the radiation is of the order of 500 thousand billion cycles per second or 500×10^{12} Hertz.[9] It is clear that the product of this number by the frequency, although that frequency may be very large, does not produce a large amount of energy. To produce 100 watts of power, like an incandescent light bulb, the number of required photons emitted per second is 3×10^{20} photons or 300 billion billion photons. The photon is thus a very tiny amount of energy. It is the large number of them that creates visible macroscopic effects. We have mentioned above that the speed of light is finite, but rather large, about 300,000 km per second. This, of course, applies to photons as well. Photons travel through space at the speed of light. This speed is so large that when we turn on a source, such as a flashlight, the light ray looks like if it appears instantaneously. The same is true of all radiation whatever the wavelength. Red light propagates at the same speed as blue light although red light has a wavelength twice as large as blue light. The same is true of waves emitted by a radio station that have much longer wavelengths. Visible radiation has a wavelength of the order of 0.5 millionth of a metre while radio waves have a wavelength in the metre range. Gamma radiation, originating from atomic nuclei, consists of electromagnetic waves having a much shorter wavelength than visible radiation and propagates at the same speed.

Although the speed of light is enormous, it is possible to sense directly its finite value during long distance telephone conversations when the signals are transmitted through communication satellites from one continent to another. These satellites are placed in so-called geosynchronous orbits, which are at a distance from the Earth's surface such that their velocity maintains them over a chosen location over the equator. This distance is of the order of 35,800 km and in the trip to and from the satellite the signals carried by the wave may be delayed by a certain amount depending on the location of the transmitter and receiver, particularly in the case of multihops. This small delay may affect communication in some circumstances, like having to wait for the partner's answer. The situation is not encountered when communication is accomplished through underwater cables since in that case the distance travelled by the radiation is only a few thousand km and the radiation originating from one continent takes only milliseconds to reach another continent. The delay, however, is rather important in space communication, for example, in the case of those robots sent to planet Mars at a distance of the order of 100 million km, depending on the relative position of the two planets, Earth and Mars, in their orbit around the Sun.

Are there other consequences caused by the finite speed of light? As will be seen in the following pages, the effect of its finite size and the absence of the aether have caused a radical change in our understanding of the working of the Universe. There are observations sometimes in physics that lead to theories that totally change our understanding of the rules governing the dynamics of the Universe. Such are observations made at the atomic level that have given rise to quantum mechanics. The finite speed of light and the conclusions reached from the Michelson-Morley results, that the speed of light is the same in all directions as measured on Earth, whatever the motion of the source or observer, are also observations that caused a radical change in our understanding of the concepts of space and time.

Who is Moving?

One simple question comes to mind: if you sit on the photon that propagates through space what do you see? How do you perceive the physical world around yourself when you travel at that speed? First, is it possible? It is said that this is the kind of question that Einstein was reflecting upon. To approach to an answer we have to go first to very basic concepts relative to motion and time. The first principle that we elaborated relative to the motion of a particle says that if no force is applied to the particle there is no acceleration of the particle and it keeps moving at the same speed in the same direction. But as was mentioned above it is legitimate to ask: the speed or motion is relative to what? We have all experienced the sensation of being at rest in a train, or another vehicle, and then looking outside at the station platform, realising that we are moving. If you are in the moving wagon and you do not look outside, you do not perceive the motion if it is smooth. Furthermore, if you look outside and see another wagon passing by, you immediately conclude that the other wagon is moving. There is no way you can tell within your world, which is the space within your wagon, if you are moving or if the other wagon is moving. How can that be? Nature is made that way. Objects in motion keep moving at the same speed and in the same direction if no force is applied to them. They act as if they were at rest. In your wagon you cannot do experiments that will tell you that you are moving. If I am in the other wagon, I am exactly in the same situation. If I drop a ball in my wagon, moving at a constant speed, the ball will fall directly on my feet under the point I let it go. The ball does not seem to know that the wagon is moving. If you are in another wagon moving relative to me and you drop a ball, you will make the same observation. However, if you look at the ball falling in my wagon you will see it falling somewhat diagonally relative to an arbitrary vertical direction. You may conclude that I am moving, but I can make the same reasoning and conclude that you are the one moving in the other direction. What does that show? The wagons form individually small Universes where the mechanical laws of physics apply as if the wagons were at rest or stationary. Motion is relative. This was the conclusion reached by Galileo and Newton: on a calm sea a sailor cannot know if he is moving or at rest by making experiments inside his ship. If a ship passes by he does not know either if his ship or the other ship is moving.

The above consideration applies to the mechanical aspects of the law of physics. Motion of objects is described in the vessel or wagon moving at constant speed as if the vessel or wagon were at rest. This question of being at rest is crucial. We often use that concept when we describe the motion of an object. We compare its movement to some point in space that we assume for convenience to be the rest point. For example a person on a bench at the railway station may be used as the rest point when talking about the wagon moving on the track at speed 'u'. The space surrounding the person on the bench is considered as attached to a reference frame at rest, while the space within the wagon can be said to be attached to a reference frame moving with the wagon at speed u. Is there, however, somewhere in the Universe a point of 'no motion', a true rest point to which motion could be referenced? For example are the stars fixed in space? Of course not, they move relative to each other and relative to us. If there were no objects in the Universe, no planets, no stars, no galaxies, and you were in a rocket could you tell if you are moving or not? No. If your friend is also on a rocket and passes by, what can you say about him? Is he moving? He will ask the same question about you. The aether filling all space, proposed at the end of the 19th century, appeared to provide a reference frame for establishing absolute motion.[10] In such a frame, he could compare his speed relative to the speed of light carried on the stationary aether canvas and determine if he is moving or not. However, as mentioned above, the experiment of Michelson and Morley showed that there was no aether or absolute reference frame. The speed of the Earth could not be determined by comparing it to the speed of light. Electromagnetic radiation propagates in free space at speed c. In my moving wagon, if I throw a ball in front of me at speed v along the direction of motion of the wagon, I will see it moving at speed v relative to the wagon and myself. However, if you sit on a bench at the railroad station you will see it going forward at speed v plus the speed of the wagon u relative to you. Thus, if I throw the ball in front of me at a speed of 1 metre per second and the wagon moves at a speed of 10 metres per second, then you, on the bench at the station, will see the ball moving at 11 metres per second in your own reference frame, which is attached to the bench you are sitting on.[11]

This is a simple transformation, called Galilean, from one frame of reference moving at constant velocity u (me in the wagon) to one at rest (you at the station). This applies to all mechanical laws. We will see, however, that the simple addition of speeds that we just made is an approximation that is valid only at low velocities. We will have a problem with it when we introduce other concepts to be developed relative to the speed of light. This may seem strange at this time, but just wait for the introduction of the laws of electromagnetism and relativity.

Thus, in a wagon moving at constant speed, all the mechanical laws of physics act as if the wagon was stationary. It is legitimate, however, to ask what happens if instead of a ball I turn on a flashlight in the direction of motion of the wagon. In doing that, I am sending a bunch of photons forward. Since we said that there was no aether, there is no specific frame of reference for light; the flashlight does not know it is in a wagon moving at speed u. Thus the photons must travel at speed c relative to the wagon. Now, assume again that you sit on the station platform and look at the beam of light going forward with the moving wagon. At what speed does the beam propagate? Assume that the red ball in Figure 3.6 is one of the photons emitted by the flashlight. Does it propagate at speed $(c + u)$, that is to say, the speed of light measured in the wagon plus the speed of the wagon relative to you? The experiment of Michelson and Morley says: no. It must still be c because the experimental apparatus in their experiment was moving with the earth at fairly large speed and no effect in connection with the propagation of light was detected.

If you measure from the station platform the speed of the photons emitted by the flashlight in the wagon your result is still c. This is a strange rule of addition. This is a most fundamental question and it took a great deal of thinking from scientists like Poincaré, Lorentz, Minkowski and Einstein to create the framework for the understanding of phenomena connected to the propagation

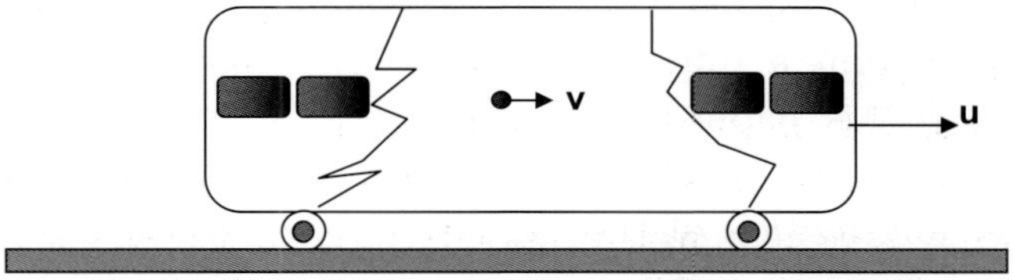

Figure 3.6. Relative motion in a moving wagon.

of electromagnetic radiation. It was found that if one tried to do a transformation from one frame of reference at rest to one in movement according to the rules set by Galileo, the laws of electricity and magnetism were not the same in the rest frame and in the moving frame. That difference seemed to provide a means of identifying who was moving by making measurements in the moving frame and this was opposite to the Galilean principle of relativity for the laws of mechanics. Consequently, a transformation that appeared to provide means of identifying motion by measurements on effects within the same reference frame did not seem to be correct. In a most elegant approach, Lorentz showed that to be the same in a moving reference frame and in the rest frame the laws of electromagnetism, unified in Maxwell's equations, needed to be transformed according to certain rules different from the Galilean rules of simple addition. These rules, strangely enough, involved the speed of light.[12] As a matter of fact, the speed c appears directly in Maxwell's equations, without a direct connection to a particular reference frame, and looks very much like an inherent property of electromagnetic radiation and space. In Lorentz transformations, the flow of time is altered as well as dimension. For example, a rod moving at great speed along the direction of its length appears to be shortened from the point of view of an observer at rest, and the shortening is the exact amount required in the Michelson-Morley experiment to have a null result. It is called the Lorentz contraction. This, of course, is a more fundamental explanation than the ad hoc reduction in length assumed to be due to interaction between the rod and the aether substance as proposed earlier. Those so-called Lorentz transformations apply evidently to all mechanical laws. However, at velocities low compared to that of light one can use, for mechanical laws, the Galilean transformation which becomes an approximation of the Lorentz transformation. Independently, Einstein addressed the question from a different point of view relying on two basic postulates. In his paper,[13] Einstein starts by making a comment on the basis of previous observations: '... the unsuccessful attempts to discover any motion of the Earth relatively to the light medium suggest that the phenomenon of electrodynamics as well as mechanics possess no properties corresponding to the idea of absolute rest'.

From there he assumes that 'the same laws of electrodynamics and optics will be valid for all frames of reference for which the equations of mechanics hold good'. Einstein raises this statement to the level of a postulate and calls it the *Principle of Relativity.* This is an extension of the Galileo principle of relativity for mechanical bodies to the laws of electromagnetism. It just says that the laws of electromagnetism have nothing special about them: they are equally valid and are the same in frames moving relative to each other.

The second postulate reads as follows: 'Light is always propagated in empty space with a definite velocity c, which is independent of the state of emission of the emitting body'. Consequently, photons emitted by a source in motion travel at the speed of light relative to the source. An observer attached to that light source sees light travelling in front of him at the speed of light. Since there is no privileged position for the observer, light appears to travel towards him or away from him at speed c whatever his position or speed.

These were bold postulates, but if experimental results agree with their predictions they must have a physical basis and they must be accepted as such. The results derived by Einstein from these two postulates are most amazing. Events that appear simultaneous in a frame of reference do not appear simultaneous in a frame in motion relative to the first frame. Time flows more slowly in a reference frame that is moving relative to one we consider stationary. An object appears to contract in the direction of its motion when observed from a stationary frame. This is the Lorentz contraction mentioned above when a reference frame transformation is made. Mass has a value that is a function of relative speed. In Einstein's and Minkowski's view, time becomes a new axis added to space: we live in a four dimensional 'space-time world'.[14]

This is a radical change from the kind of absolute space and absolute time that was expected to exist previous to Einstein's considerations. To illustrate these results we will imagine some fictitious situations that could very well represent real situations if we could achieve in daily activities speeds that become a good fraction of the speed of light. The results obtained are in terms of ratio of the velocity we can achieve to that of light. Even with super-fast rockets achieving velocities of 40,000 km per hour, this corresponds to a speed of only 11 km per second and it is readily realized that this speed is a very small fraction of the speed of light, which we recall is 300,000 km per second. Only small effects are to be observed even at that speed. Consequently the effects to be described are not visible in occupations that fill our days. Nevertheless they are clearly observed with atomic particles like electrons and nuclei and they are important in the implementation of navigation systems based on satellites. They are essential for the understanding of the functioning of the Universe and it is of the utmost importance to examine them in detail. It should be mentioned that in some cases it is possible to accumulate the effects over a certain period of time and with the instruments that are available nowadays we can make measurements that are precise enough to verify most of the results predicted.

Joe and Mary are Confused about Time Flow

To illustrate the effects of relative motion on time and space we will tell short stories about two young friends Joe and Mary, university graduate students in biology and medical science respectively. They will be our experimenters, they will go through various situations and will observe surprising phenomena along the lines of those described above.

Joe and Mary's university is on the east coast. Joe is a graduate student in plant biology and works on a research project that could be important for long space travel. He has a contract with his thesis advisor. It concerns the speed at which plants grow under certain special conditions. He needs to visit another laboratory on the west coast to get information. Since all expenses are paid he uses the fast train service, first class. However, he has a problem. One of his projects on a very special plant is at a turning point. He cannot leave the plant for a long period because the system he has implemented to water it is fragile and needs continuous monitoring. The train station manager tells him that he can take with him his flower and his watering contraption because they have an experimental wagon whose function is simply to accommodate crackpots like him. Furthermore, there is a whole section of the wagon that is mostly glass and, consequently, his plant will be exposed to sunlight most of the day. They will reserve a corner of the wagon for him for a decent price. Joe accepts and decides to bring with him the plant and the watering basin with a small adjustable tap on one side to water his plant while he would be working on his thesis in his own cabin, sleeping, or having his meal in the saloon. Joe is so concerned with his project that he had acquired from his fellow physicist Bob a very precise clock to time the watering rate of his system. He did not want things to overflow, or worse, to run out of water. He brings the whole system, clock, flower and watering basin in the experimental wagon.

Joe's clock works as follows.[15] It is made of two perfect mirrors that reflect light completely without any losses. The mirrors are set as in Figure 3.7 separated by a rigid metal rod of length L. The clock uses a single photon that is trapped between the mirrors. The clock would also work with a pulse of light. The photon or pulse travelling at the speed of light c bounces back and forth between the two mirrors, and Joe, as time passes, can count the number of times the photon or pulse hits the top mirror making a full cycle. He uses this cycling process as a clock pendulum with the proper gears attached to it. He starts the basin watering system and very confident, he checks with his clock that the watering rate that he just set is the same as the one he had calibrated at home to be correct. In Joe's arrangement, the rod holding the mirrors apart is 0.6 m long (L). The photon, travelling at the speed of light, which is 300,000 km per second, takes about 2 billionth of a second (2 nanoseconds or 2 ns) to travel from one mirror to the other thus making 250 million full cycles in one of our seconds. It is indeed a rather fine chronometer

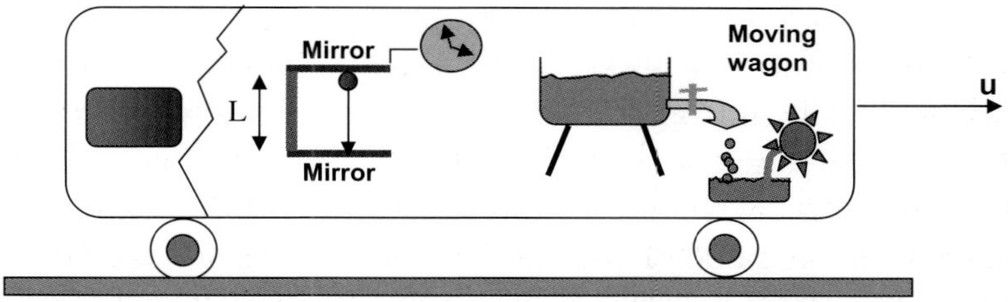

Figure 3.7. Joe's arrangement in the experimental wagon.

and Joe was able to make measurements at home on the emptying of the basin with an extremely high resolution. When the train sets in motion Joe does not see anything happening in his wagon. Everything stays the same. He has no way of knowing he is moving and it could be at a great speed even approaching the speed of light. A real TGV! Everything for him works the normal way; the photon travels between the two mirrors at its speed of 300,000 km per second and his chronometer works the same as it did at home. Everything seems to be right and he feels confident that his flower is secure. He can leave the whole contraption alone for several days and goes to his studies while the train travels across the country.

Now his girlfriend Mary has profited from his absence by visiting her parents in another town. She is aware of what Joe is doing. There were leaks. Joe had obtained for her another clock from his friend, which he had given her as a gift. The clock is identical to the one Joe has in his experimental wagon. She knows that Joe's train passes through her parents' town and she goes to the station, sits on a bench at the railroad platform waiting to wave. He arrives. Waving at him passing by, she notices right away that there is something strange about Joe's behaviour. He looks sleepy, turns his head and waves his hand rather slowly. She concludes that it must be the wine. However, she has a glance at the experimental wagon. What does she see? The TGV passes by at a fantastic speed with the clock in the wagon. She knew that she would not see much due to the speed of the train, and consequently had prepared herself with a fast movie camera to show to Joe that she was there and had seen him. She takes a movie while the train passes by. She goes home, looks at the movie and observes that during the rapid transit of the train she did actually capture Joe's clock in full action. She analyses the pictures one by one and reconstitutes the scene by following the path of the photon. She observes that the photon follows a path as in Figure 3.8.[16] The mirrors in Joe's wagon are moving with the wagon along the track at speed **u** and the photon appears to make a zigzag path to follow them. In Joe's experimental wagon the photon never misses the mirrors and always hits them at the same spot. This is what Joe saw when he checked his setup. This is also what happens in Mary's pictures. The photon never misses the mirror and hits them

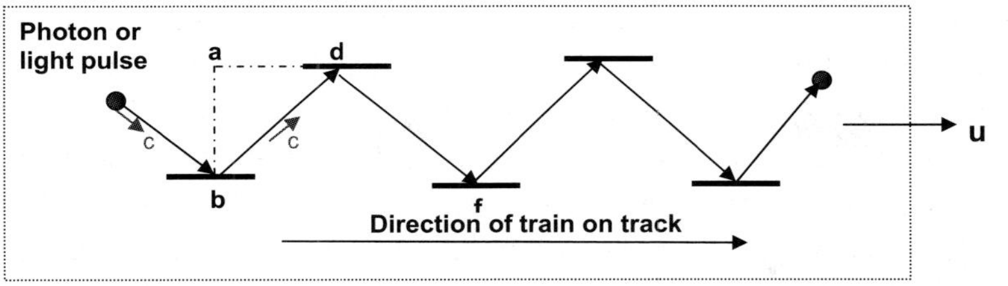

Figure 3.8. Fictitious movie made by Mary of Joe's photon clock.

in the centre. Mary immediately sees a problem because the photon has to travel a longer path and, according to a basic principle she is aware of, the speed of light is the same throughout space unaltered by movement of the source or the observer. It remains the same in the travelling wagon as it is at the railroad station. Mary makes the following calculation. She evaluates the distance travelled by the mirror centre while the photon goes from b to d in the figure at speed **c**. The distance from a to b is L, and during the time of travel of the photon the mirror has moved the distance a to d, a distance she can evaluate from the speed of the wagon and the time of travel of the photon. Remembering Pythagoras' theorem, she calculates the extra distance that the photon has to travel from b to c to reach again the centre of the mirror. She concludes that the photon takes more time than it would take to travel from a to b if the clock was not moving. Mary finds obviously that a full cycle or the period of the clock in Joe's experimental wagon is longer that the period of her own clock.[17] She calls Joe on his cell phone and tells him that according to her calculation, his clock is running slow, that he better check his basin of water because he may not have calibrated things properly when he was at rest at home, that his basin may empty soon and his flower may run out of water. Joe runs to the experimental wagon and check things out. Everything is normal, his clock seems to be running well, the water basin level is as expected by the calibration he had made before he left, and the plant is growing at the rate he had measured at home.

Joe does not believe Mary's calculation about the slowing down of his clock as told to him on the phone and later comes back to her after his trip is completed. They compare their chronometers and he finds that his photon chronometer is not losing time and has exactly the same rate as that of Mary's photon chronometer. They are more puzzled than ever. They decide to do an experiment. They synchronize the chronometers so that they give exactly the same time. Both become a clock indicating exactly the same time. Joe decides to go on another trip to verify the results claimed by Mary. If his clock were running slow in the moving wagon it should lose time, a thing he could not check before. They were not able to synchronize their clock since they were not at the same place when Joe started his trip. Now starting at the same station and coming back to the same station they can compare the clocks before and after the trip. Joe goes on his trip and comes back. He is very puzzled: he finds that his clock is late relative to Mary's clock. She was right. They repeat the experiment now with Joe staying at the station and Mary going on a trip, to make sure it is not a peculiarity of Joe's clock. They now find that it is Mary's clock that is late.

This is an astonishing result. A clock in motion loses time when it is compared to one at rest. But the word 'rest' is troubling, because motion is relative. Joe reasons that in the train at constant speed he could have said that Mary on the railway platform was the one moving in the other direction. If he had taken a movie of Mary's clock he would have found also that Mary's clock was running slow. His reasoning is right. Two clocks side by side travelling in the same direction at the same speed will indicate the same time at the end of the trip but two clocks travelling in the opposite direction will never meet again and there is no way to compare them. However, Einstein made a bold statement regarding the slowing down of a moving clock.[18] 'The results hold good if the clock moves from A to B in any polygonal line and also when points A and B coincide'. This path is shown in Figure 3.9. The results just presented are generally given the popular name 'twin paradox' and still raise questions today. The main question is in connection with the relativity of motion. When Joe was travelling he could have said that it was Mary who was moving the other way. So both Joe's and Mary's clock would be delayed and in joining them they would be reading the same time.

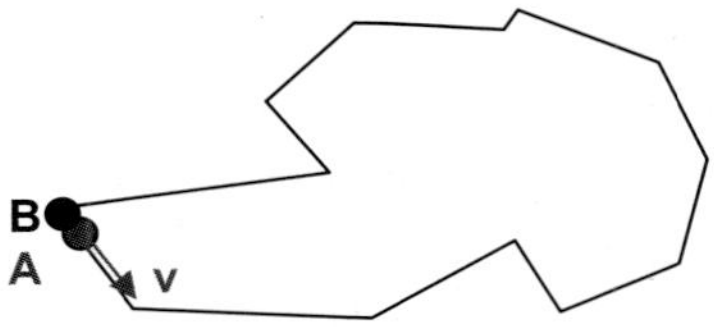

Figure 3.9. If clocks are synchronized at A and B in the same rest frame and clock at A goes on a trip, it will lag behind the clock which has remained at B. '... this result holds good if the clock moves from A to B in any polygonal line, and also when the points A and B coincide' (adapted from Einstein, *op.cit.*).

However, if the clocks join after the trip there is no ambiguity in identifying the partner that travelled. For example in the illustration of Figure 3.9, A has to come back. For doing that, A has to stop or be accelerated somehow for turning around. A will feel it as Joe would feel it in his train when either it stops or turns around. This is the idea behind distinguishing the clocks. The clock that comes back will experience some acceleration at one point in its course and that can be perceived even though that acceleration is small. Assume that there are three clocks available and that two of them are sent on a long trip in opposite directions at the same speed and return to their point of origin at the same time. They will indicate the same time but will be late relative to the clock that stayed at the point of departure as a witness. The whole arrangement of three clocks could have flown through space at high constant velocity. The results would be the same.

There may be some objection raised concerning the effect of acceleration on a clock. Since this is the way we differentiate between the clock at rest and the one travelling, it is appropriate to raise questions about this effect. In the equations of special relativity there is no explicit dependence on the acceleration but only on instantaneous velocity. As mentioned by Alley this is sometimes referred to as the 'the clock hypothesis'.[19] However, special relativity deals with uniform motion and it is not surprising for acceleration to be absent from the final result. We will see below that, indeed, acceleration has an effect on clock rate. In fact we will introduce a principle that points to the equivalence of gravity and acceleration and we will find also that gravity has an effect on clocks. Nevertheless, it is sufficient for the moment to assume that the effect is small and that it is possible to imagine situations where Joe changes directions at a very slow rate to come back towards Mary.[20]

There are many real phenomena that confirm the description we have just made using the fictitious travels of our friends Mary and Joe. For example some elementary particles have an ephemeral life and disintegrate after a known time. It is found, for example, that a muon, a heavy electron we will look at later, sometimes created by cosmic rays in interaction with the nuclei of molecules in the upper atmosphere, may reach the surface of the earth even though its lifetime at rest, as measured in the laboratory, is $\sim$2 microseconds. Even if it moved at the speed of light, it could travel only 200 metres during that time. The distance travelled is of the order of tens of km and consequently in our frame of reference its lifetime must be greatly increased. In the story told above about Mary and Joe, the time delay observed would be extremely small because of the slow speed of the train relative to the speed of light. However, a direct experiment can be done with real clocks in aircraft travelling for a sufficient time to accumulate measurable time delay. Such an experiment was done in the 1970s by J.C. Hafele and R.E. Keating[21] using four Caesium atomic clocks that can keep time very accurately over long periods. We will describe such atomic clocks in a later chapter. Those atomic clocks were flown on commercial jet aircraft around the globe twice, once eastward and once westward and their time was compared to other atomic clocks that were left on the ground at the point of departure and synchronized to them. This looks very much like Joe and Mary's experiment. Both sets of clocks in the plane and on the ground as seen by a stationary observer, for example looking down at the North Pole, are in motion due to the rotation of the Earth. What is calculated and measured is thus the difference in elapsed time between the two set of clocks as seen by that observer. However, since the clocks join back together, it amounts to the time difference observed at the end of the trip between the two sets of clocks. The results obtained are in agreement with the predictions of relativity. For example the time delay calculated between the clocks on the ground and the flying clocks in a westward direction was 273 nanoseconds while the measurement gave 275 nanoseconds. In the eastward direction the two sets of clocks travel in the same direction due to the rotation of the Earth itself and the time delay was evaluated as -59 ns while the measurement gave -40 ns. The calculations involved for the whole trip were complex since, as we just mentioned, a clock rate is also slightly dependent on the gravitational potential that the clock is in. Thus various corrections had to be taken care of in the calculation. The agreement between the experimental results and the predictions is rather astonishing. Consequently, macroscopic objects are responding as expected to time dilation when in relative motion. Does the phenomenon described apply to everything, material objects and living beings as well? After all we made a very special clock, with photons that propagate at a speed that is constant whatever the velocity of the observer or the source. One could say also that atomic clocks, although built differently from the photon clock, are driven by atoms and their internal properties. They could react in the same way. The answer to that question is relatively simple because if the effect were different for two clocks, for example an atomic clock and a quartz clock, one would be able to see a time difference between the two when they travel at the same speed. This effect would tell that the reference frame in which the clocks are situated is moving without any reference to external objects or phenomena. This is in contradiction with the first principle of relativity. Is it also true for living beings? It has to, because heart beats are clocks and that precious clock would provide in principle a means for deciding on absolute motion, which is not possible. This is why, in our story, when Joe in his train was waving slowly at Mary, it was not the wine he had in the saloon. This is why also he did not see any problem with the basin dripping water for the plant and the plant itself growing as usual. In the moving wagon everything operates normally. A twin in motion in a rocket is seen as aging slowly by his brother that stays at rest on the launch pad. There is nothing wrong with the idea of travel through space at speeds not achieved yet that would allow long trips in which the travellers would age so slowly as to allow them to reach a celestial body when they are young enough to enjoy their newly discovered world. However, it would not be very pleasant for them in returning to their home base since all their family and friends would either be old or dead.

Joe and Mary do not see Things Happening at the Same Time

We have examined the effect of the constancy of the speed of light on the flow of time in a system moving at constant velocity. We have concluded that it was a real effect, measured experimentally. For objects moving at common velocities such as those that we encounter in our daily activities the effect is small. Nevertheless, it has a great impact on the understanding of the functioning of the laws of the Universe. Consequently, it seems that we have to be very careful in thinking about physical laws and the conclusions we can draw from observations on the flow of time. Let us go back to our friends to examine a few other effects.

Mary and Joe are very puzzled about time flow and are ready to do another experiment. Since time does not flow at the same rate for Joe in the train and Mary on the railway platform, they wonder about events in succession as seen in the wagon and as seen on the platform. Certain things like a lightning striking the ground are quick and unique and should be seen as happening at the same time for Joe and Mary. They wonder if they could not check if something strange is happening on the moving wagon regarding this question also. This would be a difficult verification to perform since they would have to rely on the exact synchronism of their clock and their maintaining of time for long periods. They cannot trust this whole affair of synchronism and stability and decide they need a third partner, their friend Bob who gave them the clocks. He would be some kind of neutral reference like the ones they use in time measurement in National Institutes charged with time keeping: they normally have at least three clocks. Joe and Bob will both go on the train each with a photon clock that they have synchronized perfectly. Bob sits in front of the wagon and Joe in the back, both with a light detector that produces a sharp sound different for both when hit by a light pulse. At the sound they both hit a red button to stop their clock, an action that Mary can see from her rest point on the station platform and record exactly with her own clock. This is illustrated in Figure 3.10. They cannot rely on the hazardous lightning timing and they install rather a light bulb exactly in the middle of the wagon. They measure the distance between the detectors and the

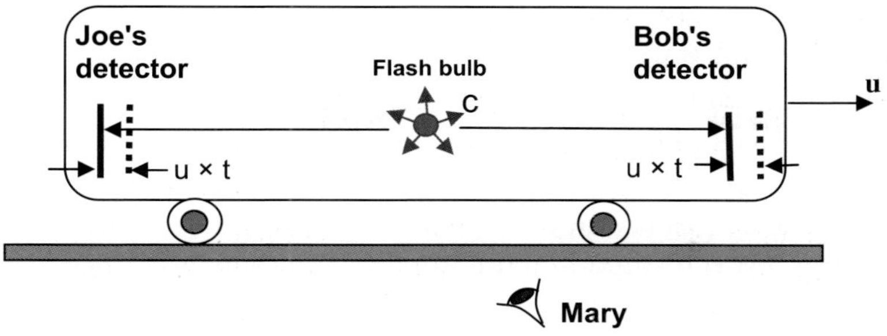

Figure 3.10. Simultaneity experiment in a moving wagon.

light bulb very accurately with a metre stick. The light bulb is turned on and off by a trigger pole on the track when they pass in front of Mary. Thus, they all have a similar clock and they can verify this whole question of time flow and simultaneity. Since Bob and Joe are at the same distance from the bulb they should receive the light pulse at the same time and the reception time on their individual photon clock should be the same. Mary outside can check the whole thing with her own photon clock. They do the experiment. Well, things went as expected in the train. Bob and Joe's detectors received the light pulse and they pressed the red button exactly at the same time as shown on their clock recording. They come out to show their result to Mary in order to prove to her that there is nothing strange in the wagon, both clock recordings being a proof of it. Mary tells them that they are wrong. She saw clearly Joe pressing the button on his chronometer to record the time of arrival of the light pulse before Bob and she has timed that with her photon clock. They check and she is right. What a strange world. What is happening?

This is a profound dilemma. But it is how nature works. Let us examine with the help of Figure 3.10 what is happening in the context of the hypothesis of the constancy of the speed of light. When the light bulb flashes in the wagon, light propagates in all directions at speed c. The light bulb does not know it is in a train travelling at speed u relative to the station platform. Consequently light reaches Bob's and Joe's detector at the same time since they are both at the same distance from the bulb. This is totally logical. And this is what

Bob and Joe observe with their precise clocks. They are going at the same speed and thus are in a common frame of reference.

However, Mary sees things moving. During the time 't' that light takes to propagate from the flashbulb to Bob's detector, the detector has moved a distance u $\times$ t since it is moving away from the source. Exactly the opposite happens to Joe's detector. It moves toward the source by the same amount u $\times$ t. Consequently there is less distance to travel for the light beam in Joe's case. Mary observes that Joe presses his button before Bob. In her frame of reference the speed of light is always c and is not affected by the motion of the source moving at speed u. This is the second postulate made by Einstein. This is a fundamental law of nature proved by Michelson and Morley in their experiment, although difficult to accept. It is thus readily concluded that simultaneity is not the same for frames moving at different velocities. In Mary's stationary frame Joe receives the light signal before Bob.

We note that if light, like sound, were carried by a medium, everything would look simultaneous. The air within the wagon would be carried at speed u and for Mary this speed would be added to the speed of sound in air. In that case, if the flash bulb is replaced by a sound source, both Joe and Bob would receive the sound pulse at the same time. Similarly Mary would see in her stationary frame that both Joe and Bob receive it at the same time. The whole effect of non simultaneity with light is due to the constancy of the speed of light and the absence of an absolute frame of reference such as the aether.

This question of simultaneity was the first item on the list of phenomena treated by Einstein in his 1905 paper. As much as the slowing down of time, the lack of simultaneity of the observation of a unique event in a moving frame as seen from a stationary frame appears to oppose common sense. But it is the way nature works.

Mary Questions the Dimension of the Travelling Clock

Mary thinks about the whole thing, simultaneity and time dilation caused by the fact that the wagon moves relative to her and she does not quite accept it. There is something that is still mysterious to her. She thinks that if time does not flow the same way when one is in motion relative to another observer at rest, something else may very well change. Since speed is distance travelled in a given time and time is affected, then length must be affected too. It makes a lot of sense. She then thinks of the distance between the mirrors in the photon clock that they experimented with. This distance is maintained by a metal rod of length L. So many things have been happening recently that it is very possible that the length of the rod is changing with speed and this is the effect they measured all the time. How are they going to verify that? The only things they have are metre bars and clocks. Mary decides to study the length of the rod spacer of the mirror. She lines up her metre bar with the mirror spacer of Joe's clock. She makes a mark on her metre bar corresponding exactly to the length of the spacer, that is 0.6 m. She keeps that metre bar and Joe goes on the train with his clock and his metre bar. He checks also the length L of the clock when in motion and finds that nothing changes. The train comes in front of Mary and passes at the same speed it did in the first experiment, but now with Joe showing her the metre bar horizontally in the window in order that Mary makes a measurement. Just when the front of Joe's metre bar passes in front of her she triggers her chronometer and stops it when the other end passes. She knows the speed of the wagon and by simple mutiplication she calculates the length of the bar in the wagon. Great news, it is shorter than her metre bar. She does not understand. Joe calls her and tells her that when he was passing by with the metre in his hand it seemed to him that she was the one moving and that her clock could have been running slow relative to his. It is thus not surprising that she measures less time elapsed and so she would see his metre shortened. She is very happy about that explanation because it may have something to do with the behaviour of clocks in motion. She tells Joe to come back at the same speed and proposes to make a measurement of the mirror spacing of Joe's clock. The mirror rod is vertical and she cannot use her chronometer to calculate its length. She decides to use directly her own metre, hold it vertically and when Joe passes with the clock in his hand, she sees them superimposed and is able to see if the mark she made on her meter still corresponds to the rod length in Joe's clock. The mark superimposes exactly on the passing rod length and the clock rod does not appear to have changed length in the moving wagon relative to the measurement she did previously when Joe's clock was nearby on the railway platform. Joe observes the same thing from his point of view. They are very confused.

The whole thing is bizarre, but reflects the real world. What happens with the measurement of the vertical mirror rod is actually a consequence of the principle of relativity, that there is no experiment you can make in the wagon that will determine your speed. If Joe had found a difference in length on the vertical rod, he could have determined who was moving by the difference observed. If Joe's clock rod had been found shorter, Joe would have said that he was moving. If Mary had found the mark she made on the metre bar shorter than the rod in Joe's clock, she would have said that she was the one moving. Since they turn out to be the same length their motion is relative and either could be moving or at rest. For the case of the metre measurement when both metre bars are parallel to direction of motion, clocks were involved. The experiment could be done in a reciprocal manner and Joe and Mary would find that the other's metre is shorter. What they observe is the contraction that Lorentz calculated in his attempt to have electromagnetic phenomena work the same way in frames of reference in relative motion. Einstein's made those transformations a principle of relativity.

In view of that finding relative to the contraction of length in the direction of motion, the orientation of the photon clock that we have studied earlier, set at right angles to the motion of the wagon, may seem to be a privileged situation. If the clock was turned by 90 degrees, with the rod separating the mirror oriented along the direction of motion, it looks as if the time dilation effect would not exist since it appears that in a full cycle the photon would travel in succession in opposite directions and the gain in time in one direction would be lost in the other direction due to the motion of the mirrors. This is not so. A simple calculation,[22] taking into account the Lorentz contraction of the rod holding the mirrors apart, shows that in that orientation the clock period is dilated exactly in the same way as in the case where the clock is oriented at right angle to the direction of the wagon motion, as in the original condition.

It appears, as Poincaré said, that the constancy of the speed of light is a conspiracy of Nature and as it is observed in all circumstrances it has to be accepted as a principle. The speed of light is the same in moving systems as in presumably stationary systems and there is no technique available within a single laboratory for the detection of motion taking place at constant velocity. The measurement must be made relative to some other object. That phenomenon causes effects that we are not familiar with and we have to accept the results predicted on the basis of the postulates of relativity.

So the Speed of Light is Independent of the Source Motion!

In the wagon when the bulb is turned on for a short time, photons are emitted in all directions at the speed of light. The source does not know the wagon is moving. It looks at rest in the wagon. Consequently, the photons forming the short light pulse emitted by the bulb expands in all directions as if they were on a globe or on a balloon expanding at the speed of light as shown in Figure 3.11. It looks as if the whole sphere moves with the wagon and Joe sees things as normal. His reference frame seems stationary to him. For Mary on the platform the speed of light is the same whatever the speed of the source and, for her, the whole sphere expands at the same rate, that

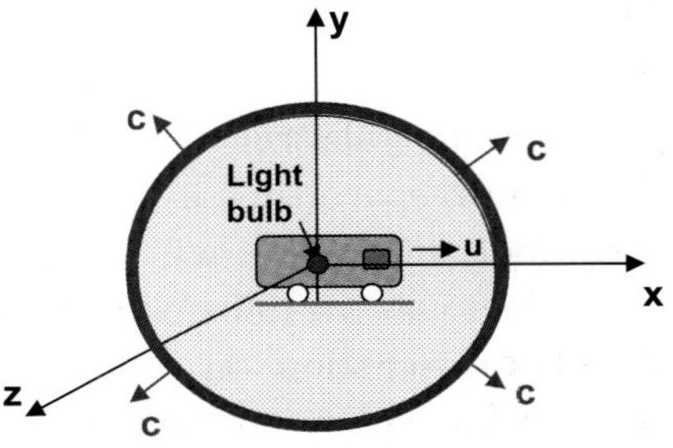

Figure 3.11. The ligthbulb sends a bunch of photons in space. These photons move away from the source as if they were on the surface of a balloon expanding at speed c.

is c. Just after the light pulse, the radiation comes straight at her, passes her, and leaves to infinity at the speed c, independent of the source speed. For such a phenomenon to take place, time and space in Joe's frame of reference must behave differently from time and space in Mary's frame. And both time and space are so entangled that they cannot be dissociated. 'Henceforth space by itself and time by itself are doomed to fade away into mere shadows and only a kind of union of the two will preserve an independent reality'[23]; it almost looks like poetry.

The Space-Time World we Live in

From the reflections of those thinkers, Einstein, Lorentz and particularly Minkowski, formerly Einstein's teacher, a four-dimensional world is born: the location of a point in the world must be given by its space location as x, y and z as well as its location in time, t. When you have a meeting with someone you need to know the address, that is to say the location x, y and z, which could mean the city, x, the name of the street and the address number of the building, y, the floor in the building (z), as well as the time, t, at which the meeting takes place. In Figure 3.11, the three axes are at right angle like lines emanating from the corner of a room to form a space. Where are we going to put time in the space-time diagram? It is a new axis and we feel we should probably put it at right angle to the other three. Well, it is clear that we are in some kind of trouble. As observed in Figure 3.1, we are not able to bring in, in a simple manner, a fourth dimension making a 90° angle with the three others. Our three-dimensional space is already occupied. Advanced mathematical tools need to be used and then the whole thing is treated in a somewhat abstract way. However, we can have some ideas about that special space-time if we limit ourselves to space motion in a plane or in one dimension, which we can do easily.

Mary's research project, also a defence research contract, is on the effect of speed and acceleration on human beings. Her thesis advisor has bought her a fast sport car to make tests. It is also well adapted to test some of the effects found on the train and she decides to be the co-investigator and do tests with Joe watching nearby. She is at the start line on the runway and waiting for the signal to move in a straight line from the point that we may call the origin or 0. She stands there, on the runway, in her one-dimensional space, with her foot on the accelerator waiting to go in gear as illustrated in Figure 3.12(a). Joe is out there, standing, and watching her, with his photon clock running. If she gets in gear he can easily get her speed in the way we described it earlier: distance travelled divided by time lapse. Now, distance can be measured in the reference frame where Joe stands in which time flows at a continuous rate, inexorably. It is not in the graphic. Can we get it in? Why not? Make time an

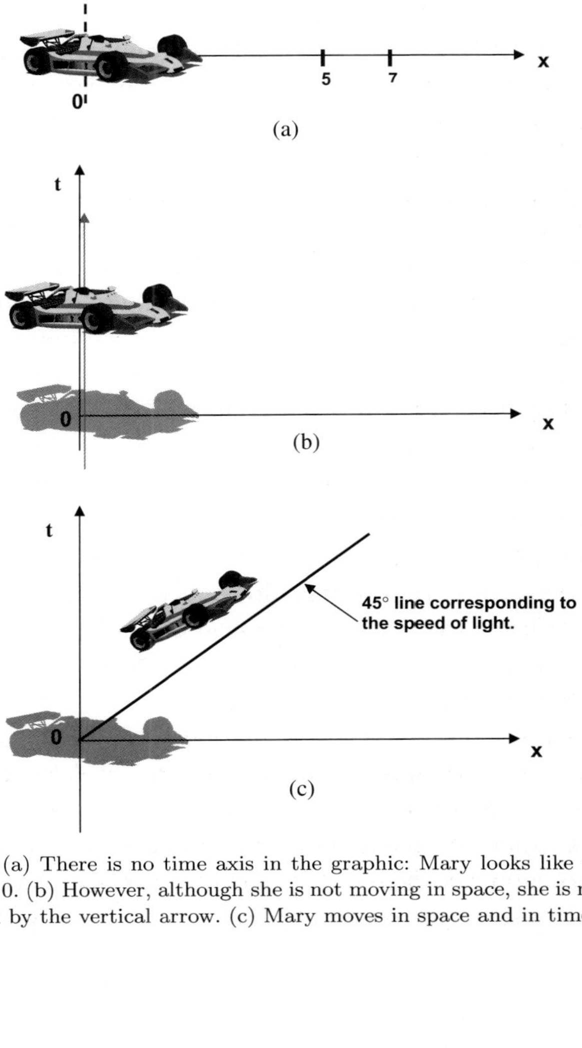

Figure 3.12. (a) There is no time axis in the graphic: Mary looks like waiting at the starting point 0. (b) However, although she is not moving in space, she is moving in time as represented by the vertical arrow. (c) Mary moves in space and in time.

axis perpendicular to the x axis thus vertical and call it 't'. What can we say about that time axis? When Mary is waiting at the start line, time is running by under her car in the same way as space would be running under the car if she was moving. So she is moving in time even though she is staying at the same point in space. A line representing her behaviour in the space-time diagram would be a straight line. We make it go up from zero, simply as a line as the t axis as shown in Figure 3.12(b). Consequently, that space-time graphic is rather special. On one axis you can be resting but on the other axis you are rushing 'all the time'.

In a space-only graphic you can get a distance between points as a difference of coordinates. For example, in the one-dimensional space of Figure 3.12(a), the difference between the point situated at $x_1 = 5$ metres and another at $x_2 = 7$ metres is simply 2 metres. In three dimensions we have to use Pythagoras' theorem and we have to deal with the sum of squares and take the square root to get the distance between points. But we know about that and it is not a major problem. On the other hand, on the time axis we need only to subtract two times t_1 and t_2 and we have the elapsed time. This is what Joe's chronometer does. But what about when Mary moves in space while she is moving in time? The units are not the same for space alone and time alone. We are now in a space diagram with an imported dimension: time. What happens, however, in the development of the theory about Mary's motion, is that the speed of light comes in as a parameter multiplying time, which gives $c \times t$; this is metre, a space unit. The speed of light plays a so important role in the passing of one frame of reference in motion to one at rest, as we have seen above, that it gets intermixed with space coordinates and the result leads to a *space-time world* instead of separated space and time worlds. In that new world we do not speak of space or of time intervals but of *space-time intervals.*[24]

If Mary stays put at the starting point on the x axis, she is travelling in time only and the space-time interval becomes a time interval. Time predominates. If she moves along the space axis 'x' she moves in a mixture of time and space since time keeps flowing. Since time appears as a product $c \times t$, and c is a constant not affected by motion of sources or observers, we could simply do a so-called graduation, or re-normalization of the axis t. We could call it *ct*, and write it in terms of metres. Then 3 nanoseconds in time becomes an equivalent 1 metre. Or we could also graduate the axis t with marks every 3 nanosecond and still call it t. Our graphic becomes regular in the sense that the spacings of the marks on the x axis and on the 't' axis correspond to similar intervals in space-time. What can we say about the concept of 'speed' in that space-time world? It was a concept rather clear up to now. In the present world representation we can talk about intervals, mixtures of time and space motion, and then an interval divided by time lapse becomes a concept that looks like speed. We will call it a rate.

Mary is tired of that reasoning and of waiting. She wants to know what acceleration and speed will do to her, what she will see. She wants results for her thesis and jumps on the accelerator. She takes off and accelerates rapidly. She has a speed car with a very powerful engine. Very soon she reaches a very high speed. Where is her trace on the graphic? At the start of her trip, she is going at a speed much less than the speed of light and her trace lies in that region of the graph, close to the 't' axis because time as an interval flows more rapidly than space is traversed. Upon accelerating she tends towards the diagonal line representing the speed of light. She can also get back in any part of the top region of the space-time world if she eases off on the accelerator. Her path in the diagram may take any shape above the diagonal but always pointing to the increasing time axis. Could she 'exist' below that diagonal line in the first quadrant? No. She would need to go faster than the speed of light and as we have seen before from the principle of relativity, she cannot. That part of the space-time world is not accessible to her. We note that she could also come back and pass in front of Joe, standing still close to x = 0, and go in the negative x region and the same rules would apply. There would be a diagonal line extending in the left quadrant with the same properties as the one in the right quadrant.

We note that the graphic can be transformed to an historical graph that we can represent as a static graphic as in Figure 3.13.

The point zero then represents a given instant; the region above the diagonal lines represent future events relative to that instant, while in the region below 0 between the diagonal lines time is

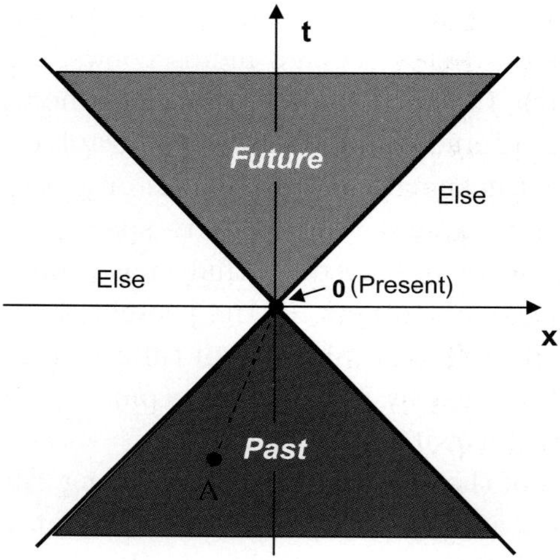

Figure 3.13. The space-time world represented as a graphic. The present moment is situated at the origin, at the crossing point of the diagonal lines that represent the Universe limiting speed, the speed of light.

negative and represents the past. On the other hand, the diagonals representing the limit set by the speed of light may be visualised in a higher dimension diagram as forming a cone. It is called the light cone. An event that happens at A has an effect at O. It is in the past of point O and is causal, that is to say it may cause effects at O. The region called future can be affected by action at O. But things happening in the other region called 'else', cannot affect present events at point O. The information would have to travel faster than light. This is not possible. We will have to wait for it to show up in the future. It is because information cannot be transferred faster than the speed of light and causal events are regulated by that law of nature.

When Mary comes back to Joe, she tells him that she kept her foot on the accelerator and that for her nothing really change in the car's behaviour. The engine was running and pushing and she thought that she had probably violated the speed laws of the World. On his side Joe tells her that when she was accelerating in sixth gear and reaching a very high speed, he observed that she was no longer gaining speed. Her car became reddish but stayed visible. It looked as if she essentially was on some cruise control. She says that she did not have the cruise control on but that very strange things also happened in the world around her. Green things in front of her became violet and when she turned her head and looked back those same green things looked reddish. Otherwise things seem normal in her speedcar. When she turned her headlights on nothing really changed. She could not see more. The green stuff still looked violet.

Speed Limit on the Space-Time World Highway

Mary could not get arrested in her crazy trip. The speed limit is that of light. She could not exceed it. Close to that speed, however, she observed strange effects, colours of objects appeared to change. Why? One of them is the *Doppler Effect*. We have talked about this effect without explaining its nature. Let us do it at this time since it will play an important role later. This effect is best understood in the case of sound waves. For example the siren of a police car going away from us has a lower pitch, or lower frequency than when the car is moving towards us. This is illustrated in Figure 3.14. This is simply because, while the car moves towards us, the emitting siren moves also towards us shortening the time between the crest of the emitted sound waves and compressing the wave: the wavelength is shortened making the frequency or pitch higher. The opposite effect takes place when the car moves away from us. The change in wavelength is simply altered by a factor equal to the ratio of the speed of the car to the speed of sound. Consequently, since the speed of sound in air is about 1,200 km per hour a police car moving at a speed of 120 km/hour will alter the perceived pitch at the observer's site by about 10%. The difference perceived by the observer will be twice that amount since

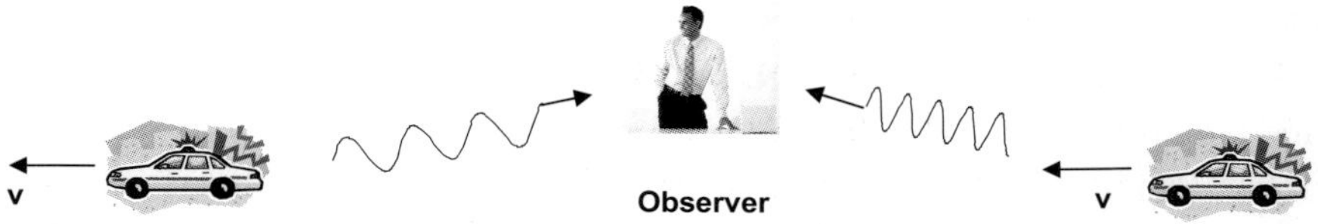

Figure 3.14. Illustration of Doppler Effect as observed in the perception of sound emitted by the siren of a police car coming towards and then going away from an observer. The pitch changes from high to low.

the effect is opposite for the car coming towards and going away from him. This is sufficiently large to be clearly felt by the ear.[25]

When Mary was approaching the speed of light, she was racing towards waves coming at her and, like with sound, the frequency of light increased, since more cycles were passing by in the same time lapse. The wavelenght of light getting to her was shorter and the color of the objects in front of her started to shift to the violet end of the spectrum. In the back, the opposite effect took place, the wavelength of the light emitted by sources she was receding from was getting longer and longer and the green leaves became reddish. When she turned the headlights on, the photons still travelled in front of her at the speed of light since she was the source and the speed of light is independent of the source speed, but Doppler effect took place also and the wavelength of the radiation emitted was shortened by her movement, catching up with the crest of the waves. The light she emitted looked like it had a short wavelength. Upon reflection on objects in front of her, the radiation came back to Mary and she saw the waves also shifted to a shorter wavelength. When she saw the world in front of her getting violet she could not determine if she was moving towards the source or if the source of light in front of her was moving towards her. The effect would have been the same. If the source was moving towards her, the speed of light being constant, the crests of the radiation emitted would appear closer together since the source would be moving in her direction: the wavelength would get shorter. Consequently the effect would be essentially the same as seen from Mary's reference frame. It is relative motion of the source and observer that determines the observed wavelength.

For Mary, in her own reference frame, her clock kept running as usual. But as seen from Joe's reference frame her clock was running slow. In the diagram of Figure 3.13, representing time and space flow in Joe's reference frame, a new axis, 't', must be introduced for Mary's time flow. It is inclined relative to the vertical axis of time. On that inclined axis time lapses are longer than on the 't' axis. On one hand, when Mary was at the starting line in space-time, her 'travel' along the time axis becomes a time rate only and 'flows' essentially at the speed of light. When she is moving at full throttle her speed limit is still the speed of light. What happens in between when she moves at a speed less than the speed of light? In that part of the graphic the limiting rate is always the rate associated with the speed of light. The interval travelled in the space-time world is invariant with the frame of reference and consequently an increase in speed in the space portion is compensated by a decrease in rate in the time portion. That simply means again that since the speed of light is constant, the time scale must change. The consequence is that time flows more slowly in Mary's moving frame as measured by Joe. An important conclusion is thus: motion in space borrows from motion in time adjusting time scales accordingly.

Mary does not quite believe that limit relative to the speed of light. She has an idea. Why not try to do something in space and have real action and freedom. With rockets she would certainly get more information on acceleration effects on human beings. She manages to get on her contract two powerful rockets[26] to see if she could not get more boost and see if she could not violate the speed limit of the galaxy. Joe and Mary strap both rockets together and lift off to space with their friend Bob as a witness. At a certain distance, far from all stars and planets, they leave their friend Bob on a capsule on his own for measuring their exploit. With the engine off, they are all moving at the same speed and thus according to the principle of inertia they may all be considered at rest in a stationary frame. After what she learned about relative motion, Mary's reasoning was that if she gets into space, she could be coasting along at fairly high speed without feeling it in the two rockets attached to each other. For example, she could accelerate, say, to 3/4 of the speed of light with Joe's rocket strapped to hers. After reaching that speed, she would be coasting along with the engine off, at 3/4 the speed of light and she would not feel it. It would be as if she were at rest. Then, un-strapping her rocket from Joe's rocket, turning on the engines again she could get a boost of another 3/4 of the speed of light, relative to Joe's rocket, and violate the law in Bob's frame of reference, which is stationary. Mary convinces Joe to try that experiment. Joe accelerates pulling Mary's rocket along. Both rockets reach the speed 3/4 c as measured by Bob in his capsule. Then, they coast along with the engine off and Joe unstraps the rockets. Mary is free to give another boost. She accelerates to 3/4 the speed of light now relative to Joe's rocket. Joe measures

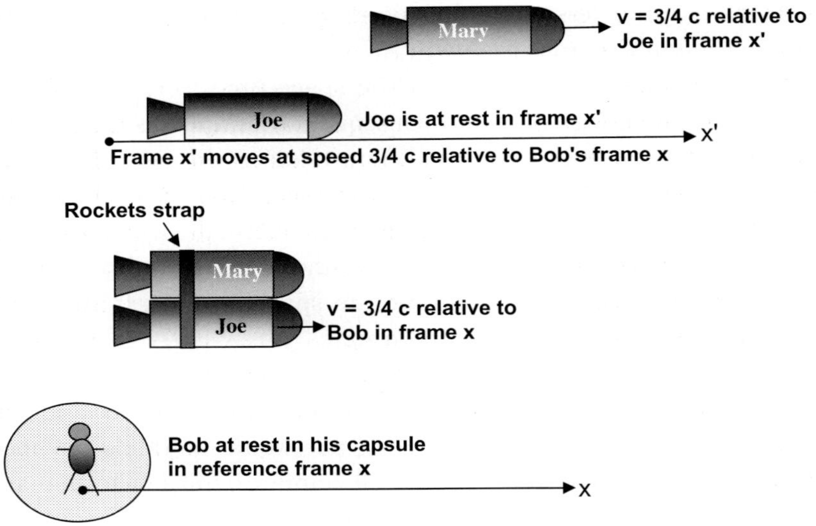

Figure 3.15. Joe and Mary verify the speed limit in the Universe.

her speed: 3/4 c. According to her, she must be going at 1.5 c relative to Bob. None of that happened. Bob, left behind as observer at rest, saw all the action. He measures Mary's speed accurately. The last boost on her rocket increased her speed to 24/25 the speed of light in his frame. This is all illustrated in Figure 3.15.

What is happening? This is the result of the law of addition of speeds as found by Einstein. One could add a new frame of reference in which Bob could be found to be moving at 1/2 the speed of light in a direction opposite to Joe's and Mary's motion; neither Joe nor Mary would still be going faster than c in Bob's frame of reference. There are no frames of reference in which they could be observed as going faster than the speed of light. In Galilean frames in relative motion, speeds add directly. If a rocket goes at speed v and boost again by an added quantity v, we conclude that its final speed should be 2v. But this is good only for low speeds. In space-time, speeds add through a special law and the sum is always less than c.[27] Furthermore, if Mary at her fantastic speed turns on headlights, the radiation or photons will leave her at the speed of light relative to her. The speed of light is independent of the source speed and the law for adding speeds is in agreement with that postulate as it should be. This situation just confirms that in space-time the speed limit is the speed of light in all situations.

Photons or electromagnetic radiation travel at the speed of light, thus on the diagonal line in the graphic of Figure 3.13. As seen from our reference frame, time slows to a halt for them. Thus, aside from an effect due to the expansion of space that we will examine later,[28] a photon emitted some billion of years ago and passing by us is identical to itself as when it was emitted: for us it does not age. It keeps travelling until it hits an obstacle, which can either absorb it or scatter it in another direction with different properties. Space is amazingly filled with photons or electromagnetic radiation travelling in all directions either from the stars, other objects' activity or from human activity, this last contribution being a rather small component of the whole.

Mary's Car had Effectively a Built in Cruise Control

To reach a high speed on the ground Mary had to accelerate her car by pressing on the accelerator; she was thus applying a force to the car. Mary was just using Newton's second principle. She was applying a continuous force and the car accelerated. The problem was that another peculiar thing followed from the predictions of relativity and the limiting speed of light. In the theory of relativity, in simple words, it may be said that the mass of the object increases with speed as observed from a stationary frame of reference and actually tends to infinity when the object approaches the speed of light. Consequently, as seen from a stationary frame, a much greater force is required to accelerate an object when its speed becomes a good fraction of that of light. In day-to-day activities we do not observe such a phenomenon, because the speeds we normally travel at are very small compared to the speed of light. The effect becomes important, like all effects caused by the relativistic principle, only at speeds close to that of light.

Can we understand a little better this phenomenon? It is best to examine it by means of the concept of *momentum*. The concept momentum is introduced as a measure of the 'quantity of motion'.

It is called '**p**' and is the product of an object speed, **v**, by its mass, m: ($\mathbf{p} = \mathrm{m} \times \mathbf{v}$). It can be illustrated by means of billiard balls on a table. The collision of two billiard balls is a common experience and we are not surprised to see the two balls in a head on collision bounce backward and retrace back their path. Taking into account direction of movement, we find that momentum is the same before and after the collision: we say that momentum is conserved as long as the collision is totally elastic, that is to say that there is no energy loss through friction on the table or as heat in the collision. According to relativity, however, the mass of the balls may be said to increase with speed.[29] On a billiard table this effect has absolutely no consequences due to the low speed of the balls. But for atomic particles accelerated under intense electromagnetic fields, as in high energy accelerators for example, this increase in mass has serious consequences concerning the design of the equipment. For example in some types of accelerators, it is desired to deviate the particles by means of a magnetic field to make them travel in a circle. It is found that due to the apparent increase in mass a much stronger field is needed to alter the trajectory of the particles than would be required if the apparent mass of the particle remained as if it was at rest.

In a linear accelerator for example, the mass of a particle travelling close to the speed of light becomes so large that no force could still accelerate it and it no longer changes speed. The momentum increases tremendously not because of its increasing speed but due to this apparent increase in mass as seen from a laboratory rest frame. It requires an extremely large force to change its momentum even though its speed does not change. Consequently, there is a trade off: instead of speed increasing, we may say that mass is increasing. Mary's rocket as observed by Bob in his stationary reference frame did not increase its speed. But Mary in her reference frame was keeping her engine on. In Bob's frame, Mary's speed was no longer increasing but the system's mass, Mary's and the rocket's mass, appeared to increase, becoming nearly infinite close the speed of light. Mary had thus a cruise control on her rocket and it was standard equipment provided by the space-time canvas. Consequently, in any reference frame, she could not exceed that speed of light whatever the power of the rocket's engine.

There is sometimes confusion relative to the interpretation of the increase of mass of an object in motion. Since motion is relative, Mary after shutting down her engine would be coasting along at a constant speed and declare that she is at rest in her own new reference frame. What is the mass of her rocket including her? Always the same. Mass stays mass. Its value would be the same as what it was before she left Bob and Joe. Her mass, however, appears to increase in Bob's and Joe's frame, reflecting an increase in her momentum without much of an increase in speed. It is in those frames that Mary's mass and that of her rocket appear to increase. In Mary's rest frame, her mass and that of her rocket are always the same. Laboratory experiments are totally in agreement with an increase in mass of an object accelerated close to the speed of light. As mentioned above, electrons in linear accelerators are accelerated to speeds very close to that of light as measured in the laboratory frame. A typical result gives an increase of mass in the laboratory frame by a factor 1.6 for electrons accelerated to 0.8 c, in complete agreement with calculation. In equipment used in accelerating particles at very high speeds, it is standard practice to speak of the energy of a particle rather than its speed since this energy appears as an increase in momentum rather than an increase in speed when that speed is close to that of light.

Let us look at this from another point of view. What is *energy*? Energy appears under many forms. For example, electromagnetic radiation is energy. That energy is said to be contained in the two oscillating fields, electric and magnetic, which constitute the wave travelling through space. When light falls on an object having a dark surface its energy is absorbed by the atoms of the object, which heats up. The energy is transferred to the surface molecules: they start vibrating more strongly and this is energy, commonly called heat. With objects, energy may take also many forms. If the objects is moving at speed v in a given reference frame we say that it has motion energy or in scientific terms, *kinetic energy*, which is proportional to the square of that speed. The constant of proportionality is simply the mass divided by 2. It is in a sense related to momentum. On the other hand, an object at rest may have *potential energy* if situated in a field with which it interacts, such as static electric, magnetic or gravitational fields. We will come back to that point.

A great principle of physics, which we will develop in some detail later, is concerned with the different kinds of energy and with the conservation of that energy. Energy is not created from nothing. It must come from somewhere. When an object is at rest in its frame of reference it obviously does not have kinetic energy in that frame: its speed is zero. Consequently, if it starts moving and reaches speed v, it gains energy and that energy has to come from somewhere.

In practice, for an object to reach speed v, a force must be applied to the object to fight its inertia caused by its mass. This is the origin of the energy: the force applied fights the opposite forces that maintain the object at rest, does some *work* in displacing the object against the forces that may be present or still in accelerating it to a given speed. The work done is energy and is the origin of the kinetic energy. Now, since kinetic energy is proportional to mass, an increase in mass, say a doubling, will double its kinetic energy. In this context, in a given frame of reference, if the mass changes with speed it is readily realized that the energy of an object moving at speed v originates not only from its speed, but also from an increase in mass. An increase in mass leads to greater energy. This reasoning leads naturally to the association of mass with energy. This is probably best expressed by the well known and most famous relation:

$$\mathbf{E} = \mathbf{mc}^2$$

In that equation, it is seen that energy is simply the product of the mass m and the square of the speed of light, c. Since we said that m depends on the speed of an object, we see that this energy depends also on the speed of the same object. A simple mathematical manipulation of the expression shows in first approximation that the part on the right, is composed of a first term representing the rest energy of the object and a second term representing its kinetic energy.[30]

This is one of the most fascinating results. We have just pointed towards a possible answer to the question we had introduced in the earlier section of this book: 'What is mass?' That equation tells us that it may be interpreted as energy. But we can also say that about many physical phenomena. For example, we have mentioned above that an electromagnetic wave consisting of oscillating electric and magnetic fields is also energy. However, in that case, we know the substance: it is electromagnetic fields. In the case of our object either in motion or at rest we just introduced the concept that mass is materialization of energy. Is it possible to make a transformation of this mass-energy into some other substance? The answer is yes. For example an electron, with negative charge −e and mass m and its clone made of antimatter with positive charge +e and same mass m, the positron, are annihilated completely when colliding. In the process, the two electric charges adding to zero disappear, the mass disappears, and two photons are created, which have an energy of 0.5 million electron-volts (eV) corresponding to electromagnetic radiation of very short wavelength.[31] The total energy of the two photons corresponds exactly to the energy of the masses before collision, mc^2.

The theory of relativity thus introduces the total equivalence of mass and energy. This is quite an achievement. By saying that, we essentially infer that energy is a fundamental 'substance' that may take various forms such as electromagnetic radiation, mass, or even motion in space. This last form, motion in space, may be vibrations of atoms in solid or random motion of atoms in a gas; in those cases the energy is generally called heat. When it is materialized as mass it gives rise to all mechanical phenomena associated with that mass. One great proof of that equivalence is in the transformation that takes place in nuclear reactors. In a fission reactor, a nucleus of uranium having a mass M may absorb a free neutron. Upon this action the nucleus becomes unstable and splits into constituents such as barium and krypton, and a few new neutrons. These constituents have mass, say m_i, where the subscript 'i' on m identifies the constituents. It is found that the sum of the masses of the particles formed, generally written Σm_i, is less than the sum of the original mass M and the neutron that initiated fission. The difference of mass has been transformed into radiation such as gamma rays or kinetic energy of the produced constituents, neutrons and other nuclei, which means an increase in temperature. The reactor is kept in continuous operation by having a sufficient number of uranium atoms in rods, the neutrons produced causing other atoms to split, creating a chain reaction that is kept under control by proper design of absorbers around the rods. This is the desired effect needed to create vast quantities of heat and boil water to drive a steam engine. When nuclear fission is not controlled as in a reactor, and the amount of uranium is sufficient (critical mass), an explosion results with an enormous increase in the temperature of the surroundings, the energy coming from the transformation of mass. It is readily calculated that the total transformation of a mass of 1 gram of matter would provide an equivalent energy of 25 million kilowatt hours, if totally transformed in energy. Since a modest house may consume about 1,000 kWh of electricity in a month, it is readily seen that 1 gram of matter totally transformed into energy would produce sufficient energy for 25,000 houses for a month. Fusion of atomic nuclei, such as the fusion of the two isotopes of hydrogen, forming a nucleus of helium, releases more energy per unit mass than the fission process in uranium. Consequently, fusion reactors would provide in principle greater energy than fission reactors. However, large scale efficient controlled fusion has not been realized yet in the laboratory. Our Sun and other stars are 'controlled' fusion reactors and the theory of relativity provides an explanation for the origin of their energy. The Sun transforms into energy 4 billion (4×10^9) kilograms of matter every second. This energy goes into space as particles and radiation, corresponding to 3.85×10^{18} megawatts. A simple calculation shows that our Sun, due to its immense size, will shine for billions of years before depleting the material it is made of. Due to its mass of 2×10^{30} kg, it is expected to last for more than 10 billion years.

Shadows of our Four Dimensional Space-Time

In the first paragraphs of this chapter we introduced the concept of multi-dimensional spaces and examined the introduction of a fourth dimension. We developed some insight by doing the exercise in the case of a two dimensional space to which we introduced a third dimension. We could appreciate the problem of interpretation of objects in three dimensions by two-dimensional beings. Can we discuss the introduction of time as a fourth dimension in our three-dimensional space in the same context? In the case of time we notice that, contrary to space in which we can fix easily an origin in a given frame of reference, the origin, as shown in Figure 3.13, is the present moment and is continuously escaping us. In fact, we may attach the frame of reference to a given moment. We thus have an instantaneous picture of the world in its various states at that moment: past, present and expected future.

On the other hand, however, we can say loosely that the introduction of time as a fourth coordinate has led us to the concept of the equivalence of mass and energy. It looks very much as if one shadow of the 'substance' energy takes form in our three-dimensional space as mass in some particular circumstances. In other cases, it may be electromagnetic fields, or still motion whose effect, in a given reference frame, is to increase mass. We are back to mass in a circular argument. In the introduction of that fourth dimension, we have also introduced a limit in the so-called 'rate' of flow or speed in the four-dimensional space-time. That limit was effectively the speed of light and in moving in space we were simply trading time for speed: time flow in a moving frame was slower than time flow in a rest frame. We are puzzled by that phenomenon, but we accept it. We know *how* it works. The mathematics leads to many effects that we can evaluate and interpret and the cause of the effects observed is the limited speed of light, constant through space whatever we do. This looks very much like the shadow of a 'substance' in another world and we interpret its behavior as a *law of nature*. We do not have access to the *why*, we accept it and we deduce consequences that we may measure.

We Stick to the Ground and Things Fall on the Floor

We are really forced to love this planet. We are 'attached' to it. It enlaces us. We try to escape its surface, and do it, but with great labour and difficulty. There is a force that keeps us on the ground; the same force that pulls objects to the ground. What is the origin of that force? Until the 18th century, the effect was observed and accepted as a way of life. Its operating mechanism was unknown. Human beings, however, must have been wondering for a long time about such a phenomenon even if they did not explicitly question its origin. It took a great mind to wake up to that phenomenon and ask the right questions with the right words. That person with such an intellect was Isaac Newton. This awakening was certainly triggered by the tremendous questioning that was taking place around that period by such minds as Brahe, Copernicus, Kepler, Galileo, Leibniz and others about the Universe they were observing. Most importantly, an answer was obtained. The answer was not perfect because it left the 'why it is that way' without answer, but it was so good that the explanation is still valid nowadays and is used at large. *Gravitation* is the engine of the phenomenon.

The principle, advanced by Newton, is that gravity creates a force of attraction between objects. In the case of two isolated objects, this force is proportional to the product of the masses of the two objects and inversely proportional to the square of the distance between them. The constant of proportionlity is the so-called gravitational constant G. If you double the mass of one of the two objects the force is twice as large. If you increase the distance between the two objects by a factor 2 the force decreases by a factor 4. This is Newton's gravitational law and is a simple law based on experimental observation. Since it agrees with measurements it must be right. It explains the motion of the objects thrown in the air, the parabolic path of balls hit with a bat, the elliptical trajectory of the planets in orbit around the Sun, the motion of the Moon around the Earth, and presumably the motion of the stars and galaxies in the Universe. This is illustrated in Figure 3.16.

In the law, the distance r is measured from the centre of the objects as if all the mass was concentrated into a single point. The gravitational constant, like the speed of light, has to be determined experimentally. It expresses the size of the force in play and is equal approximately to $6.67\times10^{-11}\,\mathrm{N\,m^2/kg^2}$. Two objects one metre apart and having a mass of one kg are attracted to each other by a force of 6.67×10^{-11} Newton.[32] It is a very small force because, if that force is applied to an object with a mass of one kg, then according to the second principle the object would be accelerated by 6.67×10^{-11} metre per squared second. After one minute with this force applied, the object would travel at a speed of 0.4 micrometer per

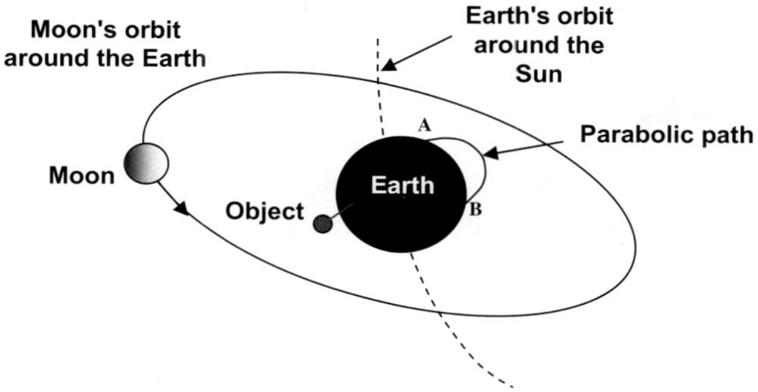

Figure 3.16. Illustration of the paths of various objects in the proximity of the Earth. An object thrown up the air without sufficient velocity falls back to the surface of the Earth and if sent at an angle to the vertical follows a parabolic path. The Moon has sufficient velocity and does not fall on Earth. It stays in orbit around the Earth. Similarly the Earth orbits the Sun. All those paths are calculated directly from the principles established by Newton.

second. We would certainly not see it move. Another example: we have all experienced the lifting from the table of a piece of iron such as a paper clip by a small magnet. A magnet that we can hold in our hands succeeds in beating the force of the whole planet Earth in attracting the paper clip. The force of gravity is thus very weak. We are in a sense lucky because if it were much stronger, similar to magnetic forces, with the size of Earth under our feet we would be stuck flat on the ground. On the other hand, there are cases that could cause us problems even if gravity is very weak. For example, if we ever reached a solid planet of Jupiter's size, life would be a little more complicated. Due to the large mass of the planet, the force of attraction would be 2.5 times larger than on Earth. We grew up on Earth and we developed as we are, compatible to some extent with the gravitational environment we live in. However, if we had appeared on Jupiter, we would probably have developed through evolution a structure and a way of life compatible with such gravity.

It is logical to examine that law in the context of the second principle. Both the second principle and the law of gravity are laws about force. The second principle tells us that a force accelerates a mass. Thus, gravity, being a force, accelerates a mass. Because of the nature of the force, objects in Earth's environment are 'attracted', tend to be as close as possible to the centre of the Earth, and thus accelerate. At first sight we could conclude that if an object is large and has more mass, the force is stronger and the acceleration is larger. No, this is wrong. If the object has more mass, it is more difficult to accelerate it because of the second principle. The two effects compensate exactly and all objects, whatever their mass or their substance, fall on the Earth surface with the same acceleration. This was demonstrated by Galileo as told in the story about his experiment from the leaning tower of Pisa. The presumed experiment could have been dropping at the same instant spheres of different materials, but having the same volume, thus having different masses. Someone on the ground at the base of the tower would verify if the masses arrived at the same time. In such experiments it is important to avoid all kinds of secondary effects such as the friction forces of the air, for example, affecting falling objects. The experiment was repeated on the Moon (free of air) by an astronaut in the Apollo program and was claimed to have been successful within the measurement accuracy of the instrument available on the mission: the eye looking at the objects falling — very low accuracy but very convincing, at least psychologically.

Due to the large diameter of our planet all objects on its surface are essentially at the same distance from the centre and consequently all objects, up to a fairly large altitude, are submitted to a similar force and experience a similar acceleration. It is about 10 metres per squared second (more precisely 9.8 metres per squared second). Consequently, after 1 second, all objects falling vertically near the surface of the Earth, if it were not for air friction, would be accelerated to a speed of 10 metres per second. This acceleration is called g, and is fairly constant over the surface of the Earth. This force, which attracts us to the centre of the Earth, is the force that we assimilate with weight. In practice, since g does not vary much from one place to another we simply use mass as a measure of our weight in kilograms, but in reality our weight should be given in Newtons. On Jupiter the mass of a person would be the same as on Earth, say 80 kg, but his weight would correspond to an equivalent mass of 200 kg on Earth.

What is the source of that force? How can an object feel the presence of another object a metre away or still, 400,000 km away, like the Moon that orbits the Earth? This was a question that puzzled Newton. How does an object with mass know its distance from another object and even worst, how does it know the mass of the other object? What is the substance that is exchanged between the two masses in order that they are aware of the companion's presence and structure? Newton did not elaborate on it and left it to the reader to speculate on. This question of action at distance is certainly troubling. It seems that there is something emanating in the totality of space from an object of mass m, a 'substance' of some kind that is felt by all other masses in the Universe. We will see that Einstein approached the subject from another point of view and obtained a different picture of the interaction. For the purpose of calculation, however, it is possible to model that substance with the help of a concept developed in the case of magnetic effects. We are familiar with the effect of magnets on the distribution of small iron filings. Figure 3.17 shows such a distribution.

In that picture a magnet was placed under the sheet of paper and iron filings were simply spread over the sheet. The iron filings appear to follow lines that emanate from the tip of the magnet and spread in space. The filings distribute themselves along those lines. We say that a magnetic field is created in space and the field follows the lines observed as drawn by the iron filings. We then make a model of the interaction of the field with the little magnets inside the iron filings. We develop a theory that the magnetic field creates forces on

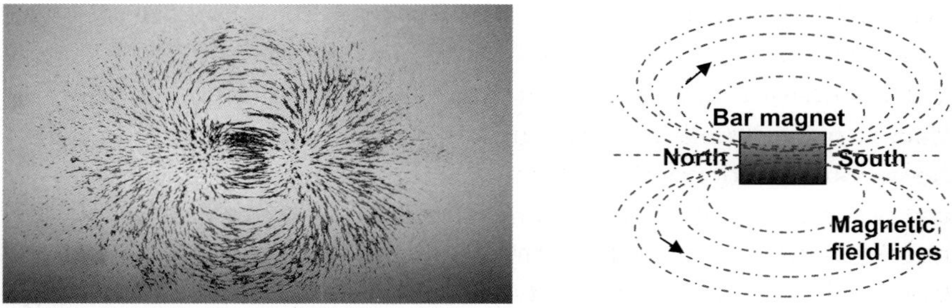

Figure 3.17. Iron filings orient themselves along so-called lines of forces that we identify as lines of a magnetic field emanating from the poles of the magnet.[33] This demonstration is rather easy to implement. The figure shown was obtained directly on my desk with a simple bar magnet bought from a toy shop. Iron filings were made from a nail with a file. They were spread over a piece of paper placed over the bar magnet.

the little magnets, and find that we can explain the essentials of the behaviour of the iron fillings. Is this magnetic field an invention of our mind? We see it through the iron filings and through its effect on the needle of the compass. We write laws for that field, it creates forces and we explain other phenomena with it, such as the concept of light propagation: there must be something real in it. Again if it works all the time there must be something right in the concept.

Why not use the idea with gravity? We then talk of a gravitational field $\mathbf{E}_{\mathrm{g}}$ extending through all of space and affecting all masses in the Universe. The force on an object of mass m in that field is simply the product of that field by the mass, or $\mathbf{F} = \mathbf{E}_{\mathrm{g}} \times m$. In some circumstances, it has been found more practical to talk about the energy of a mass m in that field. We said that if we apply a force on an object and displace it, we do some work on it. Work is effectively energy that we spend. Let us see how we can introduce that concept in the case of gravity. Since the gravitational field introduces a force there must be a way for expressing that energy. Suppose you have in your hand an object of mass, m, at the surface of the Earth. To raise it to a certain height H you must use your strength and thus apply some force on it to counteract the force of gravity. You thus effectively do work or, in other words, spend some energy on it. Because energy is conserved, the object stores that energy internally and we call it *potential energy*. The force you had to apply is nothing else than the opposite of the gravitational force of attraction. If the object is 1 kg, this force is thus $1 \times \mathrm{g}$. Consequently if you raise it to the height H above the ground, the work you have done is $1 \times \mathrm{g} \times \mathrm{H}$ and for a mass, m, it is $\mathrm{m} \times \mathrm{g} \times \mathrm{H}$. We say that the potential energy of the object of mass m is that value. This is an approximate value that is rather good near the surface of the Earth. When the height becomes a good fraction of the Earth's radius, the expression for the potential energy is a little more complicated but can be handled almost as easily. If you drop the object from height H to the ground, it will accelerate and change its potential energy into kinetic energy. Upon hitting the ground the kinetic energy will be transformed into heat: as said previously, energy is conserved. Thus the gravitational potential energy of a mass m at height H close to the surface of the Earth is just $\mathrm{m} \times \mathrm{g} \times \mathrm{H}$. We will use this concept later. It is illustrated in Figure 3.18. With the two laws, the second principle and gravitation, calculations can be made on the kinematics of all objects on the surface and around the Earth. For example, the trajectory of an object thrown in the air at an angle can be calculated exactly and the trajectory is predicted to be a parabola, which of course is what is observed experimentally. Another trajectory is rather interesting, that of an object of mass m thrown horizontally at great speed, such as the trajectory of a bullet sent at high speed from the muzzle of a rifle. Suppose the rifle is held at the window of the second floor of a house, about 5 m above ground as illustrated in Figure 3.19. If the bullet were simply dropped from the height of the rifle without being shot, it would fall to the ground right under the rifle because of gravity.

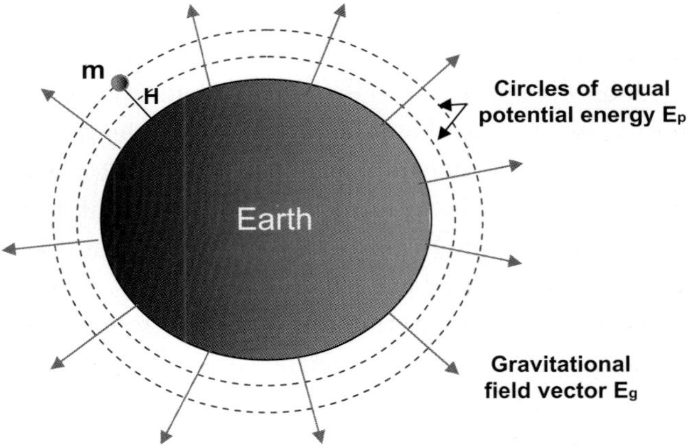

Figure 3.18. Representation of the Earth's gravitational field E_g and the potential energy E_p of a mass m at height H in that field. (Earth radius $= 6.4 \times 10^6$ metres, Earth mass $= 6 \times 10^{24}$ kilograms).

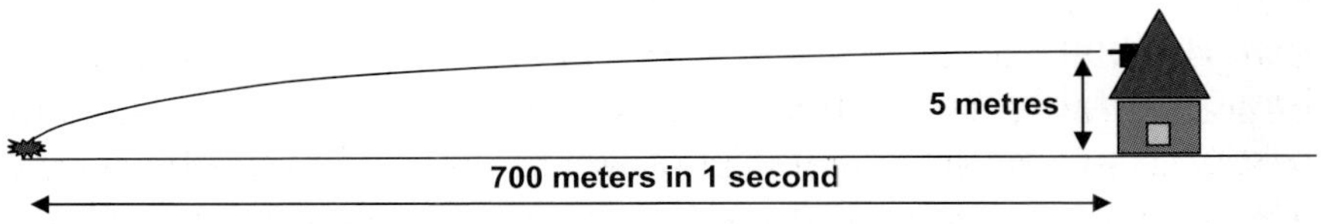

Figure 3.19. A bullet simply dropped from a height of 5 meters without being shot by the rifle would take a second to hit the ground. If shot, its vertical speed under gravity is not altered and hits the ground in the same time but at a distance crossed during the same time, that is 700 metres if travelling at 2,500 Km per hour.

How fast? Exactly as discussed previously. The bullet would hit ground at a speed of ~10 metres per second after fall duration of ~1 second. We have to remember that gravity is applied on a continuous basis and the bullet is continuously accelerated. Its speed increases continuously in its descent to the surface of the soil. Now, if it is shot from the rifle horizontally at great speed, how long will it stay in the air? If we neglect air friction, it will stay in the air exactly the same time, 1 second, before it hits the ground. Its horizontal speed does not alter the law of gravity and the bullet, being fired horizontally, does not have any vertical velocity to start with. The only difference is that it will fall at a great distance, which is the distance it should travel in one second with the speed it has at the muzzle of the rifle. With a muzzle velocity of 2,500 km per hour, and no air resistance the bullet should travel about 700 metres before hitting the ground.

A nice exercise can be done with that experiment. We know the Earth is a globe and its surface is curved. Consequently, in practice the bullet will have to travel a little longer to hit the ground, because the ground at that distance is a little lower than it would be if the Earth surface were flat. If we increased the speed at the muzzle of the rifle it will fall further. If its speed is high enough, the bullet may fall so far that it will not hit the ground any longer. It will always be 5 metres above ground or in other words it will follow the curvature of the Earth and always miss the surface of the Earth. What is the speed required? The two laws mentioned, gravity and the second law provide the tools and the calculation gives a speed of about 28,000 km per hour. At that speed, the bullet would never hit the ground and, were it not for air friction and hills in its trajectory; it would become a satellite of the Earth at an altitude of 5 metres. It would be in a low altitude orbit! Another way of saying the same thing is to realize that the bullet is in free fall in the gravitational field, but due to its horizontal speed it misses the Earth continuously. Of course air friction is present and the bullet is slowed down. It would fall to the ground before even a small fraction of an orbit. Earth satellites are placed in orbit at altitudes of several hundreds of km where there is practically no air. Artificial satellites travel in the vacuum of space like the Moon. Consequently there is no air friction and, in principle, the orbit is maintained forever. What happens if the speed exceeds that value? At a given speed above that so-called orbiting value it may get into an elliptical orbit like a comet. However, if the speed is large enough it may escape from Earth's attraction and never come back. This velocity is 40,320 km per hour.

We Cheated A Little about the Concept of Mass

We have a problem with the introduction of mass. We found that it was the substance required to obtain a description of motion of objects in agreement with observations under the action of forces. Just now, we have introduced a new concept, that of the force of gravity and we have said that the force was proportional to the masses involved and inversely proportional to the square of their distances. We found that when this is done we can explain the common behaviour of objects around the Earth and we believe it applies to all objects in the Universe. But we do not know how such a force could spread in space to infinity. We invented the concept of the gravitational field, something like an 'emanation' from an object with mass into all of space, and we assumed that another object with mass had the property of detecting that field and act according to its strength. As usual we displaced the problem to a higher level. Then the question becomes: what is the nature, the essence of that gravitational field? We have invented a nice tool to explain the mechanism that regulates the behaviour of objects with mass but we do not know the real nature of the field creating the force.

Without mentioning it explicitly, however, we created a crucial union of these two laws, the second law about acceleration under the influence of a force and the law of attraction due to gravity. We used the same value of mass in both cases and this is why we could do calculations: we assumed that the mass from which emanates the gravitational field and causes attraction between objects is the same as the mass that opposes acceleration: this is a rather major step. How do we know that it is the same 'substance' that we call mass that is at the origin of the two laws? The assumption became very interesting at the surface of the Earth because it allowed us to define a very useful parameter, the g factor, the acceleration of gravity, 9.8 metres per squared second.

We notice also that in this context the expression for gravity results in the same units as acceleration and is thus similar to acceleration. This is the concept that Einstein used to develop a model in which gravity became a property of the space-time continuum that we introduced previously. In that context, the properties of the speed of light forced us to reconsider completely our notion about space and time thought previously to be independent. We had to think in terms of a new concept that of space-time in which space and time were forever joined together into a single framework. To understand more of the consequences of that joining of time and space in the context of gravity, let us call back Mary and Joe and ask them to participate in some new experiments.

We ask Mary and Joe to go in space in one of their rockets. In their rocket, far away from any planet's gravity, they are cruising along with the engine off, there is no acceleration; they do not feel any force, thus have no weight, and simply float around inside the rocket as illustrated in Figure 3.20(a). They are in a rest frame attached to the rocket with engine off. They turn the engine on and accelerate at a rate of 9.8 metres per squared second. What do they feel? They feel like being projected to the back of the rocket and find that they can stand up on the wall opposite to the direction of acceleration, as on a floor on Earth. This is illustrated in Figure 3.20(b).

Mary concludes that it looks exactly as if they were on the surface of fair planet Earth. She observes that if she lets go an apple from her outstretched hand it accelerates to the back of the rocket at the rate 9.8 metres per squared second: it appears to fall as on the surface of the Earth. And if she lets go her hat it does the same thing as the apple: 'I wonder', she says, 'how Newton and Galileo would explain that. They would be rather puzzled, I suppose'. And if Einstein had been in the rocket with them he would have answered that there were no problems at all because gravity and acceleration are similar. The apple and the hat left free in space are reached by the back floor of the rocket that is accelerating at 9.8 metres per squared second towards them. The apple and the hat are not accelerated, they are floating in free space within the rocket and the back of the

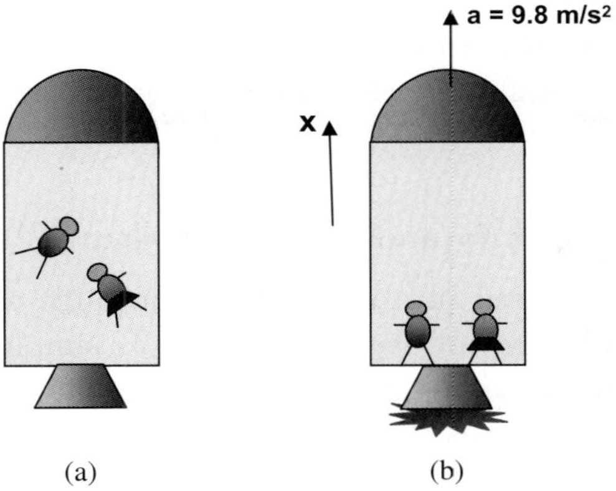

Figure 3.20. Mary and Joe go on a trip in one rocket. In (a) the engine is off and the rocket drifts in space at constant speed. They float around inside the rocket. In (b) the rocket is accelerated at a rate of 9.8 metres per second per second. They can stand on the back of the rocket as if they were on the suface of planet Earth.

rocket reaches them in the same way as it reaches Joe and Mary. It is clear that, looking at the context that way, their motion relative to the rocket is independent of their mass: they do not accelerate, the rocket does.

When Joe and Mary are coming back to Earth in their rocket, they feel the effect of gravity and are attracted by the Earth. To save fuel, they turn the engine off and let themselves pulled by the planet: they are thus in free fall. What do they feel? It is exactly as when they were in space, some time ago, a long way from Earth, not feeling any gravitational attraction. With the engine off, they were floating in their rocket. This is the same situation as when they let themselves be attracted by planet Earth and approach it in free fall: they are in a so-called free float system. And they cannot tell if they are moving or not. The same thing would happen if Mary and Joe were on Earth in an elevator falling freely. They would be accelerated to the surface of the Earth and fall at the same rate as the elevator. They would float inside the elevator and any object carried with them would float around, motionless relative to them and the elevator, whatever their size and composition. They would not be able to tell the difference between being in a rest frame in space or the situation in the elevator. The law of physics would be the same as in a rest frame. We have mentioned above the fact that we can represent the movement of a satellite as a free fall. An astronaut floats inside the satellite: he is essentially in free fall with the satellite, like Joe and Mary in the elevator.

Acceleration, Gravity and Space Curvature

Since being accelerated in a rocket feels the same as being on the surface of Earth why not say that gravity is simply equivalent to acceleration? At the same time, the concept of mass that enters the second law is the same as the concept of mass that is accelerated by gravity. This is what Einstein called his happiest thought and made it a basic principle. It is called the Principle of equivalence.[34]

The conclusions that are reached in assuming this equivalence are far reaching. They transform completely our understanding of the behaviour of objects in the space-time canvas we introduced earlier in the context of the special theory of relativity, which dealt with objects moving at constant velocities. It leads to a new theory called the General Theory of Relativity which now deals with objects under acceleration. Let us look at a few of the consequences that follow the adoption of such a principle.

Gravitational redshift

Let us ask Joe and Mary to do another experiment with their precise clocks in their rocket. Let us send Mary to the front of the rocket with her clock and ask Joe to stay in the back as illustrated in Figure 3.21. We ask them to compare their clocks' behaviour. Since they are not at the same location they have to use a means to communicate. In the rocket, however, they see each other and it is simple: they send light pulses to provide information on the tick of their clocks. Assume that Mary in the front of the rocket is the one sending to Joe the information about the ticks of her clock. At every tick, she sends a light pulse. Surprisingly, Joe receives the ticks faster than the one given by his own clock. He finds that Mary's clock is running fast although they were running at exactly the same rate before Mary went to the front of the rocket. How can that be, since they are in the same environment and at every instant going at the sme speed?

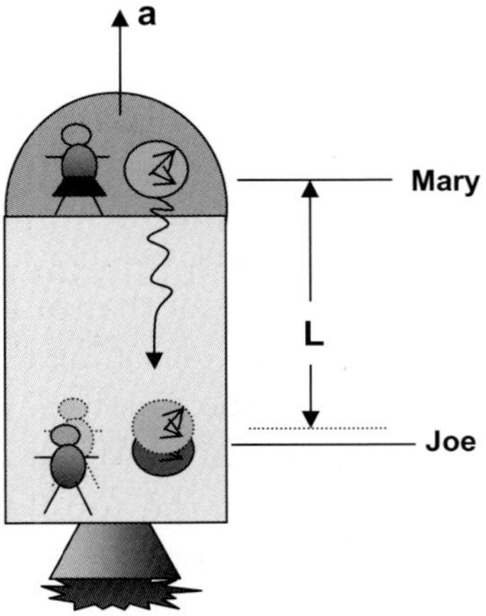

Figure 3.21. Mary and Joe compare their clock in an accelerating rocket. Comparisons of Joe's and Mary's clocks rate show that Mary's clock is running faster than Joe's clock.

Let us examine it in some detail. Assume that Mary sends a light pulse at one tick of her clock. Before it arrives at Joe's location, Joe has already moved a certain distance and is now at distance L from the position of Mary's clock when the pulse was sent. He receives the pulse earlier than he would have if he had not moved. Then, Mary sends another pulse at the second tick of her clock. If the rocket were moving at constant speed the difference in time of arrival of the two pulses at Joe's position would be exactly the difference in time of emission of the two ticks of Mary's clock. Furthermore, all subsequent pulses emitted by Mary would arrive with the same elapsed time. This is a direct consequence of the fact that if the rocket were moving at constant speed it would look to Joe and Mary as a stationary frame of reference. The two clocks would appear to run at the same rate. However, since the rocket is accelerating, Joe travels at a greater speed at the time of arrival of the second pulse that represents the second tick of Mary's clock. Consequently he receives the second pulse earlier than if he was cruising at a constant speed. At the time of the third pulse he is going still faster and the situation repeats itself. This process continues for the subsequent pulses and the time between pulses received by Joe is shorter than if he and Mary were cruising at constant speed. What does that mean to Joe? He concludes that Mary's clock is running faster than his own, although they have checked them carefully before they left on that trip. It is readily observed that the conclusion would be the same if it were Joe that sent the light pulses to Mary.

If the principle of equivalence is right, the same thing should happen in a gravitational field. Thus, two clocks situated at different height in a gravitational field, such as that of the Earth, should run at different rates. According to this reasoning the clock rate increases with height above the surface of the Earth or a clock rate decreases in a gravitational field. This is called the *gravitational redshift.*[35] If you are an employee working at the top floor of a tall building, your watch, when you reach the top floor, runs faster than on the ground. This looks almost like nonsense, but this is the way the world functions. The effect is very small, however, and even if your time at work is shorter than on the administration floor below you will receive the same pay.

A similar result is arrived at by reasoning directly about the characteristics of a photon. We have said earlier that the energy of the photon was $h\nu$, that is to say, Planck's constant h multiplied by its frequency ν. We have also said that energy could be materialized as mass through the famous relation $E = mc^2$. Then, it is very tempting to say that E in that expression is equal to the energy of the photon, which leads us to the concept of photon equivalent mass, $h\nu/c^2$. In the field theory of gravity we have introduced the potential energy of an object of mass m as mgH, where H if the height above ground. Consequently, a photon at height H in the Earth's gravitational field has an energy $h\nu gH/c^2$ added to its value $h\nu$ on Earth's surface. The difference in frequency between the two clocks is thus simply gH/hc^2. The photon clock, as characterised by its own frequency at height H, is running faster than when it is on the ground.

It turns out that the effect is about 10^{-16} per metre of elevation, thus very small. However, with the precision and accuracy of atomic clocks maintained nowadays in national laboratories, this effect has to be taken into account when scientists at the International Bureau of Weights and Measure (BIPM) in Sèvres, France, analyze the data originating from those laboratories. The altitude of the laboratories that provide the data has to be known. For example, in time keeping laboratories located in Boulder Colorado at an altitude of a few thousand metres, primary clocks are running parts in 10^{13} faster than those running at a surface close to sea level. In the context of timekeeping, which can be done to the order of nanosecond per day, or the equivalent of 10^{-14} per day, the effect is important.

In 1960, R. Pound and G. Rebka did an experiment in which gamma rays were propagated down towards the surface of the Earth from the top of a 45 m high tower. The gamma rays were photons falling in the gravitational field of the Earth. Their frequency was measured by means of a detector very sensitive to frequency shifts, a detector using Mössbauer effect, which provides extremely narrow nuclear absorption lines in certain solids. They found that the results agreed with the calculation within about 0.8%.[36] The gravitational effect on the rate of clocks was also measured in a very clean direct experiment performed in 1975 by C.O Alley and colleagues.[37] In that experiment Cs atomic clocks were flown on aircraft high above ground. The planes climbed in steps to 10,500 metres (35,000 feet) and stayed at that altitude for five hours. The experiment was repeated five times. The flying clocks at the various altitudes were running faster and the difference in time was measured against clocks that stayed on the ground. The agreement with the calculated time gained was well within 1%. Another very precise experiment was done by R. Vessot and colleagues[38] in the 1970s. A Hydrogen Maser, one of the most stable atomic clocks (to be described later) was flown on a rocket to an altitude of 10,000 km. The clock frequency was monitored all the way and an agreement with calculation based on the gravitational shift in the clock rate described above was within 0.01%. These experiments are probably some of the most convincing proofs of the validity of the equivalence principle.

Deviation of a light ray by gravity

Another effect that is predicted in assuming the principle of equivalence is the deviation of light rays in a gravitational field. This may be illustrated as follows. Suppose Mary and Joe are in their rocket and that a light ray is made to penetrate the rocket transversely through a small hole. If the speed of the rocket is constant the light ray will cross the space in a straight line and hit the wall opposite to the source at a point somewhat below the point expected if the rocket were not moving relative to the transverse trajectory of the light beam. However, if the rocket is accelerating, the light ray will hit the opposite wall underneath that point. Since they accelerate, Mary and Joe observe the trace of the ray across the rocket not as a straight line as in the case of constant speed but as a curved path since they are moving faster and faster as the light ray progresses across the rocket. Consequently, if the principle of equivalence is right, a light ray propagating close to a heavy celestial body would be deviated by the body's gravitational field. This was a prediction made by Einstein,[39] and it was readily verified by Eddington through observation of the apparent position of a star during an eclipse of the Sun, the star radiation passing close to the surface of the Sun. The deviation observed was in agreement with the predicted value.

Space-time curvature

The previous results appear to be explained by assuming that gravity is equivalent to acceleration. However, if you are asked what gravity is and you reply that it is acceleration, you will raise a lot of eyebrows. If you give the example of the accelerated rocket with its passengers Mary and Joe, you may interest more people, but questions will follow as to how the Earth navigates its way around the Sun as it does, on the basis that gravity is acceleration. You may find it easier to come back to Newton's gravitation theory, use the gravitational field concept, and answer the question using Newton's mechanical laws. But then the similitude to acceleration is not used at all. We are then back to the question as to what is that substance that spreads through all of space from an object of a given mass, a so-called field that creates forces on other object having mass.

In Figure 3.3(b), we have introduced the concept of curvature for a two-dimensional being, Flatant, living on the surface of a globe. We concluded that the radius of a circle on the sphere as measured by Flatant would be larger than the true radius of the circle within the sphere, which when multiplied by 2π gives the true perimeter of that circle as measured on a flat surface. The difference between the two radii is effectively a measure of the curvature of Flatland. We have also examined the possibility of drawing a square on that curved surface. When we used the rule of having 90° angles at each corner, we found that the lines did not close the square at the origin of the path followed. We may thus use these geometrical properties as a way of characterising surface curvature.[40] Can we do a similar reasoning in the case of three-dimensional space? It is clear that the advantage we had in the two-dimensional Flatland of seeing the curvature from outside in a three-dimensional environment no longer exists. We have to use some imagination. We can use a small sphere and measure its area. This area should be equal to 4π times its radius squared. Then we can measure its radius by digging a hole in it. We can do that in many directions and the measured radius can be compared to the radius deduced from a measurement of the surface. The difference would be a measure of curvature as in the case of the surface examined above. If the sphere contains mass, responsible for gravity, what should we expect? The answer may be obtained by imagining a fictitious experiment as we did previously. On the basis of the equivalence principle, we need something in acceleration and we need to make measurements of its dimension. A good example is a disk of a given radius set in rotation. Points on the outskirts of the perimeter are moving at a constant speed and consequently if a measurement of the perimeter is made it is expected, according to special relativity, that its dimension be altered. We note that those points on the perimeter are under acceleration since their direction of motion is changing. On the other hand the radius of the disk is not altered since its motion, as in the case of the photon clock in the train, is at right angles to its length. We thus have a curvature somewhat similar to that of a circle on a sphere: the perimeter is not equal to the expected value 2π times the radius.[41] We have curvature caused by acceleration and according to the equivalence principle we should expect similar curvature caused by gravity or the presence of mass. This is what is found from the General theory of relativity. The presence of mass causes a curvature of the surrounding space.

How about time? Referring again to the curved surface we examined in Figure 3.3(b), we found that attempts to draw a square

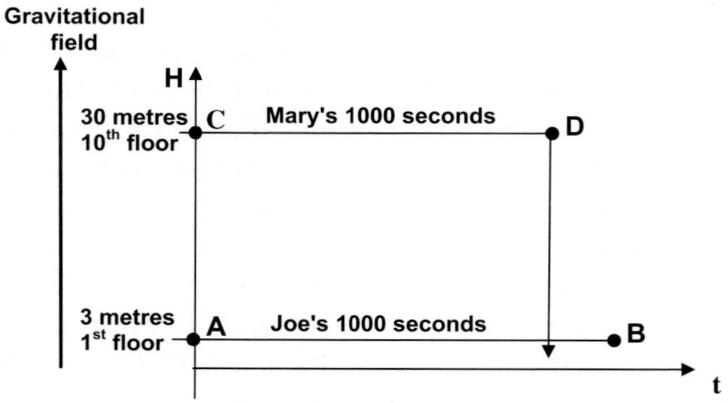

Figure 3.22. Illustration of curvature in the space-time diagram.

or a rectangle on such a surface led to one that did not close on itself. Can we imagine such a situation in our four-dimensional space including time? To obtain an idea about the possibility of curvature, we simplify the situation to a two-dimensional surface with only one space dimension on one axis and time on the other. In a gravitational field, if we try to make a rectangle on that plane we get into trouble as we did on the curved surface forming a sphere. This is illustrated in Figure 3.22. In the present case we have turned the two axes by 90° in order to have height, H or altitude up, and time horizontal.

The surface made by the two axes, t and H, is intrinsically curved. How can we see that? Suppose Joe is on the 1st floor of a building, identified as point A in the diagram. This may be 3 m above ground. Mary is on the 10th floor at 30 m altitude. If we let time pass by, say 1,000 seconds, we have a line that extends from A to B in Figure 3.22. However, Mary's clock is higher from the ground than Joe's clock and consequently as shown previously is running faster, the gravitational potential being weaker. The seconds of Mary's clock are shorter than the seconds of Joe's clock. Consequently 1,000 seconds in Mary's clock is a smaller time interval. This is shown by the horizontal line extending from C to D. If we try to close the rectangle from Mary's frame the lines overlap as in the case of the two-dimensional example of a surface on a sphere. We may thus say, as in the case of the curved two-dimensional surface, that we have a curved space-time, the curvature being caused by the presence of mass.

We thus conclude from these two fictitious illustrations that the equivalence of gravity and acceleration leads to the concept of a curvature of the space-time canvas caused by the presence of mass. What is the effect of such a curvature on the motion of objects?

We can imagine the path followed by objects in such a space by looking back at the path of objects attempting to go in a straight line on the curved two-dimensional surface formed by a sphere. In that context let us ask Mary and Joe to do another experiment for us. Let us ask them to walk on such a surface, in a path parallel to each other from a given straight line that we have drawn on the surface. We do not tell them that the surface is that of a large sphere and the starting line is the equator. Without knowing the actual geometry, they thus start at 90° from the equator of the sphere at points separated by 1,000 metres. Thus, their starting trajectories are parallel. They walk, as shown in Figure 3.23, keeping the angle of their path at 90° from the lines circling the globe all parallel to the equator. What do they find? They find that when they measure the distance between themselves with an accurate laser range finder they approach each other, or that the distance 'd' between them decreases. Since this means that, in their presumed flat space, they change direction, they conclude that there is a force present that pushes them towards each other: they believe that they are accelerated towards each other. An observer seeing the scene from outside will simply conclude that there is no force applied at all and that the inward motion is caused by the curvature of the surface they are travelling

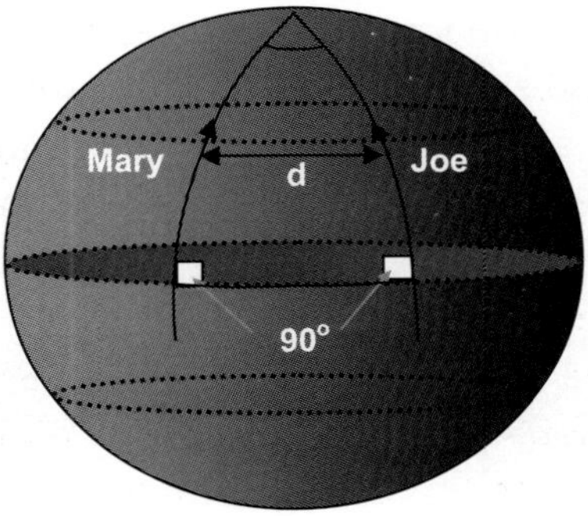

Figure 3.23. Mary and Joe feel something similar to acceleration when travelling on parallel paths on a curved surface, as if they were attracted towards each other.

on. They are guided by an imaginary grid that reflects the curvature of the surface.

Similarly, in the curved space-time canvas, in the presence of curvature caused by mass, objects travel in trajectories that look in our three-dimensional space as if acceleration were present. We may thus interpret the origin of gravity as the result of motion of objects on a curved space-time fabric. The trajectory of an object in that frame appears to us as deviating from a straight line since our point of view is limited to observation in the three-dimensional world we inhabit. This is what we interpret as acceleration towards a mass. Thus, in Newton's point of view the objects are attracted by a mass producing gravitational attraction, while in Einstein's point of view the space-time canvas is distorted by the presence of mass and objects travelling on that canvas appear to us as being accelerated towards the mass that creates the curvature. In Newton's point of view, the effect of mass at a distance takes place through gravitational force, while in Einstein's point of view, the local curvature of space-time influences the trajectory of the object. In other words, Einstein approach may be said as having transformed the concept of gravity from an action at distance interpretation to a geometrical interpretation.

The effect of the curvature of space-time on trajectories of objects is also often illustrated with the help of the motion of a glass marble on a flat plane with smooth shallow holes in it, like on a damaged parking lot with pot-holes as shown in Figure 3.24. A marble sent with some speed will be deviated by the pot-holes as if attracted to their bottom by a pseudo force. If the marble's speed is high enough, its trajectory will be simply deviated as in the case of pot-hole A; pot-hole B is very small and does not affect much the trajectory, while pot-hole C is so large that it reverses the trajectory of the

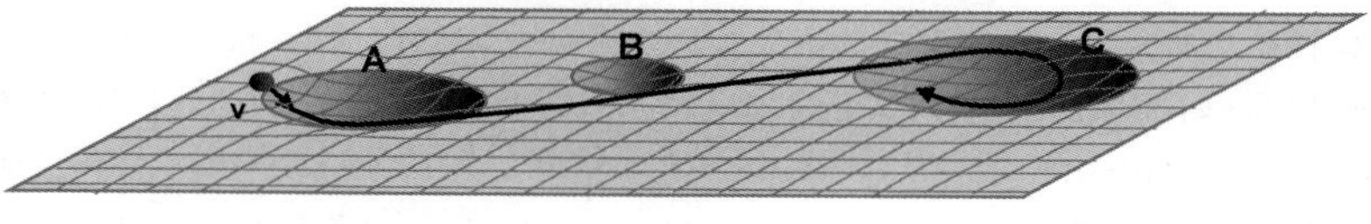

Figure 3.24. Representation of the curvature of space-time by a flat plane with pot-holes.

marble, which may then circle totally the hole before finally falling to its bottom. An observer very high above ground, very distant from the surface, will not see the potholes, and the marble will appear to travel as if attracted by some invisible objects at the bottom of the holes. This illustration provides a nice visualization of the situation. In the second chapter we made a trip in space and used exactly the phenomenon to project us in various intended directions.

We have examined two phenomena caused by the presence of gravity, supporting in a sense the equivalence principle: those were the slowing down of a clock and the deflection of light rays by the presence of a mass. A third phenomenon is explained by the presence of space-time curvature, that of the precession of Mercury's elliptical orbit around the Sun. The orbits of planets around the Sun are generally elliptical. However, because of its closeness to the Sun, Mercury experiences a stronger space curvature than the other planets and the effect is more pronounced than is the case with other planets. How is that shown in its orbit? Owing to this curvature, the relative dimension of Mercury's orbit is affected and the trajectory overshoots slightly the point of closest approach to the Sun. This point, called the perihilion, appears to move slightly relative to its position in space in the previous orbit. It is said that the orbit precesses in space. The precession, which is very small — 43 seconds of arc every 100 years — had been observed for some time but had remained unexplained by Newtonian mechanics. This is the third effect that was predicted by Einstein in the application of his theory of general relativity. The observation of those effects has given great credibility to the theory.

Deeper Thoughts and Relativistic Laws of Motion

An interesting property of the trajectory of an object in space under the effect of gravity is that the object acts as a lazy being: it takes what is called the path of least action. This concept is of such importance that it has been made a principle. In fact it is possible to deduce Newton's second principle from it. We have mentioned above that the energy of a given mass is made up of its kinetic energy, which is the energy associated with its motion and its potential energy, the energy that the object accumulates while it is transported, for example, from a given height in the Earth's gravitational field to a greater height. As amazing as it sounds, if you throw a ball up from the surface of the Earth, it will go up to such a height and take such a path that the difference between those two energies, kinetic and potential, will be minimised. This means that if you tried on paper, by means of a calculation to find a path that requires less action (kinetic energy minus potential energy), you will find that this difference is always larger than that calculated for the path that the object has taken. For example, a direct calculation shows that a ball sent in the air will follow a parabolic trajectory as the one shown in Figure 3.16 as predicted by Newton's equations. It looks very much as if the object evaluates the various possible trajectories by making a quick calculation and then takes the one of least action. It is a most interesting principle and shows that the physical world certainly hides subtleties just waiting to be discovered.

This principle is easy to apply to situations when the relativistic effects are small and can be neglected, which is the case in most experimental situations. However, the principle is still valid in cases when the velocity of the object in question becomes a good fraction of the speed of light, creates time dilation, and mass causes curvature of space-time. However, it becomes more subtle since now the rate of a clock depends on the speed of the object as well as the strength of the gravitational field and is intermixed with space. The basic idea is that an object sent up above the Earth surface will travel in a parabolic path at various speeds and will sense various gravitational potentials. The object rising above the Earth's surface senses less gravitational field and a wrist watch attached to it runs faster, while the object speed slows it down. It then turns out that, for a parabolic path with the end points fixed, a path in a gravitational field from point A to point B as in Figure 3.16, the principle of least action corresponds to maximising the travel time, as read on the object's wrist watch (proper time). In practice, the object appears to evaluate the optimum path in space-time in comparison to all other possible paths. The object in a sense fishes its way through space-time with all its bumps and shallows and follows a trajectory that appears to be of least action in the context of relativistic effects, that is, time dilation and curvature. (See reference 6 for more details on that phenomenon.)

The conclusion may be summarised in two simple laws: the space-time geometry is affected by the presence of matter and the trajectory of objects in that curved space is such as to maximise their travel time. Except for the three cases examined above, these laws lead in general for low velocities to the same results as those obtained with Newton's mechanical laws. This is not surprising since we know that Newton's mechanical laws have been used successfully for centuries and are still used today even in very complex situations.

The Speed at which Space Curvature Propagates

A basic particularity of the theory of relativity is the use of a characteristic of electromagnetic radiation without really saying or knowing why it is that way: in space, light travels at speed c. It is a ploy of nature. If light is created somewhere it immediately starts travelling through space at speed c for everyone in its path. Most fascinating! Is it a property of space or of electromagnetic radiation? Is there anything else that travels at speed c through space? Interesting questions! An object of mass m cannot, in principle, travel at the speed of light; in the reference frame in which its speed is evaluated, its mass tend to infinity and this is a barrier that cannot be overcome. We can accelerate particles very close to the speed of light, but the particle never reaches speed c. It looks as if the speed of light were an entity that belongs to a world without mass. How about gravity? In Newton's days it was thought that the effect of gravitation was instantaneous. Let us assume that something happens for example to the Sun, say it disappears instantaneously. The Earth would feel no longer any attraction and would be flying away in space immediately, in a straight line. However, we would still see light coming from the Sun for eight or more minutes. Our situation would be quite confusing. In Einstein's theory, gravitational effects propagate at the speed of light: it seems to be an absolute speed that characterises the space-time canvas.

How about actual gravitational perturbations passing by, caused by collision or collapse of large structures in space? Those perturbations would look like waves. Is there energy transported in such gravitational radiation? Could we detect those waves? It is possible to show how such a wave can be generated in space. In an elegant cartoon chosen by Taylor and Wheeler,[42] a great giant Atlas plays a game in space with two large masses. He throws the two masses away in opposite directions. Their mutual gravitational attraction slows them down and finally reverses their direction of motion. They come back as if they were attached by elastic bands, are caught by Atlas on their way back and sent again at such a speed that they reach the same distance again; they then come back again because of their attraction. This activity is repeated at will. A simple argument shows that Atlas must spend some energy in the process because he does not get back as much energy when he captures them as he spends when he throws them out. The reason is that when the masses travel, the information on their position or on the curvature of the space-time canvas is transmitted at the speed of light. Thus, when the masses approach Atlas, the force of attraction reflects a greater distance than the actual one due to the delay in the information transmitted between the masses because of their distance. The force of attraction at any instant is less and Atlas gets back at any instant less energy from the masses than he would get if the information was transmitted instantaneously. On the other hand when he sends them back the force that attracts the two masses reflects a shorter distance again owing to the delay in information transmitted. Thus, more work is done in the motion outward than in the motion inward. Atlas does some net work and since energy is conserved, some energy is lost in space as gravitational radiation. It leads to what is called gravitational waves. These waves would propagate in space at the speed of light. As suspected, since the force of gravity is so small, the intensity of such waves is very small or, in other words, their effect on surrounding objects is extremely small. We may be traversed by these waves, but we do not feel them because gravity is a very weak effect.

How could such waves be detected? We have seen that gravity changes a clock's rate and consequently changes length. The idea is then to detect the changes in length caused in an object of large dimension. This may be realised by means of a very large interferometer, with extremely long arms, kilometres long. If a gravitational wave passes by, the position of the mirrors within the interferometer would be changed and the length of the arms of the interferometer would be altered creating a visible effect on the observed interference pattern. It is thus a wait-and-see experiment, out of the control of the experimenter. How big is the effect? The change in length is extremely small and calculated to be of the order of parts in 10^{21}. This is at the limit of sensitivity of present technology, but progress is being made.[43]

However, gravitational radiation emitted by a rather exotic celestial object has been inferred by means of precise time measurements. The celestial object of interest is a so-called pulsar, which is essentially an object having a diameter of about 10 to 30 km, totally made of neutrons and turning rapidly on itself. It is the relic of a star explosion known as a supernova. It is called a neutron star and possesses an extremely strong magnetic field that may be oriented in a direction different from its axis of rotation. Electromagnetic radiation originating from various causes such as decay of the magnetic field or transformation of gravitational potential energy of accreted matter is emitted along the magnetic field axis. Since the axes of rotation and the magnetic field are misaligned the radiation sweeps through space as a beacon and may be visible at planet Earth arriving as pulses with a period equivalent to the rotation period of the pulsar.[44] The period may be as small as milliseconds and is found to be extremely stable for short measurement times, the angular momentum of such a massive object being very large. In fact, the stability of the period of such emission is comparable to that of atomic frequency standards. In a particular case, it was found that one such pulsar existed in a binary system in orbit with another star with individual mass of about 1.4 solar mass.[45] The two stars travel on elliptical orbits around a common centre of mass. Space curvature is thus extremely pronounced at their location and an analysis of the pulse period showed that their orbit was precessing as in the case of Mercury but much more rapidly, at a rate of 4.2 degrees per year. Furthermore, it was also found that the orbital period, as reflected in the behaviour of the measured pulse period, was decreasing at the rate of about 75 microseconds per year. This corresponds to a loss of energy. It was concluded that the energy was lost as gravitational energy or gravitational waves in a way similar to the Atlas cartoon described previously. It is a rather fascinating subject since it makes explicit an observation that allows conclusions that confirm predictions of the theory of relativity. No gravitational waves, however, have been detected directly yet. The direct observation of such waves would be a great achievement in support of the theory.

Is this Whole Thing Right?

On our planet, we live in a three-dimensional world and we see time passing by independently of that space. Both appear absolutely independent in our daily activities as inferred in the application of Newton's principle. However, here comes a theory supported by measurements that tells us to abandon totally that point of view and consider time and space as totally relative and intimately interlaced. Their marriage is indissoluble. The basic idea is that uniform motion is relative, that the speed of light is the same in all directions independent of the speed of the source or of the observer, that the effect of gravity is the same as acceleration and that all laws of physics are the same in so-called free-float frames. The results that follow from those principles are most fascinating: mass becomes a materialization of energy, time flows at different rates in frames in relative motion, measured lengths are not the same in those frames. On the other hand, gravity appears to be equivalent to acceleration; gravitational attraction between objects with mass is a result of the curvature of the space-time world we live in; objects with mass sense the curvature of the space-time canvas and follow in their displacement a trajectory that is independent of their mass. This has led to the statement that gravitational mass is the same as inertial mass. In the study of the Universe at large, that is to say cosmology, what does space-time curvature mean? This brings us back to the question I was asking myself as a kid about the size of the Universe. Is it finite or infinite? With the presence of all those galaxies with mass the curvature of space could very well be such that the whole aggregate of celestial bodies is finite but without ends, without borders. That sounds like nonsense. Let us look at it through the example of two-dimensional surfaces. In a flat Universe

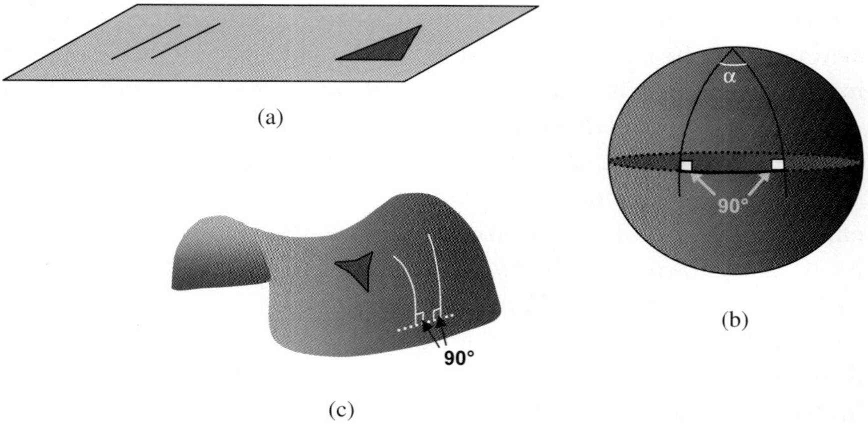

Figure 3.25. Two-dimensional spaces of various curvatures: (a) flat surface, no curvature. (b) a sphere, surface of a globe, closed space, (c) surface of a saddle, opened space.

shown in Figure 3.25(a), Euclidean geometry applies directly. Parallel lines don't meet. They extend to infinity and never join, never cross each other. The sum of the internal angles of triangles is always 180°. On the curved surface of a sphere, two lines starting parallel at the equator, thus making an angle of 90° with the equator, meet at the poles. The sum of the internal angles of a triangle on that surface is larger than 180°. This is obvious in Figure 3.25(b). The angle α is added to the two angles of 90° at the equator, which already add to 180°. This is true for any triangle drawn on that surface. There is another type of surface that we have not considered; it is that of a saddle. In that case, shown in Figure 3.25(c), the sum of angles of a triangle is less than 180° and parallel lines diverge.

It is possible to characterise those geometries by a parameter K related to their so-called radius of curvature. In the case of the sphere, for example, the radius is R and we define the curvature as $K = 1/R^2$. We say that the curvature is positive and the equivalent space is spherical and closed. We conclude immediately that the plane of Figure 3.25(a) may be said to be the surface of a sphere that has an infinite radius and the curvature is $K = 0$. In that case space is said to be flat and is open. In the case of the saddle we are in presence of two radii of curvature, one for the curving up geometry and the other at right angle for the curving down geometry. If the saddle is symmetrical the radii in the middle, at the saddle point, are identical but of opposite signs and the curvature is said to be negative, $K = -1/R^2$. The equivalent space is said to be hyperbolic and is open. These characteristics are easily visualised for two-dimensional spaces. What is the curvature of the whole Universe? If we assimilate its curvature to that of the surface of a sphere, the Universe is closed on itself, like the world of the two-dimensional Flatant of Figure 3.3(a). In other words, if you walk in the same direction you will come back to your point of origin. A light ray sent in one direction will come back to the source that has created it. But if the Universe is open as in Figure 3.25(c) the Universe is infinite and there is no coming back. At present there is no answer to the question although, as we will see, all measurements tend to lead to the conclusion that the Universe is so large that for us it is essentially flat. Consequently, there is no answer to the question that was haunting me. However, in any case, closed or infinite, one cannot fall off the edge of the Universe: there is no edge.

There is another way of summarizing the picture of the world we have just presented. It appears that the special theory of relativity has joined space and time into a single indissoluble concept or entity, a canvas over which the world evolves: this was done by accepting a conspiracy of Mother Nature, the constancy of the speed of light. On the other hand, the general theory of relativity has joined gravity and geometry together into a single concept over that canvas: this was done by accepting the principle of equivalence. These were radical steps in the evolution of our understanding of the basic laws governing the functioning of our Universe. We are still living with them today and there are no serious alternatives to that point of view.

However, the mind of the physicist is always open to questions. In physics, scepticism is as basic a principle as the others we have just outlined. Is this whole picture of the Universe at large a true representation of reality? All experiments up to this date have confirmed the predictions that follow from the theory. Nothing wrong has been found yet. As we said: 'If it works it must be right'. But does the theory answer the fundamental questions? It seems to displace the problem to a higher level. We asked what is mass. The answer was that it is energy that materializes. How is this done? It is a great question! Present theories, although not yet demonstrated experimentally, point to an answer. We will re-examine this later in the context of our description of the world of elementary particles.

We may consider at this point the evolution of physics to the level of understanding achieved in the models described above. It shows us how progress is made. In Newton and Einstein's approaches, reflection on observations and experimental facts led to the formulation of a few great principles. Those principles could be formulated in words and basic equations were written. The theory developed led in turn to phenomena directly observable in nature. Conclusions were transparent. We will see that in the case of electromagnetism and atomic physics, the same affirmation can be made. Observations led to the formulation of great principles such as the wave nature of matter and the identity of the source of electric and magnetic phenomena. In the present state of development, however, it appears that new physics theories become so complex that it is very difficult to express in simple words the principles at their foundation. The theories are based on concepts with a meaning not evident in common language. Is this the end of physics as we have known it? Will we find further simple basic principles that are still hidden from us? Will such concepts as force, energy, matter, space-time, basic to our present understanding, be sufficient? Or are we destined to be lost in complex mathematical constructions that appear to be fascinating for their symmetry and elegance but disconnected from the possibility of verifying them, their predictions being at present out of the range of any possible experimental measurements? We should not be discouraged. At the time that Einstein made his theory, there was no atomic clock to verify some of the conclusions he arrived at; we have achieved that very recently. At present, technology develops rather rapidly. As we will see in a later chapter, measurements of the Universe at large allow a verification of predictions made by new, tentative unifying theories. We may be able to extract from the combined theories and experiments basic simple principles that were hidden from us because of our limited understanding in earlier times. In the future we may very well be able to continue on that route traced by those giants.

IV

Conservatism and the Universe

Our Universe is acting in a baffling way. Although it is continuously evolving and looks very dynamic with objects moving around, changing themselves into other entities, there are certain traditions and possessions that it wants to keep, like an old greedy person. When you have strong roots it is hard to break the bonds and you act behind the scenes, attempting to keep things the way they have always been. The Universe acts that way in many regards.

Energy It Has and Energy It Wants to Keep

A most basic principle of the Universe is that in isolated systems, *energy is conserved.* We have used that principle in previous chapters. Let us look at it in more details. What is energy? We already asked that question. We gave examples of the form that energy can take. Does it come up in small pockets of some kind of substance? Can we identify it somewhere, isolated as so-called 'pure energy', waiting to be used? The reality is that, although we can describe many situations where energy can take form, we do not quite know what it is. Energy, for example, can be associated with heat and we can say for example that a hot object contains more energy than a cold one. At one time it was thought that in the exchange of heat between bodies at different temperatures a substance was flowing from a hot object to a cold one. For example, in getting water to boil with a wood fire, some substance would flow from the fire to the water. The substance was given the name 'caloric'. But there is no such thing as a flowing substance between a hot object and a cold one, and this was recognised by Lord Thomson who, in the study of the boring of cannons, observed that the temperature of the bored metal cylinder was rising without any contact with a hotter object. Just the boring was causing the rise in temperature. There was no 'caloric' flow.

We said that we did not know exactly what energy was. However, we can get more familiar with the concept if we describe more situations where it takes form. In simple language we might say that energy is the name we have given to a specific property of an object and this property can take many forms. We feel the Sun's radiation on our skin and warming our body: *radiation* is one form of energy. A car moves at a certain speed: it has some form of energy that we call *kinetic energy*. We talked earlier about the *potential energy* of an object raised a certain height in the gravitational potential of the Earth. In that particular case, we said that the potential energy was equivalent to the work that we had to do to raise the object from the surface of the Earth to that given height. An object at a certain height above the Earth has more potential energy than when it is on the ground because we had to do work to bring it up and that work seems to be stored in the object. If we let the object fall from that height it accelerates under the force of gravity, it acquires speed. When it reaches the ground level, the potential energy is smaller. Where is the potential energy gone, what is left of all that work we did to bring the object up? It has been transformed totally in motion energy, kinetic energy. When it hits the ground the object stops. Where does the energy go? Into heat. Deformations, vibrations of the particles of the object and of the ground at the time of collision create heat and are some form of energy that adds to the *internal energy* of the object and the ground. The process is illustrated in Figure 4.1. The temperature of the object and the surrounding ground is raised in the process. This is a situation that happens very often. A bullet from a rifle travels at very high speed and has tremendous kinetic energy. When it hits the target, that energy is transformed mostly into heat in the target and the residue in kinetic energy of the target that moves slightly. When we step on the brakes in our car, the kinetic energy of the car is transformed into heat in the discs and pads of the braking system.

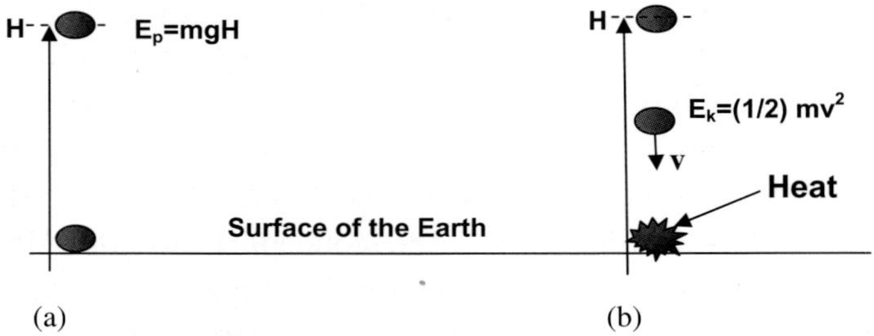

Figure 4.1. Illustration of energy conservation in a typical situation where an object is first raised to a height H in Earth's gravitational field (a), and then dropped from that height (b). In (a) the object has potential energy mgH which is transformed into kinetic energy in (b) and finally may be destroyed in hitting the ground. The kinetic energy is transformed essentially in heat raising the temperature of the object residues and the surrounding ground.

Energy is thus an entity that may take many forms and may transform into one form or another relatively easily. We are comfortable enough with the concept of heat as a form of energy. We are familiar with the use of heat to do work: in an old technology, heat in boilers produced hot water vapour that activated pistons and moved a locomotive, becoming kinetic energy. But what is in practice kinetic energy? We have an expression for it: an object with mass m moving at speed **v** has kinetic energy equal to one half the product of the mass and the square of its velocity ($(1/2)\,\mathrm{mv}^2$), as illustrated in Figure 4.1. How do we know that expression to be true? How do we know that this is the quantity that, in a collision, will be transformed into heat, and not something else, with v at the third power for example, with a constant in front to keep units right? The definition comes directly from Newton's second principle relating force to acceleration and Galileo's principle of inertia. The work done in doing the acceleration is transformed into kinetic energy and that expression comes out directly from simple mathematical logic. Consequently we can rely on that concept and apply it to many circumstances where kinetic energy is transformed into other forms of energy or vice-versa.

We said earlier that mass is actually the materialization of energy. The expression $\mathrm{E} = \mathrm{mc}^2$ is a magisterial expression of the equivalence of mass and energy. Since, as we have shown earlier, mass is a function of speed, it thus contains also the kinetic energy of the object of mass m. When the kinetic energy is taken out of the equation by means of a simple mathematical trick, that we call expansion in series, one term becomes m_oc^2 where m_o is called the rest mass of the object, that is to say the mass of the object at rest in a given reference frame. This mass in turn may be transformed into heat energy as in nuclear reactors. An example in reverse is the possible transformation of a gamma ray photon into an electron and its antiparticle, a positron. In that case, the energy of the photon is transformed into two resulting masses and their kinetic energy. The statement that energy is conserved includes all forms of energy: mass, potential energy, kinetic energy, electrical energy, magnetic energy, chemical energy, light, heat, which is molecular motion or vibration, and so on. In analysing a particular situation, all these forms of energy must be taken into account when the principle is used. It appears that the law is obeyed at all levels. Energy is a concept that we represent with mathematical formulas. In a given system, we get numbers that we add together and when we change things around, the sum remains the same, although the energy may be distributed in a different way. We thus know how it works, but we do not know why. It is very logical and has never been proved wrong. The whole of physics is based on it.

We should mention that in giving units and numbers to energy, we come to the point, when considering gravity, that it is characterised by negative potential energy. That is to say, if we let two objects start from very far apart get close to each other we get some positive energy from them. When we separate them we have to do work and we spend energy. It goes into the objects. When they get close they give that energy back. It is subtracted from the internal energy of the objects. We can thus say that their energy of interaction, their potential energy, is negative. This is most interesting, because if matter is created somehow in vacuum, as we will see later in the study of the expansion of the Universe, that matter can be considered as positive energy through its mass, but gravity, which enters into action immediately at the creation of the objects has negative energy compensating for that creation of energy through mass. The inventory may be zero. It is a free lunch as is sometimes said. We will come back to that point later.

It thus appears as a fact of nature that energy is conserved. We started to look at the concept in isolated systems. We must realize that we can make the system as large as we wish. Can we expand it to the Universe? There is no formal proof of it, but the total energy in the Universe seems to be constant.[1]

Don't Get in the Way: I Move and I have Momentum

Another important quality of motion is also concerned with its speed. We have introduced it earlier. It is called the quantity of motion or momentum. As with energy it appears to be attached to the object in motion. It is just the product of mass, m, by velocity, **v** of the object. We write it m**v** and we usually call it '**p**'. It seems to be a very reasonable and simple property. Is it of any use? It is very different from energy, which is proportional to the square of the velocity. Momentum has direction while energy does not. How does it behave when an object is in interaction with its surroundings, for example when it hits another object and bounces in another direction? The result is not evident at all. However, it is shown directly from the second and third laws that in a collision between two objects, momentum is conserved. This property is also embedded in the principle of relativity.[2] This is a most interesting property since it gives us a new equation with velocity at the first power.

Let us look at it more carefully. Suppose the objects are two balls of equal mass at rest on a billiard table. Since they are not interacting strongly with the rest of the world we can consider our two balls as an isolated system. We send one ball towards the other head-on. We have given speed to the first ball and it has acquired a momentum equal to the product of its mass by its velocity. The collision takes place. We assume that the collision is totally elastic. By this we mean that during the collision the kinetic energy is not transformed into heat through friction or vibration within the balls. After the collision, the velocity of the first ball is zero while the second ball has acquired a velocity equal to the velocity that the first ball had before collision. The momentum has been totally exchanged from one ball to the other. The momentum is thus the same before and after the collision. We can do the same experiment with balls with different masses. We find that the velocity of the two balls is different after the collision, but again we find that the sum of the product of their respective masses and velocities, that is to say, the sum of their respective momentum, is the same before and after collision. It is also true when the two balls are in motion relative to each other and hit each other at any angle. It is fascinating. These examples illustrate what is called the principle of *Conservation of momentum.* It is the principle that applied in Rutherford's experiments that led to the discovery of the atomic nucleus. In that experiment one of the billiard balls was an alpha particle consisting of two neutrons and two protons, in fact a helium nucleus, and the target was a thin film of metal, gold. He found that most of the alpha particles went through as if there was nothing in the foil but some of them deviated at large angles and some were even bounced back in a total reversal of their motion as if they had hit a heavy target. That target was the heavy nucleus of the gold atoms in the metal film. From that experiment Rutherford concluded that matter is essentially empty space but contains very small massive particles. In his case the α particles were sometimes hitting a heavy gold nucleus and were bouncing back with a speed compatible with the conservation of momentum. The principle applies also to sticky collisions, when a ball impacts for example on another ball made of putty. After collision, the ensemble moves with a momentum equal to the momentum of the first object. The mass of the ensemble is larger but the speed is less.

Such a law finds a very practical use in some of the most elementary applications. The operation of the sail of a yacht, for example, is based partly on the conservation of momentum: the molecules of air moving with the wind hit the sail and transfer their momentum to the sail and the boat. It is the same thing with light and radiation in general. Photons do not have mass, but carry energy and momentum. Upon collision with a reflective surface they reverse direction, their momentum becomes negative, and since the total momentum is conserved the surface itself must acquire positive momentum. Some artificial satellites might use this phenomenon to provide small corrections to their direction. It is not impossible that such an effect could be used to acquire speed in space travel, the photons in space being cheap (aside from the cost of getting there!).

However, owing to the smallness of the photon momentum, very large sails would be needed to reach any substantial speed.

The billiard balls that we have looked at above have kinetic energy. In the head-on elastic collisions we considered, the momentum was completely exchanged. In that particular case it meant that the balls had exchanged velocity. If so, the kinetic energy was also totally exchanged and was conserved as well as the momentum. It is a most interesting situation: in an isolated system, thus, both momentum and energy — although energy can change form — are conserved.

A very nice toy that can be used to illustrate that situation is the so-called Newton's cradle. It is available in specialist stores. One consisting of five suspended balls is shown in Figure 4.2. I am always amazed by its simplicity. The principles just evoked are best demonstrated with just two balls in motion, the other three being raised and removed from action. With all the balls in the game the system is more complex relying on the exchange of energy and momentum through the various balls, but still demonstrates principles just introduced.

The principles of conservation of energy and momentum are fundamental principles of physics and are at the basis of its

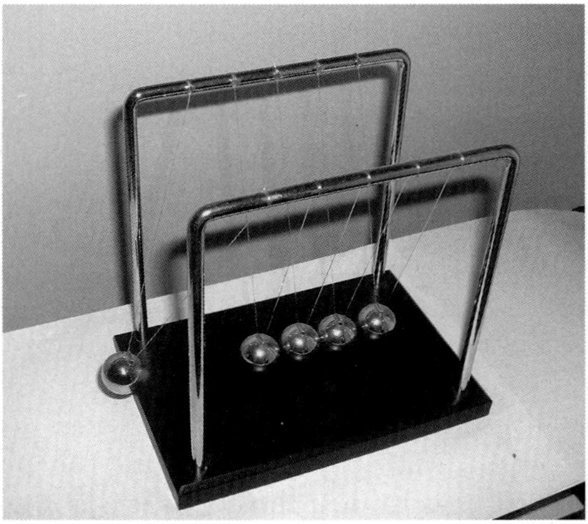

Figure 4.2. Newton's cradle. It is a rather interesting toy for illustrating the principles of conservation of momentum and energy.

development and of our present knowledge of the functioning of the Universe. For example, they are at the core of the research done on particle physics with large scale accelerators. Particles such as protons or electrons are accelerated to great velocities close to that of light. In the laboratory frame, at that speed, they acquire extra mass through the effect of relativity. They have a tremendous momentum and energy. When they collide with similar particles travelling in opposite directions, their energy is transformed into other entities, photons and other material particles with their own kinetic energy. It is possible to withdraw a substantial amount of unique information from the energy and direction of motion of the resulting particles and photons as well as to speculate on the internal parts of the particles that collided. It is essentially like throwing two watches at great speed against each other and from the pieces determining the properties of the internal gears of the two watches. This is the way most experiments are done in laboratories on particle physics research. For example, the new LHC instrument in CERN[3] in Europe will accelerate protons to a speed close to that of light with an energy greater than 10 tera-electron-volt ($1\,\text{TeV} = 10^{12}\,\text{eV}$) and the by-products are expected to provide new information on particles that are predicted in presently developed theoretical models.

The principle of conservation of energy plays also an important role in our organisation as a modern society relying on sources of energy to maintain our way of life. Electricity is at the heart of the progress made in our living standard over the last century. How is that electrical energy produced? We have all been astonished at one time or another at the sight of these large water reservoirs maintained by skinny concrete dams, a marvel of modern engineering. In a typical case, the water above the dam may be at a height, say 100 metres, above the water below the dam. What is the potential energy of a volume of one cubic metre (m^3) of water at that height relative to the energy it would have if it was below the dam? It is very simple to calculate. A cubic metre of water has a mass, m, equal to 1,000 kilograms (kg). Its potential energy above the dam is simply $\text{m} \times \text{g} \times \text{H}$, where g is the acceleration of gravity that we have introduced earlier. It is about 10 metres per squared second (m/s^2). The potential energy of that mass of water is thus 1,000,000

($kg\,m/s^2$). This combination of units, kilogram, mass and squared second is called a 'joule'.[4] We thus have a potential energy of 1 million joules stored in that cube of water one metre on its sides.

How do we transform that energy into useful easily transportable electrical energy? The metre cube of water it let down through a tunnel and flows through turbines. If it goes through in 1 second, it will develop a power of 1 million joules per second or 1 million watts, forcing the rotation of the turbine and transforming its potential energy into mechanical energy. In turn, the turbine forces an electrical generator into rotation producing electricity. Since the efficiency of the turbine and the electrical generator is not 100%, a small amount of the energy is lost as heat in the whole system. A big reservoir at a height of 100 m, delivering 1,000 m^3 of water per second to efficient turbines, would thus develop a power of nearly 1,000 megawatts. This is approximately the power delivered by a good-size hydroelectric plant. Energy is conserved all the way through the transformation from potential energy, to mechanical energy, to electrical energy, and a small amount of heat that is considered a loss. It is one of the most efficient ways of producing clean energy.

Another nice example is that of a nuclear reactor. In that case, the potential energy is not contained in a mass at a given height in the gravitational field of the Earth, but in the mass itself through the relativity connection between mass and energy, $E = mc^2$. A nuclear fission reactor transforms mass into heat and the heat is used to boil water, producing steam, driving a turbine, which finally produces electricity by means of a generator. The steps involved are a little more involved than in the hydroelectric plant, but the principle of conservation of energy applies over the whole process of transformation. The energy comes from the splitting of uranium atoms into smaller atomic constituents whose total mass is less than the mass of the original uranium atom. This difference is minuscule but the numbers involved in the transformation are so large (mc^2) that the quantity of energy becomes astronomical. In fact a single gram of matter transformed totally into heat energy would produce a total of almost 10^{14} joules. With an efficiency of about 30% in transforming heat into mechanical energy in the turbine, the transformation over a period of one year would produce a continuous power of 1 megawatt, about the power delivered by a large windmill tower, but coming from a quantity of matter barely filling a teaspoon.

Energy, Heat, Power, Perpetual Motion and all that Stuff

This whole subject of energy transfer and conservation has given rise to a most fundamental and elegant physical theory that we call Thermodynamics. As its title suggests it is concerned with heat, but is essentially a science of energy transfer or heat dynamics. It can almost be thought of as a kind of philosophy. It is fundamental to the understanding of the whole of physics and in particular to the processes taking place at large in the whole Universe since its beginning. Thermodynamics is based on principles that are most satisfying to the mind. For example, one law deals with equilibrium and may be written as follows: 'two objects isolated from each other but at the same temperature as a third one are themselves at the same temperature'. It is quite logical, almost obvious. Actually, it can be understood as a definition of temperature. For example, we could state the following: two objects in thermal equilibrium with a third one are in thermal equilibrium with each other. We then say that, in such a case, the three objects are all at the same temperature. It is sometimes called the 'zeroth' law of thermodynamics. The numbering incites us to think that in the development of thermodynamics science, it had almost been forgotten as too obvious. The first law is the principle of conservation of energy that we have discussed. Again, its logic is very satisfying to the mind. When we reflect on that law, we find that it has a deeper meaning than just conservation: it also says that the Universe does not allow us to create energy from nothing; *perpetual motion is not allowed by nature.* The quantity of energy you get out of a machine is at most the energy that was stored in or that you have put in; you only change its nature, electrical, thermal, mechanical, nuclear, etc. When you fill your car's reservoir with petrol, you pour in joules of potential chemical energy that will become mechanical energy when reacting with oxygen in the cylinders of the engine. We hear often of that 'genius' who has invented a 'black box' that when connected to the electrical company entrance box of their house reduces consumption of energy and their electric bill. No way. Energy is conserved and the only thing a box like that can do is dissipate heat, changing the quality of the energy that was available in the wires and increase the energy bill. In fact, as far as we are concerned as consumers, there is quality in the type of energy available. For example, electricity is very high quality energy, because we can transform it in many other types, such as mechanical, light, heat, with great efficiency. It is not the same with heat itself: if it is used for heating a house then it is used directly and is thus very efficient. But if we wish to produce mechanical work with heat, the processes we know as those used in the steam engine are rather inefficient.

To produce mechanical energy from heat, experience shows that we really need two baths, one at a higher temperature than the other. In the transport of heat energy from the hotter bath to the colder bath one can extract some of that energy and do mechanical work such as lifting an elevator or running a turbine. This is what a steam engine does. It extracts heat from a hot furnace, produces steam, activates a piston, and rejects the heat that was not transformed to the cold bath, which is the environment. This process has been found in practice to have a maximum efficiency, which is simply the difference of absolute temperature between the two baths divided by the absolute temperature of the hotter bath. This efficiency is generally not larger than 40% due to the limits of temperatures possible with the materials used. In practice this can be stated rather simply: *it is impossible to build a machine whose only function is to extract heat from a reservoir and produce mechanical energy.* This statement is actually one of the representations of the second law of thermodynamics. No one has succeeded in building a ship that pumps water from the ocean, extracts heat from it, produces mechanical energy for turning the propellers, and transforms the water it has pumped into blocks of ice thrown overboard. Such a contraption would be a perpetual motion machine, not violating the conservation of energy principle since no energy is created, but violating the second principle. It would be rather interesting, but it does not exist. We need a bath colder than the ocean on which the ship floats to make the machine possible. This is a second law that nature has put in our way in order to keep us reasonable and make us work and be imaginative. If someone comes to you and tells you that he has invented a machine that transforms heat into mechanical energy, ask him about the temperature of the baths he is using and the efficiency of the machine. If the efficiency he proclaims is larger than the efficiency you calculate from the temperatures of the baths, do not invest in his company. When I was active in teaching I often received messages from people that claimed to have invented a gadget that does exactly what is forbidden by the second law. An interesting one amounted to extracting energy from the motion of electrons in solids. Since the electrons are in motion they have energy. Why not use it since it is free? What the device would do is cool the solid without external intervention and produce electricity. This is not possible. The idea is similar to that used in a device described by Feynman.[5] It consists of a paddle wheel attached to a ratchet; thus the wheel can turn only in one direction. If you leave it in air, under a glass bell jar to avoid air currents, there will be at some instant, by chance, some differences in the random speed of the molecules of air hitting the two faces of one of the paddles. A net momentum will be transferred from the molecules to the paddle and a greater pressure on one side of the paddle than on the other side will result. Since the wheel can turn only in one direction, it will start rotating in that allowed direction. A fantastic machine, but it does not work, essentially because the random motion of the molecules would also act on the ratchet itself and the wheel would turn equally well in both directions. I have received similar proposals with magnets arranged in a contraption that can baffle the most lucid and objective physicist. They were all supposed to produce energy without input. The second principle tells us that this is not possible. However, it is extremely difficult to convince those people that it will not work. Often, when you are too harsh in refuting the arguments, the inventor will make a threat that he will tell an enemy country about the idea. It is almost a question of faith: you cannot argue rationally with faith in the background. However, physicists must be vigilant and honest: it is neither fair nor ethical to let those people think that their invention may work. Although I knew that the persons would be frustrated by my analysis, I never hesitated to tell the truth to the ambitious inventors, who of course were always very disappointed by my conclusions.

The second law introduces the concept of absolute temperature. We are familiar with the Fahrenheit and Celsius scale based on some easily accessible fixed temperature points, such as the freezing and boiling points of water. In the Celsius scale the number 100 is set at the boiling point of water and 0 at its freezing point. One then uses a convenient substance that changes dimension as linearly as possible when temperature changes: for example, the thermal expansion of a small volume of mercury in a small bulb forces it into a capillary tube, the height of which can be read in the column. Then, assuming that the expansion is linear, it is relatively easy to calibrate the column of mercury at the two points in question and interpolate between the two fixed points. The device can then be used to some extent outside that range by extrapolation. In the theory of thermodynamics, an absolute scale of temperature, also called the Kelvin scale, is introduced. This scale is based on the efficiency of heat engines that cannot exceed 100%. It leads to the absolute zero of temperature where it is no longer possible to extract heat from an object cooled at that temperature. It is not possible to cool objects below that absolute zero. The graduation of that scale has been made such that the distance between the rungs is the same as in the Celsius scale. This situates the absolute zero as −273.15 degrees Celsius. At that temperature, all activities on a classical scale cease. At that temperature the world is frozen solid, whatever that means. But what really is temperature? It is a measure of the level of excitation of atoms forming an object. In a gas, it is connected to the speed of the molecules travelling around and hitting each other. In a solid the atoms or molecules cannot move around. Temperature is a measure of the amplitude of vibration of those atoms or molecules. In a liquid, motion is inhibited, but atoms are still moving around somewhat and temperature is a measure of the random motion or vibration of the atoms. Then what is zero Kelvin? It is the temperature of an object that is cooled to such a state that all atoms and molecules have stopped moving, vibrating, or turning on themselves. Can it be achieved experimentally? Almost, but not quite. Quantum mechanics tells us that there is a minimum energy that cannot be extracted from a system in vibration. But that energy is very small and for all practical purposes we can have atoms almost at rest in special cooling traps that have been implemented recently. Such a cooling trap is a fascinating device in which photons are made to enter into interaction or collide with an atom. During the collision, due to special conditions regarding the wavelength of the photon and the resonance frequency of the atom, the photons gain momentum and since momentum is conserved, the atom on its side loses the same momentum. This means that since the atom has less speed it has lost kinetic energy and is thus cooled. The energy is gone into the scattered photon that has a higher frequency. It takes about 10^5 photon-atom collisions to stop an atom initially moving at about three hundred metres per second at room temperature. It is now standard practice in some special devices, such as atomic fountain clocks, to form an aggregate of more than a million atoms as a small ball about one cm in diameter and with very little motion within the ball. Since these atoms are essentially at rest we say that their temperature is close to zero Kelvin and their properties can be studied at will in that state.

We are all Guilty of Vandalism

A concept of thermodynamics embedded in the second law is more subtle, but can be expressed in words as well. It says that whatever we do in the Universe, we increase disorder. We are all guilty of vandalism. All our actions lead to greater disorder whatever we do. In the science of Thermodynamics disorder is called *Entropy*. Long ago, I had bought an old used sail boat, a so-called cat-boat that a colleague of mine, Bob Vessot, the physicist who implemented that beautiful experiment on relativity with a Hydrogen Maser, had recommended to me. Even after that venture, we are still friends. The boat was rather slow, inefficient and was mostly creating waves. It was baptized *Entropy* by the previous owner and I understood the reason rapidly: the disorder that it created added considerable entropy to the Universe. It also created arguments between people onboard about why it would not sail well; that also added to the entropy of the Universe. Growing a crystal in a saturated liquid or building a house seems to create order. But the activities connected to manufacturing the required material and creating the elegant house produce more disorder in the surrounding environment than the order that is created. When we mix two baths of cold and hot water together, a system that has a certain order before mixing, the resulting ensemble takes an average temperature. It results in a state of lower order. Everything is mixed up. The entropy of the mixture is higher than the sum of the entropy of the separate baths. If we separate the water to re-establish the original state, say we remove some of the water, heat it again, and cool the rest in a refrigerator, we re-establish the order, but the energy needed is such that we created more disorder in the Universe in so doing. The entropy of the Universe has increased. And it goes on like that with all the practical activities we are engaged in. It is a lost battle to start with.

The whole question of entropy increase is connected to the concept of reversibility, which is another topic that thermodynamics addresses. In a gas, according to kinetic theory, molecules are moving at random, they collide, change speed, change direction: it looks like chaos, but it is very well organised. If we could take a slow motion film of one molecule and observe its behaviour we would find that it behaves in a normal way, obeys Newton's laws of motion, and furthermore conserve energy and momentum. We could run the film in reverse and our mind would not be able to tell if the film runs forwards or backwards. The motion of the molecule appears to be totally reversible in time. However, at the macroscopic level things are very different.

Suppose you drop a bottle of beer on the kitchen ceramic floor. It shatters in pieces and the liquid spreads around. Take a movie of that accident and show it to your friends in reverse. The bottle will reassemble and the beer will re-enter the bottle. The bottle will jump back to the counter. Everyone will laugh and see that the process is not possible. They will conclude that the film is run backwards. It is an evident fact of life. The process is not reversible. Why is that so? Why is it that when a bottle shatters on the ground, a short moment later all the molecules suddenly do not re-assemble and jump back to the counter? One way of answering that question is to observe that in the process the order of the Universe would be decreased without any work done that would increase the entropy somewhere else. The entropy, as we have seen, increases in all processes into which, as human beings, we get involved. Another particularity is the fact that the Universe seeks always the lowest energy state possible. We could make it a principle. If you drop an object it falls down on the ground, never up in the air. The potential energy on the ground is smaller than above. Thus the bottle falls down: it finds a lower state of energy on the ground than on the counter. But it shatters and falls apart. These pieces are all at ground level with the same potential energy as when the bottle was assembled as a whole. Why did they not join together to form another bottle? In the case mentioned, the bottle, seeking the lowest possible energy, fell down and broke in pieces. However, nature offers an infinite number of possibilities for the pieces of glass to scatter around, all possibilities with essentially the same total energy. Any possibility is equally as likely as any other. These possibilities are so numerous that the one with a new bottle reconstituted with the pieces is as negligible as any other distribution on the floor. If we tried a great number of times, the probability of getting a new bottle will not increase with time, but it is possible that on one occasion the pieces will re-assemble to form a new bottle. In practice that probability is so low that we will not see it happen during our lifetime... or the Universe's lifetime.

That property of thermodynamics, related to the behaviour of entropy thus has important implications in the Universe at large. The present state of our Universe is very much dependent on those principles of lowest energy, irreversible processes, and entropy growing with time. The entropy of the Universe is continuously increasing. We cannot escape it and whatever we do reduces the order in the Universe. It thus seems that at the origin of the Universe, the order was a maximum and has decreased with 'time' since then. We have just introduced again the concept of time. It thus looks like there is a direction of evolution for the Universe and that the concept of time is involved with it: we call it the *arrow of time*;[6] and it seems to be oriented inexorably in one direction to a state of greater disorder. We are heading towards a dull state where order has vanished and nothing can be done about it. We should not worry too much, however. Such a state will not appear tomorrow.

The Skater Physicist and Angular Momentum

Have you ever noticed that elegant skater on ice, turning around at various speeds seemingly without pushing herself with her skates? In a typical exercise, the skater reaches high speed, stops suddenly on the tip of one of her skates and start turning on herself with her arms fully extended. She brings her arms in and sticks them close to her body. What happens? Her revolving speed, number of turns per second, increases in a substantial way. Interesting! Why is that happening? Why is she turning faster? With her arms extended the skater had what is called a given *moment of inertia*. It measures the extent of mass distribution of the skater in space. The greater the distance of the mass from the axis of rotation the greater is the moment of inertia. Thus the skater with her arms extended has a greater moment of inertia than when she has her arms close to her body. When she rotates on herself at angular velocity ω, which is simply 2π times the number of turns she does per second, we say that she has '*angular momentum*' equal to *moment of inertia* $\times\ \omega$. It is very much like the linear momentum except that mass is replaced by moment of inertia and linear velocity by angular velocity. When the skater brings her arms in, she reduces the moment of inertia, and her revolving speed increases. This is illustrated in Figure 4.3.

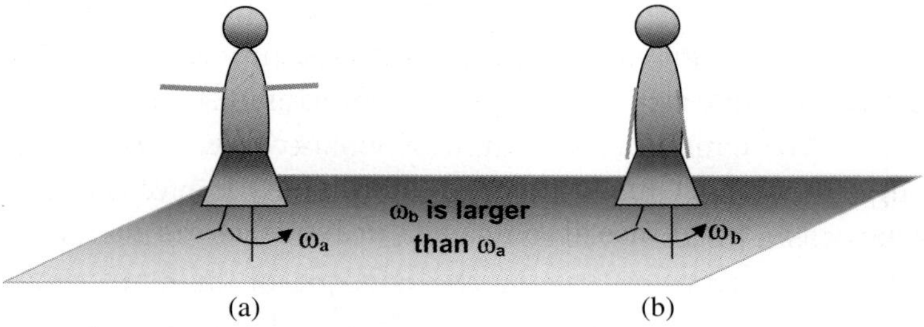

Figure 4.3. Illustration of the principle of conservation of angular momentum for a skater stretching or bringing her arms along her body. In case (b) the moment of inertia of the skater is smaller than in (a) and her rotating speed must increase to conserve angular momentum.

It is a living demonstration of the conservation of the angular momentum and it is another facet of the conservatism of Mother Nature. When a certain quantity of angular momentum is created somehow in an object and no other forces are applied to the object, that quantity of angular momentum is conserved. The Earth keeps rotating, conserving its angular momentum. The Sun and the Moon, however, have an effect on it. They create tides, consequently friction. It causes some loss in angular momentum. Its revolving rate decreases with time but very little. It has been measured to be of the order of 1.6 milliseconds per century.[7] Atomic clocks could certainly improve the accuracy of the measurement, but due to various other causes, the stability of rotation of the Earth is poor and fluctuations inhibit better accuracy. Objects in interaction may exchange angular momentum but the sum of all the components will stay constant with time. It is another trick of Mother Nature to keep us on guard, but it is also a gift because it introduces a parameter that stays constant, a property that can be very useful.

In atomic and particle physics, this property is extremely important. For example, a photon may have positive or negative angular momentum, corresponding in a classical picture to its polarization to the right or to the left. We call this 'circular polarization' to the right or to the left. When a photon with right circular polarization is absorbed by an atom, it disappears. According to the principle of conservation of energy and angular momentum, the atom has to gain energy and angular momentum with values equal to those attached to the photon to start with. In practice this is exactly what is observed. The atom finds itself in a state with more energy and with an angular momentum larger than it had before absorption of the photon. The process works in reverse: if an atom emits a photon with circular polarization, the atom must lose energy and angular momentum. These properties are extremely useful in many experiments and devices.

Other Conservation Laws: How about Electric Charges?

Are there other quantities that are conserved in the Universe? The answer is yes. There are several others but some are approximate and some apply to specific circumstances such as mass at low speed, in non-relativistic situations. However, a few are fundamental. Electric charge is one of those.

When an object becomes charged by static electricity, it accumulates a net charge, but another object close by becomes charged with the opposite polarity. How can that be? The atoms have lost one electron and they now become ions with the nuclear charge not being compensated by the appropriate number of electrons. The whole ensemble is still neutral. Similarly, in a storm, parts of clouds above our heads become charged with a given polarity while other parts become charged with the opposite polarity. The charges originate from a division of the original elements into charged ensembles. The elements are neutral to start with having net zero charge and the sum of the charges after separation is till zero.

Processes encountered in radioactivity illustrate very well the conservation of charges. For example, the emission of alpha particles, consisting of two protons and two neutrons, seems to create two elementary charges +e. However, the nucleus left behind has lost two positive charges, and thus the atom, neutral to start with, finds itself with two electrons with charge −e that are not compensated for by the positive charges of the nucleus. They escape rapidly. The sum of all the charges is still zero. Beta decay in which an electron is emitted is even more illustrative. As we mentioned previously, beta decay may be thought of as the transformation of a neutron into a proton and an electron. Thus the net charge before and after the emission of the electron is zero. In both cases, alpha and beta decays, the total number of charges is conserved.

We could list many other situations. Consequently, extrapolating to the limit, it appears that the total quantity of charges in the whole Universe is constant. Going further, we could formulate the hypothesis that the universe is neutral with exactly the same number of positive and negative electric charge[8] but we have no way of checking it. However, if the Universe were not neutral, there could be regions with a given surplus of charges, and the electric force being so large compared to the gravitational force, it would be a very different world. The smallest fractional difference in charge between stars or galaxies, for example, would have an important consequence on attraction or repulsion, because of the number of particles involved and the strength of the force involved.

This brings us to deeper thoughts. Quarks are rather exotic particles with charge $(-1/3)\,e$ for the down quark and $(+2/3)\,e$ for the up quark. They join to give birth to protons and neutrons forming nuclei. In turn, nuclei join with electrons, having a charge $-e$, to form the visible matter of our Universe. How did those particles appear? As we will see later, according to the Standard cosmology scenario, matter in the Universe was produced by means of a materialization of an original intense field into matter and antimatter, quarks and antiquarks, electrons and antielectrons. The process looks rather symmetrical. There was no net charge created in it. Due to the high temperature existing at that time, particles and antiparticles were continuously created. They also annihilated through collisions between themselves. At a very early period of the evolution, however, it appears that there was a break in the symmetry and matter won over antimatter by a small amount, parts per billion. Not all matter disappeared and we now live in a world made of matter. However, if the universe is neutral, it appears also that the number of negative charges belonging to the electrons that survive was compensated for by an equal number of positive charges belonging to the quarks forming protons and neutrons, all left over from the annihilation process. It is a real exploit. We will come back on that point, as we will look further into the transformation of a neutron into a proton and an electron, in radioactive beta decay. That process appears to provide a connection between nuclear and electronic charges.

Nevertheless, we can assume with some conviction that in the Universe electric charge is conserved, at least locally. We can apply the principle and use it with absolute confidence in the determination of parameters in various experiments for example in nuclear fission and fusion reactors as in accelerators of particles.

Reflections on Conservation of Energy and Entropy

There are many quantities that are not conserved. For example, relativistic mass is not conserved: it can be transformed into radiation. Speed by itself is not conserved. A combination of speed and mass, momentum and energy, is conserved. This is how Mother Nature distinguishes herself: conservatism in certain aspects of her properties showing up her character. Energy may transform itself into many entities. It thus appears to be more fundamental than those quantities that are somewhat ephemeral. If an object is moving at velocity $\mathbf{v}$ in a given coordinate system its mass increases, but if you are moving at the same speed as the object, it appears to be at rest and in that particular system its mass is not altered. This property is thus a property of the space-time canvas carrying the mass and the laws of physics, as well as the conservation laws, are the same in all reference systems.

Mother Nature provides also a direction for the evolution of time. Entropy is getting larger with time and basic physics shows that this is the direction that time flows. We are heading to a state with maximum entropy or total disorder. This does not look very promising but the scale of time in that evolution is so large that we do not need to worry about it for our own existence. Nevertheless, this phenomenon exists and shows an inner property of the Universe. We do not know why this is going in that direction. If we assume that the Universe one day will start to contract under the influence of gravity, will entropy start to decrease and will time reverse? This question is part of the debate on the nature of the evolution of the Universe.

The arrow of time is fundamental to an understanding of the Universe evolution and is coupled strongly to entropy. Unfortunately, it appears also that it is one of the least understood concepts. We have made an axis of it in the space-time canvas and deduced many properties associated with it, all supported by experimental data. We will leave it at that for the moment and will come back to the matter in a future chapter on the evolution of the Universe.

Conservation Laws and Symmetry

In the following chapters we will often refer to the concept of symmetry among objects, ensemble of such objects and even physical laws. We will see in later chapters that our Universe behaves in a rather strange way, such as differentiating left from right and showing a preference towards one of the two in the case of beta decay, for example. Let us examine now the concept of symmetry and its connection to some of the conservation laws that we have observed in nature and have examined in the previous paragraphs.

In general we say that we are in the presence of symmetry if an object, a concept or a law looks the same from a different point of view or perspective, for example, a different direction. As an example, use a sphere: it looks the same from whatever direction we look at it. We can rotate it by any angle on an axis and it looks the same. We say that it has spherical symmetry. If we look at a cube, we find that it appears the same when we look at it from particular directions. If we turn it by 90 degrees, it looks the same as in the previous view. We can enumerate many objects like those and by means of geometrical arguments associate with them particular symmetries, such as spherical, axial, cubic, hexagonal, etc. In our trip through the Universe small-scale we looked at a snow flake. We said that it had symmetry. Furthermore, that symmetry reflected some internal symmetry of the crystal cells forming that snow flake down to the molecules that assemble to form such crystal. Symmetry is thus a very useful tool to explore internal structures.

The concept of symmetry, on the other hand, becomes most interesting when we look at it in the context of physical laws. We may well ask if the laws of nature have some symmetry or in other words, are the laws of nature symmetrical when viewed from different perspectives. A different perspective might very well be simply a translation in space. We may ask if physical laws are equally valid in this room as in another room. Are the laws equally valid on the Moon as on planet Earth? This in fact is something we discussed and assumed earlier as valid without being too explicit about it. If it were not so, our physical laws would not be very useful, being local, and would be different in various locations. We would need to have laws for every location we wish to study. We then say that translation in space is a symmetry of our physical laws: these laws are invariant under translation in space. We have also introduced another particular symmetry without referring to it as a symmetry: it is the invariance of the laws in frames of reference moving at different constant velocities. In Joe's wagon, physical laws were the same as on the railway platform where Mary was at rest. We called it the principle of relativity. However, it is a symmetry since the laws stay the same from a different perspective: every constant speed is a different perspective. These are known as Poincaré symmetries. How about translation in time? For example, we may ask if Newton's second principle remains the same with time. From what we said previously it is quite obvious that it does not change with time otherwise we would not be able to use it. This applies to all physical laws. This is thus another symmetry of the physical laws since they remain the same if we look at them from a different perspective, such as a different time. We then say that the laws of physics are invariant under translation in time. How about rotation in space? Space is isotropic and physical laws are invariant if observed from different angles.

If it were only for those observations, we would simply conclude that nature is fine and that we can continue calculating and speculating with our physics the way we have always done, not worrying about rotations and translations in space, constant speeds and time. However, it can be proved that conservation laws that we have described earlier are directly connected with those symmetries. This was shown in 1918 by Emmy Noerther. She proved mathematically that invariance of physical laws under an operation that makes a symmetry explicit leads to a corresponding conservation law. The connections are:

Invariance under space translation leads to the conservation of momentum.
Invariance under time translation leads to conservation of energy.
Invariance under rotation leads to conservation of angular momentum.

As pointed out by Feynman, those relations between symmetry and conservation laws are some of the most beautiful and profound things in physics. They apply to the large-scale world as well as the atomic world that we introduced in the previous chapter.

Are there other symmetries or hidden invariance from which we could deduce other conservation laws? The answer is yes. We will use them later on occasions when we will introduce the subject of particle physics. Let us examine a few of those symmetries and their consequences on conservation laws. One of them is connected with a local symmetry found when we examine the laws of electricity in their general form. This is the theory of electromagnetism as developed by Maxwell, a theory that unites the concepts of electricity and magnetism under a single formalism. That theory is based on the existence of electric and magnetic fields that can be expressed in terms of potentials. It turns out that the laws of electricity are invariant when those potentials are altered according to certain rules like adding a term to them or changing their phase. The alterations and the rules form what is called a *gauge* transformation and the invariance under such transformation leads to symmetry in the laws. Electromagnetic laws are invariant under a gauge change. We say that they have gauge symmetry. The word gauge is not really appealing, does not tell us much about the nature of the operation done and is probably not well chosen. However, the physics community uses it at large and we have to live with it. We may think of gauge symmetry in electromagnetism, although this explanation is imperfect and somewhat naive, as a symmetry that persists when the zero of the potentials suffers a translation in a specific way. The laws are functions of those potentials and are not altered when such translation is made. It is something like measuring the voltage developed by a battery with a digital voltmeter whose zero is offset. On measuring the voltage of each pole in reference to ground or a common point we find that those voltages are unsymmetrical. It really does not matter in practice since what is important is the difference of voltage between the two poles of the battery. What is the physical implication of such symmetry? Although not obvious, it turns out that gauge symmetry in electromagnetism leads to *conservation of electric charge*. It is a most fascinating result and it turns out that gauge symmetry is also important in advanced theories that describe the behaviour of elementary particles in the case of strong and weak forces.

The symmetries considered above were connected to invariance under continuous transformation such as rotation and translation in space and time. There are, however, important symmetries that are connected to discrete transformation such as inversion of time and space coordinates. For example, we may make a transformation that consists in reversing time. Then in all physical laws t becomes −t. Time appears to run backward. Similarly, we may invert space coordinates. For example, coordinate x becomes −x. This last operation is called a parity transformation. It is like looking at an object in a mirror set in the y-z plane. What happens to the laws of physics under such transformations? You may think, for example, of the motion of billiards balls on a table when they are looked at in a mirror. The balls will appear to move from left to right in the mirror rather than right to left on the real table. Nevertheless, everything looks fine and we do not see anything abnormal about the motion of the balls. It will be the same thing if we take a film of the motion of the balls and project it in reverse, so that time is inverted. We will not be able to tell the difference from the situation when the film is projected in the proper direction. The motion observed appears to obey the laws of physics. These laws are thus invariant under time reversal and parity transformation.[11] This kind of discrete transformation may also be applied to electric charges in the context of matter-antimatter transformation, in which positive charge is replaced by negative charge and vice-versa. This transformation is called *charge conjugation*. We may very well ask what would happen if we were transforming all matter in our Universe into antimatter, thus inverting the sign of all electric charges. It appears that we would not be able to tell the difference.

Those discrete transformations have been given operators' names, C for charge conjugation, P for parity transformation, and T for time reversal. The whole set is normally grouped under the notation CPT. It is a most exciting contemporary subject of research since it deals directly with the soul of our Universe. We would expect that the laws of nature are symmetrical under any of those transformations. For example, we certainly do not expect that some phenomena would prefer left to right (or vice-versa) or in other words would violate parity. In fact, the laws of electromagnetism, gravity and strong forces stay the same under parity transformation. However, the weak interaction, responsible for beta decay, the phenomenon by which an electron or a positron is spat out by a nucleus, violates parity. It appears that only particles with left-handed characteristics enter into weak interactions. This is a most intigruing observation. This whole subject touches also upon the very sensitive question of the predominance of matter in our Universe. It appears that C and P symmetry violation is required for our existence. We will come back to it when we examine our present understanding of the so-called standard model that describes the characteristics of elementary particles.

We just introduced the concept of symmetry violation. This phenomenon is sometimes referred to as symmetry breaking that took place at a given time under a particular circumstance. As an example, imagine a round table where a committee meeting is expected to take place with six participants. The person who prepared the room distributed the six chairs evenly around the table and placed on the table six blank pads right in front of the chairs. At exactly mid-distance between the pads she placed a useful tool, a pen. The setting is symmetrical. Five of the members arrive and sit down. Each individual picks up without hesitation the pad in front of the chair that he/she has selected. It is obvious that it belongs to the chair in front of the pad. How about the pen? The situation is totally symmetrical. Since there are no conventions, a person does not know which pen to use. If a first person decides to choose the left pen then every one else will have to do the same. The original symmetry no longer exists and this is readily visible for the sixth member of the committee that arrived late. He observes that the only pen he can use is on the left side: the original symmetry no longer exists for him: it was broken arbitrarily at the arrival of the first participants. For him the situation appears to lack symmetry and it looks natural. We may say that the symmetry that existed before the first committee member picked up a given pen is now hidden. We will find situations similar to that in the behaviour of some physical phenomena. For example, some particles of matter, called ferromagnetic, become spontaneously magnetized upon reducing their temperature. Above a given temperature the microscopic magnetic moments are oriented at random. The situation is homogeneous and is spatially symmetrical. Upon cooling, the magnetic moments tend to orient themselves all together in the same direction and under a given temperature, called the Curie temperature, the material becomes a small macroscopic magnet. The original symmetry is broken. We will see later that in order to explain some phenomena, particularly in the context of particle physics, we will have to assume that symmetry was broken at a given time of the Universe evolution. That symmetry is now hidden and it is our task to analyse the situation and an attempt to make that symmetry explicit.

V

There is Light and We See!

So the world is energy. Radiation is energy and objects have mass that we may interpret as a materialization of energy. Mass obeys specific laws in the canvas of space-time. All that is needed to explain the behaviour of objects possessing mass or to land an astronaut on the moon is Newton's principles and the concept of gravitational attraction. Relativity provides a higher level of understanding of our perception; it makes explicit a geometrical model with basic principles and concepts that explain the how and seems to touch the why it works the way it does. Mass creates a curvature of the space-time grid and objects moving on the grid follow that curvature. In our three-dimensional space they appear to be attracted by those objects with mass: the result is then essentially a confirmation of Newton's law of gravitational attraction. The model shows also that if we look more carefully at the predictions of the model, we will see phenomena that are quite uncommon in our day-to-day activities, phenomena that were not predicted in Newton's formalism. These phenomena were observed and consequently confirmed the model made by Einstein. But an important observation remains unexplained at this stage of our description: what is the origin of electromagnetic radiation, of light, and what is that strange behaviour of electromagnetic radiation, being either a wave or a particle or both depending on circumstances?

What is that substance that enters our eyes and by which we 'see' objects? We call it light and we have discussed it only as saying that it was travelling at speed c. What is that concept of duality, wave and particle, which we introduced without much explanation?

Newton thought that light consisted of corpuscles. Huygens and others claimed that light had a wave nature. These two claims are such contradictory concepts that one is led to conclude that one of the two must be wrong. Well, it turns out that both are right. It looks like another of nature's conspiracies to baffle the human mind. Not only, light travels at a speed that is independent of the motion of its source and its observer, it has two personalities. Depending on the technique used to observe it, light may be interpreted as waves or as corpuscles that we now call photons. When light falls on a detecting crystal such as a photoelectric cell, (one of those devices used in the construction of solar panels), it clearly interacts with the crystal as if it was made of particles. At very low light level, one can detect the arrival of photons one by one as distinct clicks in the detecting circuit, like rain drops on a metal roof. When a beam of light is sent on an opaque barrier that contains two narrow slits that are very close to each other, light passing through the slits produces, on a screen behind, a pattern that is characteristic of interferences produced by waves. The phenomenon is similar to interferences of water waves around obstacles. Our eye is made of a lens that is well suited for wave manipulation, and of a retina containing many cells, cones and rods well suited for detecting individual particles. Our eyes look very much like one of those digital cameras equipped with a lens that focus the image on a detector made of millions of little cells detecting the incident photons. The retina of our eyes even contains detector cells that are sensitive to different photon energies or wavelengths allowing a differentiation of colours. Evolution was aware of the nature of light and also made our eyes most sensitive to the wavelength at the maximum of emissivity of the Sun, which is in the yellow part of the spectrum. Our eyes are thus sensitive to colour through wavelength and intensity through photons: light behaves as if it had properties of both. But, aside from saying it is wave and particle, what is light physically and what is its origin? A scientific description of the phenomenon was elaborated in the 19th century and is entangled with the development of theories describing the behaviour of electrical and magnetic phenomena. The basic laws of electricity and magnetism were discovered during the 18th and 19th centuries, by Coulomb, Ampere, Faraday, Hertz, to name a few.

Maxwell at the end of the 19th century provided a full description of the nature of light using the various laws found by his predecessors and by joining in a genial way the two phenomena.

It is difficult to follow logically the historical development of scientific knowledge. Developments made at different dates in various sectors are entangled. For this reason, phenomena discovered at a given epoch are sometimes much easier to understand in the light of discoveries made later. Consequently, in order to simplify the description, the following paragraphs will not attempt to follow exactly the historical succession of developments but will rather insist on clarity, introducing sometime in the description of a concept discovered at a given date ideas that were elaborated later. For example, in the 19th century, development took place on the understanding of the constituents of objects around us: the atomic nature of matter was unravelled in a scientific manner by scientists such as Prout, Dalton, Gay-Lussac, Millikan, Avogadro and Thomson. At the beginning of the 20th century, the discovery of the atomic nature of the world provided a framework for Rutherford and Bohr to develop a model of the internal constituents of the atom and at the same time understand the origin of that radiation that we call light. However, it is Maxwell that developed the classical framework describing the nature of that radiation at the end of the 19th century. It is only later that the origin of radiation found a basic description by means of quantum physics in an exchange of energy within atoms.

Electricity

The first observation of the presence of electric charges in nature, several thousand years ago, was through the phenomenon called 'static electricity'. Walking on a carpet on a dry day liberates electric charges that flow through our body leading to a relatively painful spark when we touch the door knob. Combing one's hair also liberates charges through the contact of the plastic comb with the hair, and, in the dark, sparks reflect their motion through air. Actually, from observations made on various materials that attract or repel each other under various circumstances, two kinds of electric charge were assumed to exist. Like charges repel each other, while unlike charges attract each other. By convention they were called *positive* and *negative* (Franklin). Positive charges developed on glass when rubbed with silk, while negative charges developed on rubber when rubbed with cat's fur. Why are there only two types of charge? This must have been a serious question at the time of observations. There could have been, for example, another type of charge that, when approached by another one, is pushed sideways instead of being attracted or repelled. We will see that such an effect exists when charges are in motion. This transverse motion, however, is due to the creation of a so-called magnetic field in the environment due to the motion of the same electric charges. When they are static, electric charges attract or repel each other. It is a choice that nature made and the vocabulary, negative (−) and positive (+), is a choice of the human being. We could have chosen a different pair of symbols, like (#) and (&). We are now familiar with the symbols (+) and (−) and it is almost like a second nature to the scientist. Furthermore, the signs associated with the electric charges are coherent with our mathematical conventions and provide a means of assigning a sign to the forces that are developed when those charges are interacting. We now know that in those electrostatic experiments, the negative charges on the rubber band are caused by an excess of electrons, the elementary discrete particles with a very small mass and a negative charge −e, discovered by J.J. Thomson at the end of the 19th century. The presence of positive charges on the glass is caused by a deficiency of electrons. Objects are neutral in normal circumstances. If negative charges are removed from a neutral object, an internal imbalance is created and the object becomes electrically positive. The charge of the electron, −e, is another peculiarity of Mother Nature. It is the minimum quantity of electricity we can isolate and is extremely small in terms of units of electricity. It is quantized, meaning that it is a quantity that exists freely only as a multiple of itself. It cannot be split to result in a quantity that would be a fraction of e. The electric current used in a kitchen appliance consists of a flow of such a quantity of electrons (of the order of 10^{19} electrons per second) that it looks continuous for standard measuring equipment and the discreteness of the electric charge is totally masked.

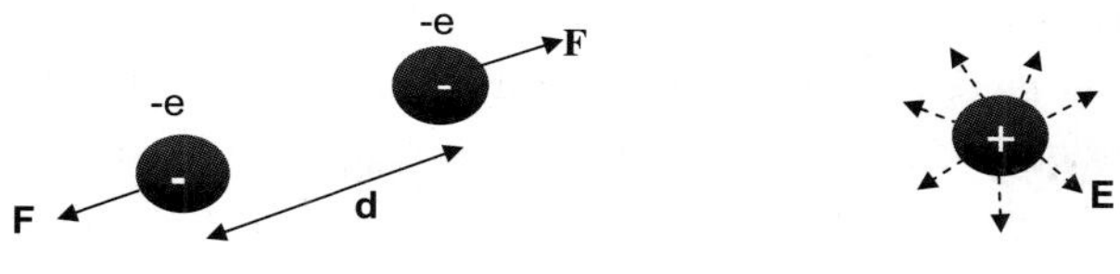

Figure 5.1. Two electrons repel each other. (a) The force of repulsion is proportional to the inverse square of the distance, d, between them (Coulomb's law). (b) Representation of the electric field E of a positive charge by arrows (vectors) pointing away from a positive charge.

What is that law of attraction or repulsion obeyed by electric charges? It was elaborated by Coulomb in the 18th century, well before the discovery of the electron, by means of experiments made on electrically charged bodies and measurements of the forces exerted between them. To illustrate and simplify things suppose we have two free, static, electrons in space as shown in Figure 5.1(a). They repel each other. What is the rule governing that repulsion?

The repulsion force is proportional to the product of the charge of the two electrons and inversely proportional to the square of the distance between them as in the case of gravity.[1] However, it is much stronger than gravity. The gravitational attraction of the two electrons through their mass is about 10^{-42} times their electric repulsion, a fantastic difference. Why does such a difference exist? We don't know; it is another mystery of nature. But if the two were of the same strength the world would be very different and we would probably not be here to ask the question. The constant of proportionality introduced to get units right in the repulsion of the electrons is given the frightening name, permittivity of free space, and is represented by epsilon zero (ε_o). It is one of those fundamental constants of nature. It characterises free space or vacuum and is appropriately called a vacuum constant. We will see that it plays a major role in nature's conspiracy relative to the speed of light. The law of repulsion between electric charges such as that carried by the electron is called Coulomb's law. Note that it is an experimental law. It is not a law deduced from a logical, abstract argument. It is how nature works when it comes to electric charges. If one does an experiment and by chance finds experimental data that does not agree with the prediction of that law, the whole field of electricity has to be revised (or the experiment is false). Coulomb's law has not been falsified yet. The repulsion seems to act with no material connection between the two charges. As in the case of gravity, we are allowed to ask the question what is exchanged for such a force to exist between objects that are not in what we understand as physical contact? In the case of gravity we introduced the concept of 'field', a very convenient tool that allows easy calculation of the force of attraction. The same concept of field can be transposed in the case of electric charges to describe the 'substance' that emanates from a given charge to the site of the other. We call it the *electric field*, and the whole theory of electricity is based on that concept. Let us imagine that one of the two electrons is a test charge and we examine the effect of the other electron on it. The force exerted on the test charge is equal to the product of that test charge with the intensity of the field produced by the other electron. The field intensity created by an electron as in Figure 5.1(b) is diluted over the surface of a sphere. If d is the distance from the charge, $4\pi d^2$ is the area of that sphere. Then, since the area increases as the square of the distance, the field intensity thus decreases as the inverse of the square of the distance as if diluted over space. In that case it seems obvious that the force being proportional to the product of the test charge by the field surrounding the charge also decreases as the square of the distance. The field has been introduced exactly to do that. In the case of gravity the theory of relativity suggests that the attraction is caused by a curvature of the four-dimensional space-time canvas, an abstract concept that appears as a shadow in our three-dimensional space: in Newton's world, we call it the law of gravitational attraction. Could a similar concept be developed for the case of electric charges? It does not appear that an explanation of the same nature as that introduced by general relativity in the case of gravity using the concept of space-time proved to be a satisfying approach. In an earlier chapter, we mentioned that the electron appears as an entity surrounded by a cloud, testing its surrounding by emitting virtual photons and reabsorbing them. This is actually an interpretation of a theory called quantum electrodynamics. In that framework, the force between electric particles is created by an exchange of virtual photons which act as mediator of the force. This is what appears to differentiate gravity from electrical effects, gravity in Einstein's model being caused by a curvature of the space-time canvas.[2]

Magnetism

Another most interesting phenomenon observed in antiquity was the mutual attraction of certain stones that we now know to be made of iron oxide. They have the property of exerting forces on each other and attracting pieces of metal containing nickel, iron, or some other elements. They even have the ability of transmitting by contact that property to another piece of material such as steel. As shown in Figure 3.17, iron filings orient themselves around a magnetised object in such a way that it looks very much as if some kind of lines are emanating from the ends of the object, called a magnet. These lines are called lines of force and it is natural to think of them as a field emanating from the magnet. Looking at the iron filings, this magnetic field appears to be real. It permeates space like the electric field and causes effects at some distance from its origin. The magnet could also be said to alter the surrounding canvas of space around itself in some kind of new dimension. This alteration is represented by the magnetic field. From a pattern such as the one shown in Figure 5.2(a), it is natural to think of the field lines as emanating from one end of the magnet to enter the other end. Similar to electric charges, it is assumed that the end of the magnet contains a magnetic charge or a magnetic pole. By convention, one end is called the North Pole and may be said to correspond to a positive magnetic charge. The lines of magnetic field are then assumed to originate from that North Pole or to emanate from it. The other end is called the South Pole and the lines of magnetic field enter into it. It is most interesting to realise that if a magnet is cut in half, one does not end up with a single North pole and a single South pole in each hand. Two symmetrical poles are created in each part and the net result is two complete magnets. It appears that a single pole or in other words a monopole does not exist in practice; although, from advanced theories, its existence is not totally banished from our Universe.

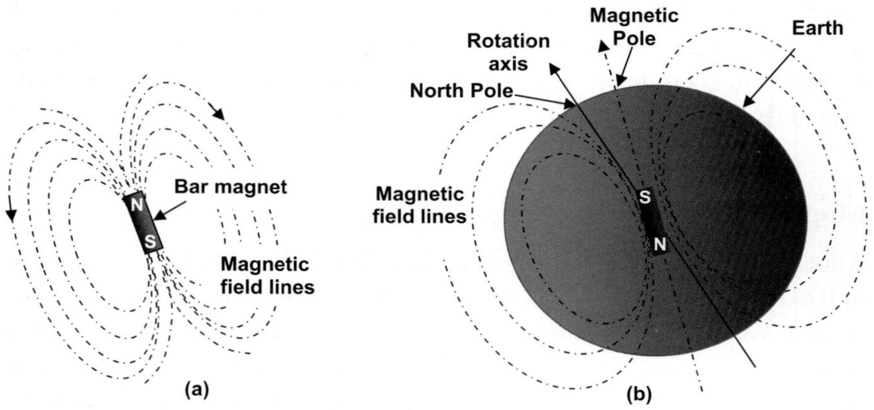

Figure 5.2. (a) A bar magnet produces closed lines of force representing the magnetic field. The lines of force or magnetic field, by convention, are said to emanate from the North Pole and to enter the South Pole. (b) The Earth's magnetic field is very similar to that of a bar magnet but, as explained in the text, its magnetic North and South poles are interchanged from the geometric poles.

A short diversion. The Earth has a small magnetic field. Its origin is not too well understood. At the surface of the Earth its value is about 1/2 gauss.[3] In order to grasp the magnitude of that field it is sufficient to know that it is a field about 1,000 times less intense than the field developed by common magnets sold in toy shops. But it is large enough to deflect a small magnetized needle free to rotate on a pivot, giving rise to a very useful navigation instrument called a compass. The identification of the South and North Poles of the Earth, however, leads to some confusion. The North Pole has already been identified through geometrical means as the convergence of the meridians. It is the geographic pole on the axis of rotation. Unfortunately, that Pole is a South Pole as far as magnetic properties are concerned. Furthermore, the location of the magnetic North pole is different from the geographic North Pole, showing their difference in origin. The end of the compass needle that points to the North Pole is actually a North pole, which means that a bar magnet that would represent the magnetic property of the Earth would be oriented as in Figure 5.2(b) in order to attract the north pole of the needle toward the North magnetic pole.

Is it possible to produce a magnetic field with something else than a magnet? The answer is yes. A magnetic field is also produced by an electron moving at a certain speed in a given frame of reference. The intensity of the field in that frame of reference is proportional to the speed of the electron and its charge $-e$. If we want a stronger magnetic field we can simply add more electrons moving together. The best way is to use a piece of copper wire containing a large quantity of quasi-free electrons ($\sim 10^{22}$ per cubic cm),[4] and set them in motion by means of a voltage applied between its ends. This effect was discovered by Oersted (1820) in experimenting on the force exerted by a magnet on a wire carrying an electric current. The field has the same property as that created by a magnet. Ampere observed that a similar force was present between two wires carrying an electric current. That process for creating a magnetic field seems far simpler than the process that is taking place within a magnet. At least we seem to identify its origin. It is produced by moving electrons. Are there in a magnet any quasi-free electrons that move like that to produce a magnetic field? The suggestion was made by Ampere that it was so. We will see below how the magnetic field of a magnet is now explained by atomic physics and quantum mechanics in connection with the mysterious spin nature of the electron.

Motion of Electrons and Radio Waves

An electron in free space may move in a straight line in accordance with Newton's first law. In a frame of reference that moves at the same speed, the electron appears to be at rest. If another electron is close by at rest in that same frame of reference it is repelled according to Coulomb's law mentioned above. But what happens to the electron if it travels in the neighbourhood of a magnet at rest? It deviates from its path in a rather complex manner. If the electron travels at right angle to the magnetic field lines it then deviates at right angle to both the magnetic field and its own direction of motion. It looks as if a force is applied to the electron at right angle to its path and to the magnetic field. That force is equal to the product of its velocity by the magnetic field intensity. The constant needed to get units of force right bears again a terrible name: *permeability of free space*. It is called mu zero, (μ_o).

We have thus two types of forces that affect the motion of electrons: one due to the presence of another electric charge or its electric field, and another one if the electron moves in a magnetic field. It looks very much that these are irreconcilable properties of the electron and that the magnetic and electric behaviour of electrons is due to independent physical phenomena. However, another property found by Faraday and Henry (1831) brings in a totally different point of view: when a magnet is moved around a loop of copper wire or when the loop is moved in and out of a magnetic field, it is found that the electrons within the wire start moving creating an electric current. This is illustrated in Figure 5.3. There is thus an electric field developed in the wire by the changing magnetic field that forces the electron to move around. We could expect such an effect since we said that electrons in a copper wire are essentially free. We can expect also that the effect of a changing magnetic field on a free electron to be somewhat like the mirror image of the motion of an electron in a magnetic field. The electric field developed within the wire is proportional to the speed of the displacement of the magnet

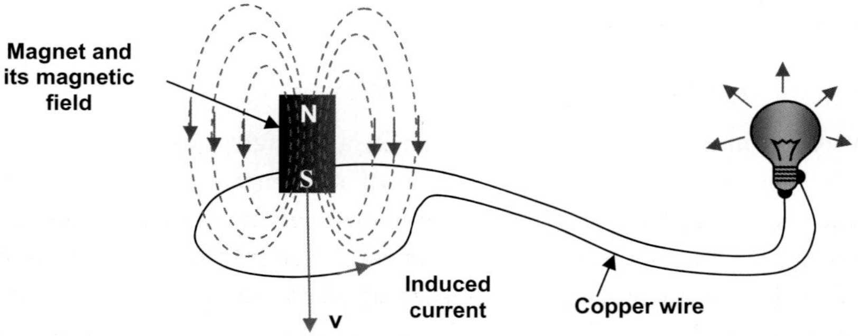

Figure 5.3. A magnet moving in the neighbourhood of a loop of wire causes a variation in magnetic field amplitude at the site of the wire. This in turn creates a current in the wire, which if the coil is made of many turns could be used to light a small light bulb. An interesting application consists of using this device to charge a small battery as is done in some flashlights sold in specialised stores. Inside the flashlight, a magnet is free to move back and forth within a coil made of several turns of copper wire. When the flashlight is shaken back and forth, the current generated in the coil by the motion of the magnet is used to charge the battery.

or the variation of the magnetic field with time as seen by the wire. It is a direct transformation of mechanical energy into electrical energy and it is at the heart of the operation of electric generators used at hydroelectric plants or the alternator used in a car to charge the battery.

That last effect appears also to connect directly time-varying magnetic effects to time-varying electric effects. Previous to that observation, magnetic and electric effects were understood as phenomena of totally independent origins. It took the genius of Maxwell to use the connection and develop a whole new theory that joined the electric and magnetic phenomena together into a single framework. He arrived at four equations that describe the behaviour of electric and magnetic fields in a most elegant way, showing the connection between them.[5] Those equations lead to the electromagnetic force that we introduced earlier as one of the four that we said existed in nature.

The most surprising result of Maxwell's theory (1864) is that it leads directly and in a beautiful way to propagating waves. These waves are made of an electric field and a magnetic field at right angle to each other, oscillating with time, and propagating at a speed equal to that given by a combination of the two constants just introduced for electric and magnetic phenomena. This is so important that it deserves to be written:

$$\mathrm{v} = 1/\sqrt{\varepsilon_o \mu_o} = \text{speed of electromagnetic waves obtained in Maxwell's theory.}$$

And guess what! This speed, v, when one uses the known values for ε_o and μ_o, as obtained independently from measurements on properties of the electric and magnetic fields in static conditions, turns out to be the same as the speed of light. From that finding and the conclusion that light is a wave made of an electric field and a magnetic field at right angle and propagating through space at the velocity given by the above expression, there is a very logical step. The wave is illustrated in Figure 5.4.

At the time Maxwell developed the theory, that step was resisted, but soon it was realized that the theory provided such a logical and

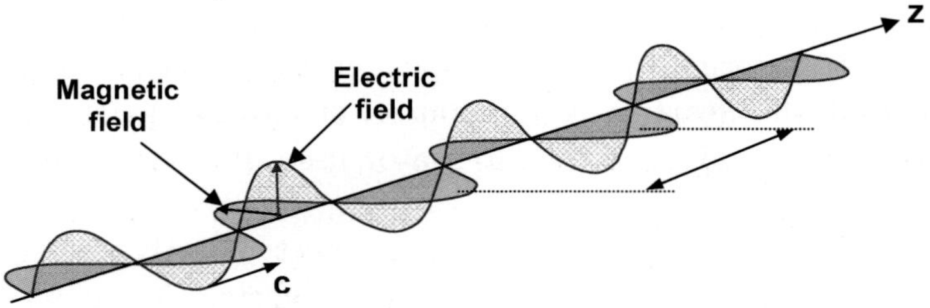

Figure 5.4. Electromagnetic waves are composed of oscillating electric and magnetic fields at right angle and propagating in free space at speed c. Individual crests are separated by one wavelength λ.

elegant framework that there could not be any better explanation. The theory is now fully accepted and shows that all radiation, gamma rays, X rays, light waves, radio waves propagating through the vacuum of space, are of the same nature: they are called electromagnetic waves.

This is another of those most fascinating properties of the vacuum of space: it supports electromagnetic fields and those fields propagate at speed c. That speed comes out naturally from the theory without a reference frame and without any assumption except for the validity of the experimental laws mentioned above and used to derive it. There is no material support for the waves: vacuum, or whatever vacuum may mean, is the support. The waves appear to exist by themselves in the same way as magnetic and electric fields appear to exist in vacuum in the static case. The wave speed is c, which Michelson and Morley found to be independent of the source's and observer's velocity. It is that property that Einstein made a principle on which he developed his theory of relativity. Consequently, it is readily realized that Maxwell's theory already includes relativity. The speed of light comes in without any hypothesis about the nature of the supporting medium. Furthermore, it is found that Maxwell's equations do not transform from one moving frame of reference to another by means of a simple Galilean transformation. The passage from one frame to another is done by means of the Lorentz transformation that includes relativistic effects.

How do we create an electromagnetic wave? Take a small magnet and let it sit static on the table. There is a magnetic field created around it, as illustrated in Figure 5.2(a), and 3.17. Everything looks stable. Move the magnet very quickly to one side by a few centimetres. What is happening in the space surrounding the magnet? The magnetic field has changed very rapidly. With your eyes, you don't see anything happening, but this change of magnetic field in time and in space has created an electric field at right angle to the lines of force of the magnetic field and the two fields, attached to each other, start immediately propagating through space at velocity c. As throwing a stone in a pond causes a perturbation of the water surface that creates a wave propagating through the whole pond, you have caused a small electromagnetic perturbation in space and the perturbation starts travelling. You have created a small wave packet of radiation at very low frequency that travels to the confines of space. Such a wave packet travels at the speed of light. The frequency of that radiation is a function of the speed at which you have moved the magnet. The radiation cannot disappear by itself. If the intensity of the magnetic field changes for some reason, the electric field changes also. If the electric field changes, the magnetic field also changes. The process is self-sustaining. Hertz (1886) was the first to identify that these particular waves with long wavelength generated by changing electric and magnetic fields had properties such as reflection and polarization similar to those of light.

An electromagnetic wave may also be created by the motion of an electron. If an electron is made to oscillate in a wire, for example going back and forth, oscillating magnetic and electric fields are created around the wire and energy is dissipated through space under the form of an electromagnetic wave having the frequency of the electron oscillation; the wave travels at speed c. This is the way radio and television stations, accelerating electrons in their antennas in a back and forth motion, transmit their signals through space. These waves excite electrons in our home antennas and we detect the information transmitted by the stations. It is like the waves created by a large ship moving on a river. The waves propagate on the surface of the water and, reaching the shore, sets in motion all the small boats anchored at their deck. Energy is transferred to the small boats. Cell phones operate the same way. A tower antenna somewhere transmits radio waves captured by the cell phone small antenna. The cell phone transmits back a signal to the tower. The thousands of radio, television and cell phone transmitters on the planet fill space with electromagnetic waves from kHz to hundreds of MHz. The whole planet is surrounded by a dense layer of radiation that spreads through space and broadcasts our presence to far distant galaxies. Anybody listening? It is amazing that we can filter with a cheap radio receiver the station we wish to listen to or still receive a very clear telephone call on our cell phone. This is done with very effective technology providing a filter at the input of the receiver of those instruments selecting the proper frequency or the appropriate code in digital communications.

Maxwell's theory explains exactly the behaviour of radiation propagation in space. We note that it is based entirely on classical concepts. What do we mean by that statement? It just means that the whole theory is based on laws derived from experimental observations at the macroscopic level. Electrons are assumed to obey some simple laws in relation to their motion in electric and magnetic field. A most important conclusion of the theory is the prediction of the speed of propagation of perturbations of electric or magnetic origins, the speed of light, c, which is just a combination of two constants introduced to describe the motion of the same electrons. This theory is one of the most powerful classical theories ever elaborated and has never been found to be wrong in any circumstances. The exploit achieved by Maxwell in the joining of separate observations relative to electric and magnetic phenomena into a single framework and the prediction of the existence of electromagnetic waves and numerous other phenomena is one of the greatest achievements of modern science. This is what makes the force of a theory: unification and predictions.

At first sight, it would appear natural to expect symmetry between the magnetic field and the electric field in Maxwell's equations. It is readily observed that it is not so. It appears that we have the presence of electric charges but not magnetic charges as such. To be entirely symmetrical the theory should make room for so-called monopoles. Dirac has shown that the existence of such monopoles is consistent with quantum mechanics and would provide a path for quantization of the electric charge, the monopole having also a minimum value. It may be mentioned that the energy involved in the creation of such an object is enormous and not available in present accelerator technology. The condition, however, could have existed at the earliest moments of the creation of the Universe, during the so-called Big Bang event that we will describe later. But the existence of such monopoles has eluded all attempts to observe them. We will see that a scenario based on the rapid expansion of the Universe in its early evolution could explain this failure because of the enormous dilution that took place at that early moment. We are left with an asymmetry in Maxwell's equations and, in view of the inclination of nature towards symmetry, the monopole remains a question of interest.[6]

A Bit of Atomic Physics

The wavelength of visible radiation is very short. While radio waves have wavelength in the range of metres, the wavelength of light is in the range of 400 to 600 nanometres, very short indeed. The corresponding frequency is in the range of 750 to 500 terahertz.[7] Getting light from an antenna by creating oscillations at that rate is not practical. Actually, light is observed all around us originating naturally from the Sun, reflecting on objects surrounding us, from the Moon at night reflecting the Sun's radiation, or from the distant stars. A hot bar of metal emits light the colour of which depends on its temperature. How is light produced by those objects?

When light, as emitted by the Sun, is observed through a glass prism, it is seen to decompose in many colours spreading from violet to deep red. The projection on a piece of white paper looks continuous, making evident a colour spread, as that of a rainbow. The pattern is called the spectrum of the radiation. If you select one colour with a slit, say, red light, and pass it through another prism, it does not decompose: it stays red. This was one of the observations made by Newton in his study of light. Thus white light can be decomposed in sub-elements and red light as other individual colours cannot. If one uses a higher definition instrument like a spectrometer equipped with a grating, a greater resolution is obtained. Each colour is further spread over a larger surface and in some sections of the observed spectrum, shows lines brighter than others, as well as dark lines and even dark regions. The spectrum is not continuous. The same observation is made when light from a house fluorescent tube is examined with such a spectrometer. How can that be? What is the mechanism that produces these so-called *spectral lines*? To understand that mechanism, we need to look at the nature of matter, that is to say the nature of the elements constituting the Universe we know.

Several thousand years ago, a Greek thinker, Democritus, addressed the basic question regarding the ultimate constituents of objects around us. By means of abstract reasoning he came to the conclusion that matter was made of indivisible particles. He called these ultimate constituents, *atoms*. However, it took millennia before that concept was given a scientific basis. This was done by means of measurements on the chemical reaction of certain elements. This was the work of Prout, Dalton, Gay-Lussac and Avogadro in the beginning of the 19th century who noted among other things that in chemical reactions gases always combine in exact proportions, either by weight or volume. The results of those experiments showed regularities that could only be explained if it were assumed that matter was made of corpuscles, what Democritus had called atoms and whose existence was assumed on the basis of general observations and abstract reasoning. Another proof of the existence of an atomic world below our visible world is in the observation of the random motion of small objects such as smoke or dust particles. First observations were made in antiquity, but the phenomenon was carefully studied only in the beginning of the 18th century by the botanist R. Brown through observations made under the microscope. The observed random motion of small particles of dust is due to collisions of the molecules or atoms of the supporting medium in random motion. A random disequilibrium in the number of collisions along a particular direction may cause a small displacement of a given dust particle in that direction. Due to the disproportion in the mass of the particles, atoms or molecules versus dust particles, the displacement is very small, but is nevertheless visible under large magnification. A mathematical formulation of the phenomenon was given by Einstein in the first decade of the 20th century. The phenomenon provides the most direct quasi-macroscopic proof of the existence of atoms and molecules and of the validity of the so-called kinetic theory that is formulated regarding their random motion and the nature of heat.[8]

The objects that surround us, existing under the three forms, gas, liquid and solid, are just an agglomeration of a large quantity of those atoms. In the case of gases, aside from colliding with each other, the atoms are free to move around, while in liquids and solids they are held in place in groups by means of interactions between themselves that are mainly of electrical origin. Sometimes in gases two atoms join together to form what is called a molecule. This is the case for example for the nitrogen molecule that exists in the atmosphere as a dumbbell N–N called N_2. The same is true also of oxygen. These atoms find a more stable state in that combination, while some other gases that we call noble gases, such as helium and neon for example, remain as individual atoms.

How big are these atoms? We had a glance at that in a previous chapter. Let us look at it from another point of view. Have you ever noticed, on the road one rainy day, how a drop of oil spreads around as a thin layer over the surface of a water pool? We can actually have an idea of the thickness of the layer by the nice colours observed depending on our angle of sight. This phenomenon shows that the layer is very thin and uniform since we see interferences in the reflected radiation. The thickness of the oil layer may also be obtained from a simple calculation. Just divide the volume of the oil droplet that created the film by the area of the film over the water pool: it shows that the thickness is of the order 10^{-8} m or less. This is thus the maximum size that an atomic layer can be. We note that the layer may still consist of a few layers of atoms and the size of the molecules forming the layer must thus be smaller than that number. So atoms or molecules are very small. Actually we know by means of other precise measurements that atoms are less than a nanometre in size, obviously not visible to the human eye. This is where we enter the atomic world. In the classical world of Newton dealing with massive objects observations are macroscopic and direct. We can make measurements using a visual approach. The atomic world is out of direct assessment by those classical means. All our conclusions on atoms have to be obtained by interpreting measurements made with instruments that are sometimes rather sophisticated.

Atoms are thus the indivisible parts of what we call elements. Elements are those constituents of matter that when divided into smaller and smaller parts stay the same until we reach a point where the last part cannot be divided anymore: we have reached the atomic level and that ultimate part cannot be divided without changing its nature. It is an atom. Water can be decomposed by electrolysis into oxygen and hydrogen, but the atoms of those substances cannot be decomposed into something smaller without changing totally their characteristics: hydrogen and oxygen are called elements. Considerable development followed these early findings and a large number of elements were identified. However, the relation between them was not clear. It appeared that some elements, although different in some aspects, had similar characteristics and properties in chemical reactions, for example. It looked as if they had internal parts that were common. Those observations culminated in the development of a table of elements by Mendeleev around 1869 making explicit some common characters of various elements by locating them in a regular arrangement of rows and columns. In particular, there were empty 'cases' in the table leading to the conclusion that some elements had not yet been discovered. It was not long before those were found in nature. Mendeleev's table was accepted showing again the strength of the concept of symmetry and the power of prediction in getting a model accepted. Mendeleev's table has been developed further over the succeeding years and today it consists of 92 elements found in nature and several others that can be 'created' artificially in instruments such as accelerators but with an ephemeral lifetime. The table of elements is a day-to-day working tool in the hands of scientists.

What are the internal components that make some atoms acting similarly in chemical reactions and cause symmetry in Mendeleev's table? The important fellow is the electron that provides the chemical properties of a given element and is at the heart of the symmetry of the table of elements. On its side, the atomic nucleus gives mass to the atom and provides a stable anchor to the electrons.

The electron, as a subatomic particle, was discovered by J.J. Thomson at the end of the 19th century in experiments using a cathode ray tube (CRT), essentially a device very similar to the traditional television tube, which is rapidly losing popularity confronted by competition from liquid crystal display devices. In that CRT, electrical particles emitted by a red hot metal plate called a cathode are accelerated under vacuum by means of a voltage and made to hit a fluorescent surface (screen) in the front of the tube where they produce visible radiation. It is readily shown that an object in the path of the particles produces a shadow on the screen showing that the particles travel in a straight line. Electric and magnetic fields may be applied to deflect the particles in a unique way and the ratio of their charge to their mass (e/m) can be determined from known laws of electricity. The particle is found to be the same as that hypothesised as responsible for the transfer of electricity in electrochemistry and had been given the name electron.

In Thomson's experiments the value of charge and mass is obtained as a ratio. The absolute value of either is not obtained. Another type of experiment is required. This was done several years later, by Millikan and Fletcher (1909) in a famous experiment in which the charge of the same particle was determined. In their experiment, the motion of electrically charged small oil droplets was studied in combined gravitational and electric fields.[9] In that experiment, in order that the oil droplets acquire definite electric charges, X rays or other radiations are used to release electrons by ionising air molecules around the oil droplets. In that case, some electrons become attached to the oil droplets. At a specific value of an opposing electric field, a given droplet can be stopped in its fall in the gravitational field. From the value of the electric field necessary to balance the gravitational force, the charge on the droplet can be determined accurately. The experiment is also carried while the oil droplet is allowed to move under the combined influence of the gravitational and electric field providing a determination of the dynamics of the motion of the droplet within the viscous substance, air or other gases. In practice the droplets are found to be charged with an integral number of elementary charges, proving without doubt the discreteness of electric charge in the Universe.

The results combined with those of Thomson on e/m provided a unique determination of both the mass and the charge of the electron.

On the other hand, experiments carried out by Rutherford showed that atoms were essentially empty structures with a very small particle acting as a central nucleus charged positively. Around the turn of the 19th century it was found that certain elements were emitting strange radiation. It was the birth of the field of radioactivity. One of the types of radiation consisted of heavy particles containing two positive charges of electricity. They were called alpha particles. Rutherford used those particles to study the nature of elements. To do so, he bombarded a foil of gold with alpha particles and made measurements on their transmission through the foil. Most of the particles were transmitted directly with very little alteration of their path. The foil looked rather transparent not offering opposition to the passing of the alpha particles. However, some were deflected at large angles and even reflected back by the foil. It looked like a bullet hitting a sheet of paper and being reflected. The only conclusion that could be reached was that the foil was essentially an empty structure that contained very small but massive electrically charged particles that changed the path of the heavy alpha particles or reflected them back when by chance a collision occurred. This was a crucial experiment that dictated the direction for making a model of the atom. Although, from Thomson's experiments, electrons were known to be part of the atomic structure, the actual configuration nucleus-electron was not known. The atom was neutral and consequently the question was: where is the electron in the atom? Is it part of the nucleus? Is it inside the nucleus or is it around the nucleus and enveloping it?

In a model imagined by Bohr, on the basis of Rutherford's findings, electrons, having a very small mass, are assumed to orbit a heavy nucleus like small satellites. Hydrogen was known to be the lightest element, and from various experiments it was concluded that its nucleus was the simplest one with only one positively charged particle. It was called the proton and was found to have a mass about 1,800 times that of the electron. The model is very much like a small solar system, the nucleus acting as the Sun and the electron as a planet as shown in Figure 2.10 of Chapter 2. The proton and the electron exert electric forces on each other. They attract each other. As in the case of the repulsion of two electrons, that attraction force is much larger than the gravitational force even if the nucleus is much more massive than the electron. Another particularity of the two particles is the exact equality of their charge. If they were not equal, hydrogen atoms would not be neutral and would repel or attract each other. The world as we know it would not exist.

What makes an element different from others? It is primarily the number of electrons and protons. However, it is not possible to simply add protons to the hydrogen nucleus to make a different atom. Protons strongly repel each other when in proximity and it would be difficult to stack protons on each other to build nuclei with many of them. They would fly apart due to their strong electric repulsive force. As just mentioned, gravitational attraction between protons is totally negligible as compared to electrical repulsion and loses the battle. However, nature invented a neutral particle, the neutron, which appears to stabilise constructions involving several protons. The neutron acts then as a kind of catalyst of a strong nuclear force that enters into play, a force much larger than that of gravity and electric forces. That strong force allows the neutron to join the proton to form a stable nucleus. That atom, the nucleus of which is formed of that combination is called deuterium, a heavy isotope of hydrogen. Simply adding a neutron to a hydrogen nucleus does not change its chemical properties much since those are mainly given by the electron. Chemical compounds formed with deuterium, however, have physical characteristics slightly different from hydrogen, such as diffusion, due to the doubling of the mass of the nucleus. This property may be important in biological systems. The presence of deuterium in nature is much diluted. There is about one atom of deuterium per 6,400 atoms of hydrogen in ordinary water.

The next element in the Mendeleev table is helium. It is made of a stable nucleus consisting of 2 neutrons and 2 protons sticking to each other by means of the strong nuclear force. Two electrons orbit the nucleus. Helium is neutral as all the elements are. Heavier atoms have nuclei made of a larger number of protons and neutrons with a number of orbiting electrons equal to the number of protons.

Electrons always provide the chemical characteristics of the atoms. The filling of orbits is done according to certain rules that create the symmetry observed in Mendeleev's table. We will develop that table later according to rules developed in the context of quantum mechanics.

The natural situation of electrons is to be attached to nuclei to form atoms. However, they can be dislodged from that stable situation by several processes. Collisions between atoms themselves in a hot gas, for example, can dislodge electrons from their mother atoms. Heating of a solid object at very high temperatures sets the atoms in violent vibration and dislodges electrons from their attaching nucleus. They may escape the surface of the solid and be sent free in space. On the other hand, electrons in metals are also quasi-free: the interactions between the atoms in a copper wire, for example, is so large that the electrons do not feel attachment to any atom in particular and circulate freely in the metal. Copper and silver are filled with a sea of quasi-free electrons. Since electricity is a flow of electrons, copper and silver are good conductors of electricity.

We mentioned above that a magnetic field can be produced by an electric current. Since static magnets produce a magnetic field we can ask what is the agent inside the magnet that creates the magnetic field? Is it a current within the atoms? It is not a real current, but something similar: electrons possess an internal property that we call spin with an associated magnetic moment. For illustration we may think of the electron as a small sphere rotating about an axis. The electric charge e being distributed on the surface of the sphere would create a magnetic moment. However, this picture is not right since we believe the electron to be a point in space. Furthermore, it gives the wrong answer, the magnetic moment measured being twice as large as that calculated. We will come back on that point later. In any case, the electron acts as a small magnet with magnetic moment $\mathbf{m}$ and if many of them are oriented in the same direction the ensemble looks like a macroscopic magnet and produces a strong magnetic field as shown in Figure 5.5.

Melted iron, when solidified in some specific manner, acts exactly that way, with all the micro-magnets oriented in the same direction. This interpretation of the origin of magnetism gives us immediately

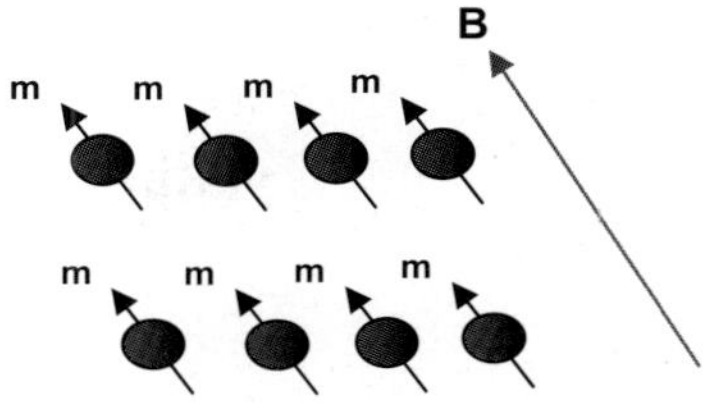

Figure 5.5. Electrons possess a small magnetic moment like a small bar magnet and when grouped together and oriented in the same direction they create a strong magnetic field.

a clue about why when we cut a magnet in half we end up with two separate identical magnets, each one with north and south poles. Since electrons with their 'spin' property are themselves little magnets with north and south poles, cutting a macroscopic magnet in half does not alter the properties of those individual magnets. It is only the number of them in each new magnet that is different from the original magnet. North and south poles are still present in the new magnets.

How about the Origin of Light?

Where is light coming from? The explanation was first given by Bohr in the 1910s by means of his famous planetary model of the atoms. This model is based on the earlier finding of Rutherford of the existence of the atomic nucleus. If such a nucleus exists, the question was then: where is the electron? As mentioned above, in the beginning it was thought to be mixed with the nucleus forming a pudding, but with Rutherford's discovery, showing that the atom was essentially vacuum, it was more natural to put the electron in some kind of orbit around that nucleus. The model made was thus an arrangement that looks very much like a planetary system, the negative electron attracted by the positive nucleus and orbiting it at great speed. One then uses classical electrical and mechanical laws as those outlined previously to determine the nature of the motion of the electron. In the case of hydrogen, for example, a neat little system is constructed consisting of a single proton and a single electron orbiting it as shown in Figure 5.6.

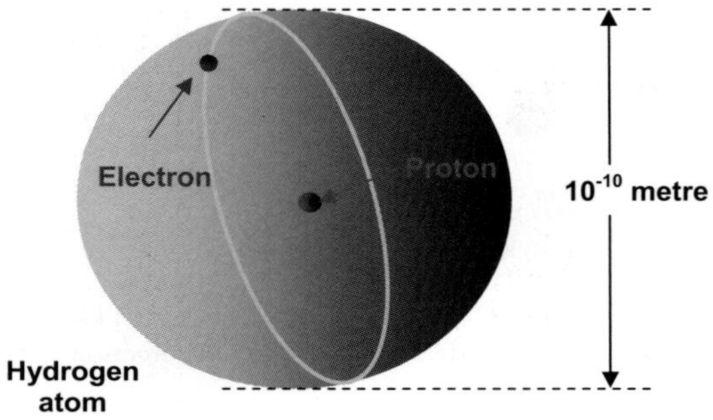

Figure 5.6. Planetary model of the hydrogen atom. The electron in its lower energy state may be visualised as orbiting the nucleus, which consists of a single proton, at a distance of approximately 0.05 nm.

Unfortunately, such a system is not stable, because an electron in its revolution around the nucleus is like an electron in an antenna. Its motion around the nucleus is similar to an oscillation back and forth and it emits radiation. That radiation is energy and must come from somewhere; as we have seen energy must be conserved. For the electron, this means that it needs to fall towards the nucleus since this is where its energy is lower. However, this means that finally the electron falls into the nucleus reducing its size to that of the nucleus. This is totally contradictory to observation. Hydrogen atoms exist in nature and have a finite size of the order of 0.1 nm, much larger than the nucleus, which is of the order of 10^{-15} m. All measurements on hydrogen or compounds of it lead to a dimension of that value. When a model does not work, one has to use imagination and, based on observations on the existence of atoms, one has to bring in new ideas and introduce ad-hoc hypotheses. Great steps were already accomplished by Planck and Einstein. Planck had already made the hypothesis that the increase of the intensity of electromagnetic radiation was discontinuous, that is to say, radiation intensity was increasing by small, finite steps. He explained totally the behaviour of the radiation emitted by a black body. On his side, Einstein had made another crucial step in assuming that light is actually formed of particles that we now call photons, explaining completely the behaviour of electrons emitted by a solid exposed to ultraviolet radiation. This phenomenon is the photoelectric effect,

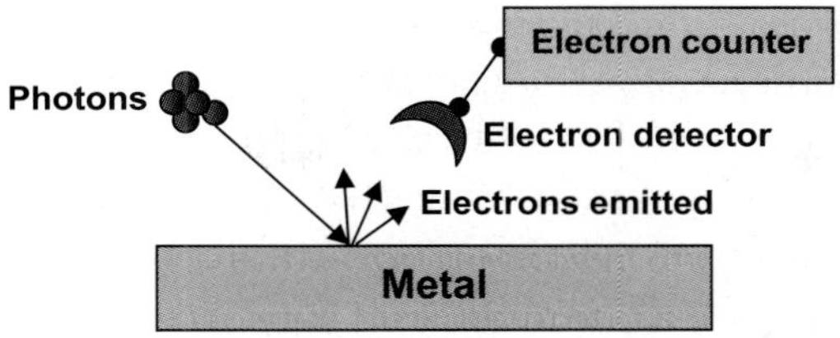

Figure 5.7. Illustration of an ideal set-up for observing the photoelectric effect in which electrons are extracted from a piece of material (electrical conductor, sodium for example) by incident photons.

discovered more than a hundred years ago. We have mentioned it previously. What does it consist of? To illustrate it let us make an ideal setup as in Figure 5.7. When light rays fall on the surface of a piece of electrical conductor they interact with the electrons within the material. The electrons are somewhat free within the material due to strong interactions between the neighbouring atoms, but they are attached to the material itself as a whole. To extract one from the material we need to do some work on it, we need a minimum amount of energy. The work to be done or energy to be spent is to fight the residual confinement force of the electron within the ensemble of atoms in the material. The energy needed to dislodge the electron from the material has been given the name *work function.*

The energy supplied must thus be larger than the work function. The energy of the incident photon is related to its frequency ν by means of the relation $E = h\nu$, in which h is Planck's constant. Thus the photon must have a minimum frequency for extracting an electron from the surface in order to count it at the detector. More photons of course carry more energy, but since the photons interact individually with each electron a greater number of photons with that minimum energy will only increase the number of electrons extracted from the surface. Below a certain frequency, whatever the numbers of photons, no electrons are extracted from the surface because the energy of an individual incident photon is not large enough to win over the work function. This explanation agrees totally with experimental data. This is a most elegant proof of the existence of light particles. Although the word photon was introduced later, the concept of light as a particle received its confirmation with Einstein's model of the interaction of light with matter.

On the other hand, Bohr made another step, but a most draconian one: he assumed that the energy of light particles is originating from electrons changing in discrete steps the radius of their orbit in travelling around the nucleus. But those orbits are very special: they are stable orbits, not decaying with time and not emitting radiation. The change in orbit size leads to a change in energy and thus, an electron seeking a state of lowest energy, falling from one stable orbit to a lower stable one, must get rid of the energy corresponding to the difference of energy of the two orbits. This is done by emitting a quantum of energy as radiation, hν. The process is called a quantum jump. It is illustrated in Figure 5.8.

Consequently, discrete light quanta, with specific wavelengths or frequencies, are coming from transitions of the electron from a higher orbit to a lower one. In normal circumstances, the electrons are all in their lowest orbit also called lowest energy state. However, it is possible to excite electrons to higher orbits by means of an electric discharge in a gas, for example. In that process the atoms of the gas in question are partly ionised. Free electrons in the discharge

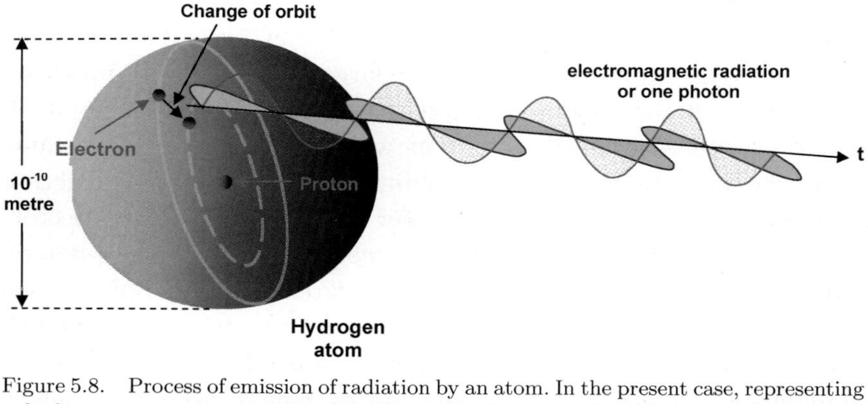

Figure 5.8. Process of emission of radiation by an atom. In the present case, representing a hydrogen atom, the electron falls from the first excited orbit to the lowest one. Radiation is emitted by the atom. If the two lowest orbits are involved, the difference of energy corresponds to emitted radiation with a wavelength of 121.6 nm. It is called Lyman α radiation and is not visible being in the vacuum ultraviolet. We have shown that energy is emitted as a wave. But we could simply have represented the process as the emission of a ball of energy, a photon with energy hν.

collide with atoms exciting the attached electrons to higher orbits from which they cascade down to the lower ones. Free electrons may also get re-attached to a nucleus and also cascade down to lower energy states.

In his model, Bohr made the hypothesis that the electron could not revolve around the nucleus at any arbitrary distance and that certain orbits followed by the electron around the nucleus were privileged and stable, they were not radiating. This is called *quantization* of the orbits in which the electron has a very specific energy. The orbits correspond to energy levels and are identified by a so-called *quantum number*, 'n', that takes only integral values, $1, 2, 3 \ldots$. The theory shows that the spacing between the energy levels is proportional to the inverse of the square of n. Since only specific orbits are allowed, the electron falling from a higher stable orbit to a lower one would emit a photon with a very specific energy $h\nu$. This energy is equal to the difference in energy of the electron in the two given orbits. The lowest possible electron stable orbit has a radius of 0.053 nm. It is called the Bohr radius. It is obtained directly from theory as a combination of several fundamental constants. This lowest orbit is identified by the quantum number $n = 1$. The number n has been given the name principal quantum number and increases by unity with each higher orbit. When n becomes very large the electron is very loosely attached to the nucleus and is easy to dislodge. One may wonder about the speed of the electron in its orbit around the nucleus or the frequency of its revolution, if such questions make any sense. The numbers come out directly from the theory and in its lowest orbit the electron makes 6,570 tera-revolutions per second (6.57×10^{15} rev per second) and travel at a speed of 2.2×10^6 m per second, a good fraction of the speed of light.

The theory thus provides rules for the wavelength of the emitted radiation between any pairs of orbits or energy states. That radiation, emitted as discrete wavelengths, forms spectral lines series in agreement with data observed experimentally. Two of such series are illustrated in Figure 5.9.

The Lyman series corresponds to electron transitions from higher orbits to the lowest one identified by the so-called quantum number

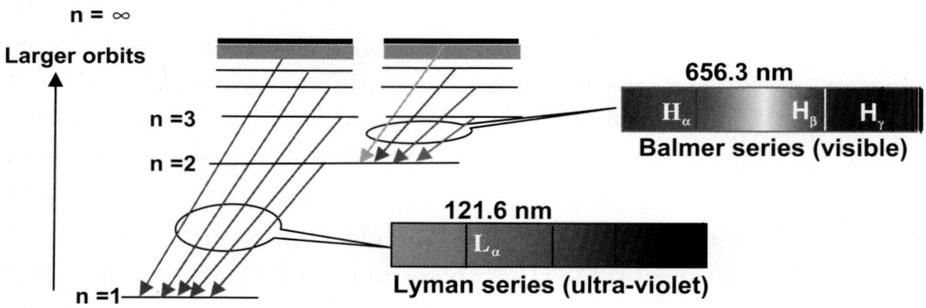

Figure 5.9. Electronic transitions between orbits or energy levels lead to series of spectral lines which may be used as evidence of the presence of a given element. The background colours in the figure are shown only to illustrate the actual colour of the spectral lines emitted by the hydrogen atom. The background is not present in the spectrum. The H_α line is the well known line of hydrogen that is present in nebulae and is visible with a modest telescope.

n = 1. The wavelength of the spectral lines included in that series fall in the ultraviolet and they are not visible. In the case that the excited electrons fall back to the level n = 2 the series of spectral lines falls in the visible. The one at 656.3 nm, called Balmer alpha (H_α) is the one that gives the reddish colour to a hydrogen electrical discharge.

It should be mentioned that the process of emission may be observed in reverse: if radiation of a wavelength that corresponds to the difference in energy between two orbits, the electron in a lower orbit may be excited to an upper orbit. The radiation is absorbed by the atom and is re-emitted at a later time by the electron seeking lowest energy and falling back to the lowest orbit.

It is quite amazing that with such a simple approach and a bit of imagination, one can get into the heart of atoms and create a possible model of the atomic world. The only ad-hoc hypothesis introduced is the peculiar stability of certain orbits compared to some others in between that are forbidden. We can apply the model to more complex atoms with many electrons orbiting a heavier nucleus. The spectrum of radiation emitted by those atoms is of course more complex, but the model, at least in principle, provides a basis for explaining qualitatively the nature and origin of the spectra emitted by any element.

In this context light reaches us in discrete packets and at very specific discrete wavelengths. How about the spectra of those standard incandescent lamps that we use commonly in our homes? If one examines the spectrum of those lamps, it looks rather continuous, without specific discrete lines. In those lamps, it is a wire of tungsten that is heated by passing a strong current through it. The electrons are forced to travel through the metal, are accelerated, collide with other atoms, are stopped in their motion, even escape partly from the metal, fall back into it and recombine with the atoms. Their motion is totally random, involving a continuum of oscillation in all directions at various velocities, and they emit a spectrum of radiation that looks continuous. The spectrum emitted is called black body radiation, the type of radiation studied by Planck. The property of that radiation is determined by the temperature of the object that emits it. The tungsten filament in a lamp is raised to a temperature between 2,000 and 3,000°C. The spectrum has a maximum of emission at a wavelength of about 650 nm, a wavelength in the reddish part of visible light. This is why objects seen under that type of radiation look somewhat reddish, because there is more light at that wavelength. The Sun's surface temperature is about 5,500°C and the maximum of its spectrum is at a wavelength of about 500 nm. Our eyes, adapted by evolution to that type of spectrum, see the mixture of wavelengths reaching us as white light. The pencil we have in our hand is at room temperature, about 25°C. It emits radiation, but due to its low temperature the maximum of its spectrum is at very long wavelength, a radiation that our eyes are not sensitive to. Why do we see it? It is light falling on it, from the Sun or from a desk lamp that excites the electrons in the pencil. These electrons re-emit that radiation at other wavelengths characteristic of the material forming that pencil: this is called reflected radiation.

In a gas, contrary to the case of a solid such as a metal, atoms do not interact very strongly. They move in space and collide among themselves. These collisions last for a very short time and they are essentially free most of the time. In such a case, the radiation emitted by the atoms when they are travelling between collisions contains discrete spectral lines characteristic of the atoms in question. For example a Neon gas, heated by an electric discharge, emits radiation which consists of sharp spectral lines with one predominant, a red line at 633 nm that we are familiar with, emitted by Neon signs on buildings. That wavelength is one of the many spectral lines whose wavelengths are determined by the particular steps that one of the Neon electrons makes from one orbit to another. The spectrum of each gas has specific characteristics of its own. The discrete spectral lines within that spectrum are determined by the inner properties of the electrons of the atom forming the gas. These lines are a representation of the personality of a given gas. They can thus be used to identify the nature of elements without having direct contact with them. Spectral lines are for atoms somewhat like the fingerprints are for a person. This is why we know so much about the Universe and its constituents. Radiation that we receive here on Earth, originating from a celestial object some billions of km away can be used to determine the substances forming that celestial object. Gases in the path of that radiation, in the space traversed to reach us, can alter the spectrum emitted by the celestial body by absorption at very specific wavelengths. Dark lines appear in the spectrum and these lines can be used to identify the intermediate gas. Most of the studies of the cosmos are based on these spectral studies from the very long wavelengths to the ultraviolet and gamma rays. These studies show us that helium and hydrogen are the major constituents of the Universe that we see. We will see later that the properties of such spectral lines can also be used to determine precisely the dynamics or the motion of celestial objects and conclude almost with certainty concerning an intriguing property of the Universe: it is expanding.

So, How Much Do We See?

Our eyes are sensitive to a very narrow band of the total electromagnetic spectrum, which actually is infinite since frequency may spread from zero to infinity. On a linear graphic scale the width of frequency that our eyes are sensitive to is very narrow. Our eyes are sensitive from the deep red to violet radiation that our brain interprets as light; our body is somewhat sensitive to infrared that we call heat. This is best represented on a log scale in which the

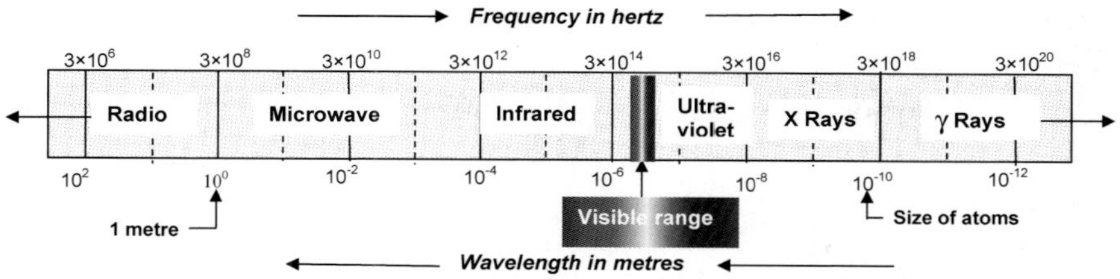

Figure 5.10. Electromagnetic radiation spectrum plotted on a log scale. It spreads from zero frequency and zero wavelength to infinity. Ours eyes are sensitive to a very narrow range of wavelengths around 500 nm. Most of the spectrum can be detected only by means of man-made instruments or through secondary effects on matter.

equal marks spacing represent an increase by a factor of ten. This is shown in Figure 5.10.

As is made obvious on the scale chosen, the part of the spectrum that our body is sensitive to, as light (visible) and heat (infrared) is very narrow. Most of the radiation exists outside the range of sensitivity of our eyes. The human being had to wait for the development of sophisticated instruments to fully appreciate the existence and properties of radiation such as X rays, gamma rays and long wavelengths such as those used to make our lives (sometimes) more interesting through radio and television.

VI

A Provocative Theory: Quantum Physics and Its Use to Improve Our Well Being

The theory elaborated by Bohr on Rutherford's model of the atom is quite a fine achievement and it provided a direct input for a possible visualization of the constituents of atoms and their behaviour. The imagination that went into it is most fascinating because it deals entirely with classical concepts such as orbits, revolution, orbital speed, and jumps between orbits, all classical concepts imported from the known laws of planetary motion and adapted to the invisible micro-cosmos. In practice, however, one does not make measurements on such concepts as orbits and electron speed within atoms. We cannot go inside the atom and measure dimensions and speed with a metre stick and a clock. This does not mean that we cannot speak about such concepts as distance of the electron from the proton or its revolution rates around the nucleus in its orbit, but we do not really make measurements of them. The experimenter observes light coming from the atoms or the behaviour of those atoms in applied magnetic and electric fields. One then deals with concepts such as atom energy, radiation wavelength, frequency, polarization, atom trajectory, path deflection and magnetic moments, parameters that can all be measured from our macroscopic framework. Another important point is that one needs to realise that an electron in orbit is under acceleration since it changes continuously its direction. Consequently, according to electromagnetic laws, that is to say Maxwell's equations, it should radiate and lose its energy continuously. It should finally fall into the positive nucleus. The world would not be what it is. To avoid that difficulty, Bohr introduced the concept of non-radiating orbits and their quantization was arbitrary, forced by experimental observations on optical spectra emitted by atoms. In the theory, there is no basic principle or fundamental reason that provides a framework in which such a situation could exist. There are also problems in the details. The observed spectra are often much more complex than those derived from the simple orbital model. That model, although pushed to its limits, through substantial refinements made, for example, by Sommerfeld who introduced relativity into it, is not satisfactory in providing a correct description of the details of the observed spectra.

A Baffling Experiment

Parallel to that exploration, experiments were made that uncovered magical properties of the atomic world. I say magical, but it is really a weak word to qualify it. In the early 1920s, two physicists, Otto Stern and Walther Gerlach, conducted an experiment to study the actual state of magnetism of silver atoms. To do this, they formed a beam of silver atoms by heating at high temperature an oven containing a small amount of silver. In such a case, the metal vaporizes and through proper collimation, a narrow beam of silver atoms is formed in vacuum as in Figure 6.1. In that arrangement the beam of silver atoms is allowed to pass between the poles of two bar magnets. One bar has a triangular cross-section and has a sharp edge forming one pole. The other one has a rectangular flat cross-section and forms the

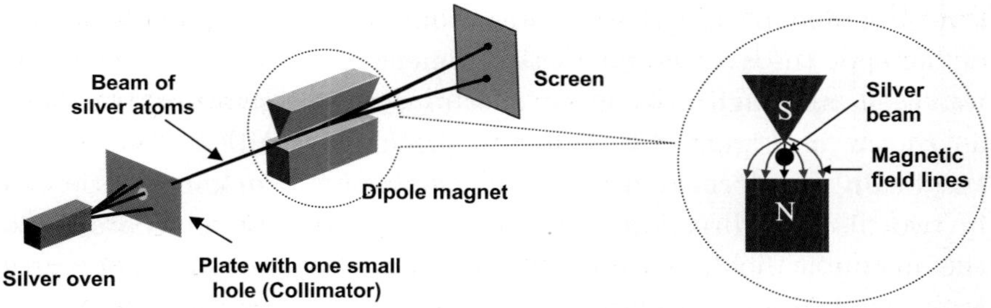

Figure 6.1. The Stern–Gerlach experiment.

other pole. The magnetic field produced is extremely inhomogeneous, very strong at the sharp edge of the triangular bar and weaker at the rectangular bar. The field spreads in space reaching the surface of the rectangular bar magnet as shown in the inset of Figure 6.1, thus decreasing in intensity. This arrangement is a special case of the general configuration called 'dipole magnet' often used in this kind of experiment. Magnetic forces are different from electric forces because in the case of magnetism we do not deal with monopoles as in the case of electric charges in an electric field. A bar magnet with its North and South poles is said to have a magnetic moment oriented along the axis joining the two poles. Since it has direction associated to it, we say that it is a vector. In the configuration shown, due to the rapid spatial change in magnetic field intensity, there is a net force acting on any object with a magnetic moment that happens to be in between the poles. The direction of the force on the object, up or down in the figure, depends on the orientation of the magnetic moment relative to the magnetic field produced by the magnets. The silver atoms possess a magnetic moment and should thus act like little bar magnets. If silver atoms behaved like classical bar magnets we should thus expect the atoms passing through the inhomogeneous field of the dipole magnet to be deviated like any small magnets, such as magnetized iron filings, depending on their orientation. Since the atoms enter the space between the magnets at random, the orientation of their magnetic moment should also be random and one would expect on the recording screen a continuous spread of atoms arriving at various angles with a maximum at a point in line with the source. The deposit of silver atoms at the screen should be blurred and one would calculate from the width of the spot the average magnetic moment of the silver atom. What was seen, although after great efforts, is the presence of two spots as shown in Figure 6.1. What a shocking result! It looked like the beam was formed of atoms with magnetic moments oriented in two discrete directions, up or down, parallel or anti-parallel to the magnetic field, nothing in other directions. It did not agree with classical laws of electromagnetism, did not seem natural, and is not.

A Question of Matter and Wave

The atomic world defies common sense. It reacts in unexpected ways. How can we explain the Stern–Gerlach result with such a simple thing as silver atoms, which were known to possess a magnetic moment as a little bar magnet? It was a challenge and it took a great deal of thinking and imagination to arrive at a model that would provide a framework capable of explaining the results and extend itself to the whole field of atomic physics observations. Sometimes, one is tempted to invent an ad-hoc mechanism to explain an observation. The explanation given then applies only to the particular observation made and lacks generality, which is essential to a sound theory. In particular, if the model does not predict a decent number of applications, verified in later experiments, the ad-hoc mechanism raises scepticism. We have seen that the explanation for the failure to find motion of the Earth through the aether was of that kind. It was assumed that the aether was reacting on the apparatus and contracting it. The explanation presumed the existence of the aether to start with and was invented for the particular result found. It raised scepticism. The principle of relativity assuming the constancy of the speed of light in general provided the basis for a theory that had a wide spectrum of applications and made predictions regarding contraction of objects at high speed of the nature that Lorentz had assumed. That principle was accepted because it explained many other observations and made predictions that were soon verified.

Before attempting to provide an explanation of Stern–Gerlach's result, which is found in the introduction of the electron spin and its so-called 'quantization' in a magnetic field, let us go back to that time, in the 1920s, where brilliant minds were struggling with hot discoveries on the properties of atoms, electrons, and light waves. In a bold manner of thinking, Louis de Broglie proposed that, based on symmetry with the electromagnetic wave to which we attach a particle, the photon, we could very well attach a wave to the electron (1925). This wave would be another property of the electron as important as its mass and charge. Similar to the case of light, the electron would have a dual personality depending on the situation in which the observation is made. This step was an audacious application of the symmetry principle that is very often used in physics to introduce new concepts, new properties and even new particles. We will see later that symmetry found in the large number of observed so-called elementary particles led to a unification of particle physics and the discovery of the presumably true elementary constituents of matter.

But what would be the wavelength of the wave to be associated to the electron? The answer to that question is rather simple and was based on symmetry with light. A photon has momentum: a simple calculation shows that the momentum is $p = h\nu/c$ or h/λ where h is Planck's constant and λ is the wavelength of the associated wave; c is always the speed of light. The wavelength is thus h/p. We can use that same expression with the electron, and use for the momentum the usual value $p = mv$, which is the product of its mass by its velocity. The electron wavelength is then simply Planck's constant divided by its momentum or $\lambda = h/p = h/mv$. The faster the electron travels, the shorter the wavelength is. Soon after that hypothesis was made, experiments were done in which the reflection of electrons from crystal planes, observed by Davisson and Germer (1927), showed that more electrons were reflected in certain directions than in others. An interference pattern was observed in the detection of the electrons as if a wave was associated with them. This was similar to diffraction in crystals as observed in the case of X-rays. The necessity of using a crystal to observe interference phenomena of the electrons was dictated by their short wavelength requiring an interacting medium having small dimensions such as the space of a few Angstroms between crystal atomic planes. The electron wavelength deduced from the experimental results was in agreement with that predicted by the theory.

How fast is the electron wave propagating? This is a most interesting question since in principle a wave extends in space and we are used to think of the particle as being localised and moving along a definite path. How can we thus represent that particle by a wave? One way would be to assume that the wave is concentrated around a given spot where the particle is. We could think of it as a pulse or a burst. Such a burst, also called a wave packet, can be constructed by adding a large number of waves with different frequencies all propagating in space in the same direction. In such a situation, the wave packet is relatively well localised. When the velocity of propagation of that burst is calculated using the expression we have assumed above for the wavelength and using mass according to relativistic principles, we find that the wave velocity is exactly that of the particle. This is the expected result because it would not be logical that the wave travels at a speed different from the electron itself.

Is this duality specific to the electron or does it exist for all material objects? Experiments were done with molecules such as H_2 and diffraction phenomena were observed upon reflection from surfaces of a crystal (Stern, 1932). Could we associate such a wave to macroscopic objects such as an apple thrown in the air at a certain velocity? In principle we could do so. However, due to the mass of those objects the wavelength would be extremely short and no interference effects of the nature of those observed with electrons would be observed. Furthermore, statistical phenomena that take place in the macroscopic world alter the coherence that is present in the atomic world and as we will see later it is not possible to transpose directly the conclusions drawn in the microscopic world to predict phenomena in the macroscopic world.

A most interesting aspect of the hypothesis concerns the wavelength of the electron travelling rapidly around the nucleus. When the calculation is made with the relation between velocity, momentum, wavelength, and size of orbits as calculated in the Bohr model of the hydrogen atom, it turns out that stable orbits are those which contain an integral number of cycles of the wave associated with the electron in those orbits. The wave propagates in an electron orbit and through interference reinforces itself at each revolution. This situation is illustrated somewhat naively in Figure 6.2. It produces a standing wave oscillating in time but with crests and troughs fixed in space.

If the orbit were not an integral number of wavelengths, the wave crests would overlap with the troughs on subsequent revolutions. After a few trip around the orbit, due to destructive interference, the wave would destroy itself, like waves coming from many directions on the surface of water. We can imagine the happy state of mind of the first scientists that made that simple calculation and arrived at that result.

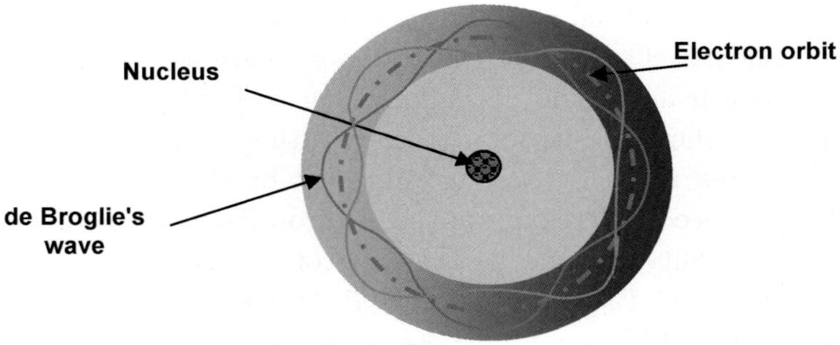

Figure 6.2. Stable orbits in Bohr's model are those that have an integral number of cycles of the matter wave calculated from de Broglie's hypothesis and the speed of the electron in the particular orbit.

Wave and Particle, is it Compatible?

Einstein in his theory of the photoelectric effect has shown without ambiguity that light behaves as a particle. We call it the photon. However, it had been known for a long time that light also behaves as a wave. How was that discovered? Similar to waves on a water surface it is relatively easy to produce interferences that prove clearly the wave nature of light. The idea is to create an arrangement in which part of the wave is delayed from another part and then to make those two parts add up to each other. We may call these two parts wavelets. If the crests of the two wavelets superimpose on each other, there is a re-enforcement of the wave. If a crest is superimposed on a trough there is cancellation. Using the description we made in Maxwell's approach, we may say that the electric field of one wavelet adds to the positive electric field of the other wavelet if it is positive or cancels it if it is negative. The relative sign of the electric field of the two wavelets at the screen depends on the length of the path they have travelled. We say that the two wavelets interfere. One way of creating the delay between the two wavelets is to let a primary wave go through two slits close to each other as shown in Figure 6.3(a) (Young, 1800). Figure 6.3(b) illustrates the difference in path lengths between rays of the wavelets.

The light wave goes through the two slits and in arriving at the screen, depending on the path length traversed by the wavelets, an interference pattern is observed. This interference pattern is caused

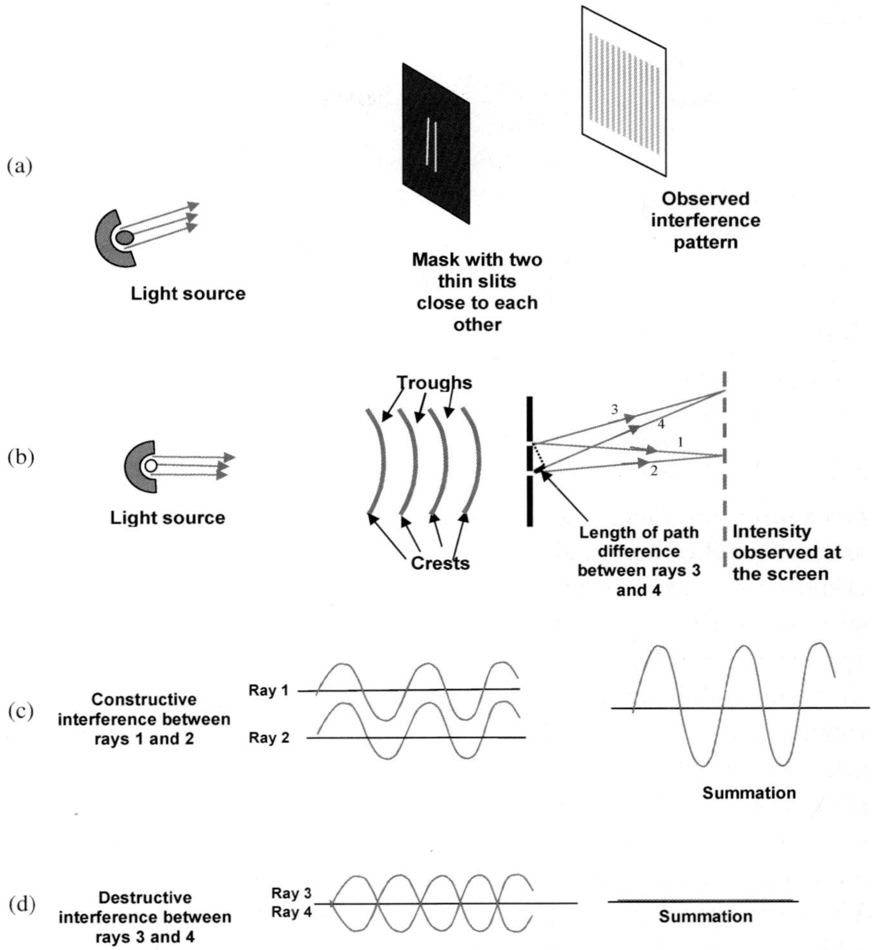

Figure 6.3. Technique for obtaining interference fringes with light. In (b), which is a bird's eye view of setup (a), it is readily observed that path 4 is longer than path 3 for rays meeting away from the centre of the screen. The path difference may be such that a crest superimposes on a trough and produces a minimum of light, as illustrated in (d). At the centre of the screen the distance travelled by the two rays, 1 and 2, is the same and the two superimposed wavelets produce a maximum of light, as illustrated in (c).

by the addition of the wavelets now coming from the two slits. When the path lengths are identical (rays 1 and 2 hitting the centre of screen) or are different by an exact multiple of wavelengths (rays on specific oblique paths), the crests of the wavelets add to cause a re-enforcement of the intensity as shown in Figure 3(c). When the path lengths are not an exact multiple of wavelengths, for example rays 3 and 4 oriented at a given angle having a path difference corresponding to 1/2 wavelength, the addition of a crest and a trough produces a null result as shown in Figure 3(d).

The two experiments, photoelectric effect and interferences with slits are most convincing regarding the dual personality of electromagnetic radiation. Can we do a similar experiment with electrons to prove clearly that they too have a dual personality? A large number of experiments have shown the corpuscular nature of the electron and many commonly used devices use the 'particle' property of electrons. The functioning of classical TV tubes is based on that side of the electron personality. In a TV tube, a large quantity of electrons is produced by heating a wire of tungsten at such a temperature that electrons are ejected from it. Those electrons are accelerated by means of an electric field, deviated by means of a magnetic field, and made to hit a fluorescent screen to produce a bright spot. The electron beam is scanned across the screen in a well structured pattern and an image is formed by varying its intensity. The whole behaviour of the electrons can be described by the classical electromagnetic laws that we have mentioned previously and the kinematics of moving bodies. Their paths is calculated as if they were particles with mass m and charge $-$e. They have all the properties of true particles.

On the other hand, we have also mentioned that if we let electrons reflect from a crystal surface we obtain interference patterns that depend on the spacing between the atoms of the crystal. The patterns obtained in such experiments do not lead to easy interpretation and the explanation of the observations requires some elaborate geometrical analysis. For a better explanation of the wave nature of the electron it is best to imagine an ideal experiment of the kind done with light as shown in Figure 6.3. The experiment is imagined as a heuristic instrument for better understanding the electron characteristics and is not done in practice. The phenomena that would be observed in such an arrangement are simply inferred from the results obtained in more complex experiments. However, it is an experiment that shines light clearly on the mysterious nature of that particle. In that ideal experiment the source of radiation in Figure 6.3

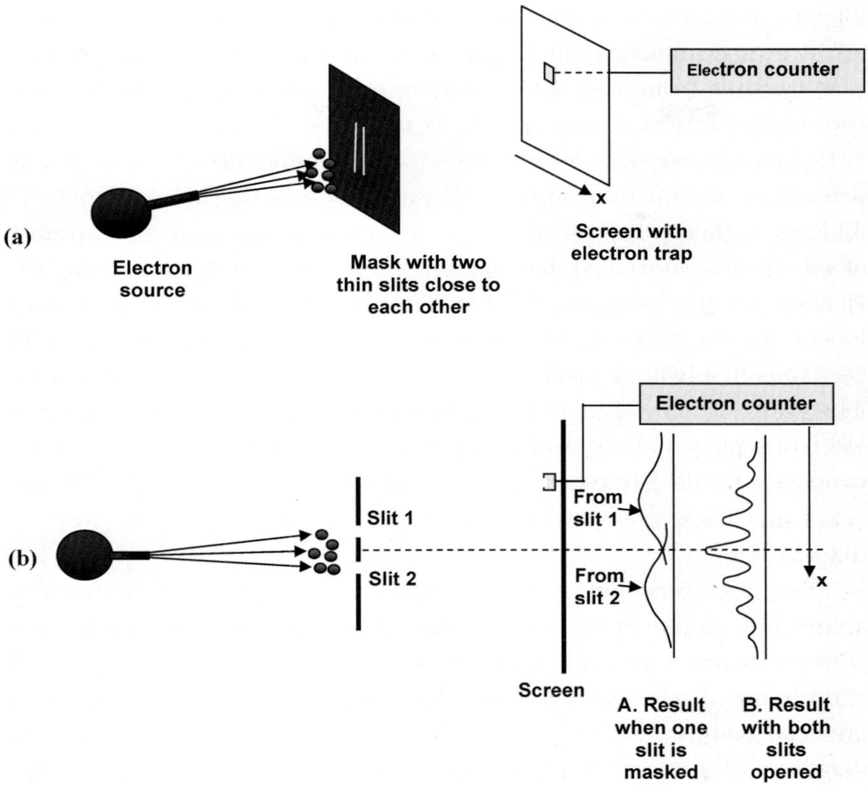

Figure 6.4. Fictitious set-up imagined for observing electron interferences. The screen has a counter of electrons that can be displaced horizontally and the results of various experiments are shown behind it. (Adapted from R.P Feynman *et al.*, Lectures on Physics, *op. cit.*)

is replaced by a source of electrons and the idea is to describe what would be observed at the screen on the basis of our knowledge of the electron properties. Let us look at such an experiment as illustrated in Figure 6.4. The screen is equipped with a counter of electrons that can be displaced horizontally in order to determine the number of electrons arriving at a given coordinate, x, by counting them one by one.

In this set-up, the electrons arrive at the slits as a flux of particles. If slit 1 is blocked the counter of electrons counts all those electrons that are traversing slit 2. The results are illustrated by means of the pattern 'From slit 2' in graph A in Figure 6.4(b). This is expected.

The probability of counting electrons arriving at the screen from slit 2 is maximum at the point on the screen that is on a straight line passing from the source through slit 2. If slit 2 is blocked, the other pattern, 'From slit 1', is observed. Now, if both slits are left open one sees an interference pattern build up in the counting process as shown by graph B, 'Result when both slits are opened'. This is rather puzzling since, in a simple reasoning, the number of electrons counted at the screen should just be an addition, the electrons being particles. If they interfere, it must mean that they have a hidden property that we do not detect when we simply count electrons in a beam or when they hit a screen as in a TV picture tube. The result is very similar to that obtained with light waves and the electrons must have a wave property: this is just the material wave proposed by de Broglie. There is thus symmetry between light and electrons: both are wave and particles, depending on the way we detect them.

To understand better the electron characteristics, let us examine a few situations. Suppose we reduce the electron flux towards the slits to such a level that an electron traversing the slits finds itself completely alone until it reaches the screen. It arrives at the detector and the counter makes a click. It should be pointed out that the detector always counts a complete electron. There is no such thing as a fraction of an electron reaching the detector. The following electron traverses the slits later and arrives after the first one has been detected. The counter makes a separate click. And it keeps going like that, no two electrons being present together at the same time between the slits and the screen. The electrons are counted one by one and make definite clicks. We might think that we should not see any interferences since electrons are particles and must go through a given slit as a whole. The electron cannot divide its mass or its charge in two parts. However, the interference pattern persists. It looks as if each electron is passing through both slits as we would expect with radiation. Suppose we put a lamp at the exit of slit 1 and try to see where the electron passes. Again this is a fictitious experiment since the electron is extremely small (a point in space!); to see it would require an extremely short wavelength not realizable in practice. Such a lamp emits photons. A photon hits the electron, is reflected and reaches our eye. We thus determine with certainty which slit it has gone through. At about the same time the detector makes a click telling us that the electron we have seen has reached the detector. If this is done and we determine which slit the electrons are going through, the interference pattern disappears. We are back to graph A. The observation of the electron has somehow disturbed it.

Before observation, the probability of the electron traversing one slit or the other is nearly the same. When one slit is blocked the electron goes through the opened slit and acts as a true particle. When both slits are opened its wave nature takes over. The experiment tells us that the electron wave extends over both slits and there is probability for the electron to be at both slits. When we detect the electron with a lamp and determine that it goes through slit 1 for example, we have localised it with certainty and the counter at the screen makes a click showing a maximum probability as if the electron comes from slit 1. The probability of finding the electron at slit 1 when detected is unity. If we reduce the light intensity in order to diminish the interaction with the electron we could expect a reduction of the perturbation. We would not perturb the electron as much as with full intensity and we should start seeing interferences. However, the lamp emits photons and reducing its intensity reduces the number of photons and not their energy which is fixed by their frequency or wavelength. What happens then is that only those electrons will be seen when a lamp photon arrives at the time an electron passes through the slit. Those not seen will not be perturbed and will produce interference fringes and those seen will produce on the screen a pattern of the nature obtained when one slit is masked. The final result will be a mixture of an interference pattern superimposed on a continuous pattern of electron arrivals when one slit or the other is blocked. You might think of a less perturbing experiment by reducing the wavelength of the radiation emitted by the lamp. If the wavelength of the light emitted by the lamp is longer, its frequency is lower and the photon is less energetic. True, but in that case, because a long wavelength does not allow precise measurements, you will not be able to determine exactly the position of the electron and which slit it has gone through. Frustrating! But this is what nature is: very subtle. It is not possible to go further in the basic explanation of the phenomenon. One can use advanced mathematics and write equations that can baffle the non-specialist, but the simple truth is what has been described: the electron behaves as a particle, the counter counts clicks, and as a wave, the counting of the arrival of electrons shows interference fringes. The whole process can only be dealt with in terms of probability. There is an *intrinsic uncertainty* in nature at the microscopic level. There is no hidden command in the source that would dictate which slit an electron should go through. This whole affair may be analysed entirely by means of reasoning on amplitudes and probability. The amplitude of the wave at slit 1 may be called a_1 and that at slit 2, a_2. The detector at the screen, depending on which slit is blocked, will detect either the electron with probability $(a_1)^2$ or $(a_2)^2$. However, if both slits are opened, the detector will detect $(a_1 + a_2)^2$, which is very different from simply the sum of the square, $(a_1)^2 + (a_2)^2$, because in the first relation, if a_1 is positive (a crest) and a_2 is negative (a trough) and both a_1 and a_2 are equal, the sum is zero. In the second expression, the two numbers are squared and the negative sign disappears upon squaring. Although we count electrons as a whole there is a property, called phase, associated with their motion in space.

This whole affair looks again like a conspiracy of Mother Nature to test our tolerance. We had duality of light that acts as a wave and a particle. We now have the same thing with the electron that acts as a particle and a wave. And all atomic particles act that way. It is a real challenge to our understanding.

Wave Mechanics

Maxwell's theory leads to the existence of radiation travelling at the speed of light. In the theory, nowhere is a photon appearing. That classical theory leads to a so-called 'wave equation', an equation from which all the properties of the wave can be described. It is like a recipe: you apply the equation to a specific experimental arrangement and solve the equation with the experimental conditions set in: you get a solution that dictates the behaviour of the wave. On the other hand, in Bohr's theory, the electron is considered to be a particle and nowhere does the concept of matter waves appear in the equations of motion when treated in a classical way as a particle orbiting a proton. If a wave can be associated with the electron, should we not have also an equation guiding the matter wave as in the case of Maxwell's theory for electromagnetic waves? Schrödinger answered that question by proposing such an equation (1927).[1] It is not derived from experimental laws as in the case of electromagnetic theory. It is more or less a principle in itself, as is the second principle in kinematics written by Newton for the connection between acceleration and force through mass. Schrödinger's equation is rather complex, has nothing to do with laws deduced from experimental observations and is not derived from first basic principles. The first hypothesis in the theory is the identification of the electron with a wave. Amongst various approaches, the equation can be obtained in a heuristic way by using a classical wave equation and using de Broglie hypothesis for the wavelength of the electron. The equation applies to a so-called *wave function* that determines the electron behaviour. The equation dictates the behaviour of that wave function for cases where the electron is placed in various circumstances, for example, in an electric field such as the one produced by the proton. It deals only with observables such as the mass of the electron, its electric charge, its energy, its angular momentum and other parameters that can be measured. It does not deal directly with orbits, speeds, or periods of the electron evolution around the proton as in Bohr's model, but situates the electron in the electric field of the proton at a given distance from it. The heuristic approach used in deriving Schrödinger's equation is not a proof of its validity, but the results obtained are in total agreement with experimental data and appear to confirm its validity.

Questions were raised about the exact nature of that wave function. In the case of Maxwell's approach for electromagnetic waves, we know that the wave is a description of the behaviour of oscillating electric and magnetic fields. It is energy in those two fields, which is transported by the wave and the electromagnetic wave is thus a form of energy. But in the case of the electron what is the wave made of? What does it represent? An interpretation was given by Max Born: he proposed that the wave when multiplied by its complex conjugate, that is to say, the square of its absolute value, represents the probability of finding the electron in the region of space where the calculation is made. We may think of the electron as an object not too well localised, as a kind of cloud represented by that wave function, whose squared absolute value represents the density of that cloud at a given point. If the wave represents a probability, it does not appear to be an actual physical substance. We may think of it simply as a number at each point of space. It is in a sense a mathematical entity attached to the electron that provides a tool for specifying its behaviour without actually knowing what it is. This kind of proposition brings back the question of wave particle duality to which we do not have a satisfying answer. But it works astoundingly well.

Schrödinger's equation is a most fascinating tool, but physics loses part of the simplicity that was introduced by Newton. Solution of the equation for the case just mentioned, that of the hydrogen atom, an electron in the electric field of the proton, actually one of the simplest cases, is a tedious mathematical task. But it can be done exactly. It is one of the most inspiring exercises because it gives all the results obtained in Bohr's planetary model and provides a description of other observations that remained unexplained in that model. In particular, stable electronic energy states appear naturally. The solution in that simple case is an inspiration for calculating and predicting the behaviour of complex atoms made of many protons and electrons.

In practice, the solution of Schrödinger's equation leads to more restrictions on the behaviour of the electron than Bohr's model and consequently it introduces new *quantum numbers*. Those numbers characterise the wave function and the associated electron energy states. In order to visualise the properties associated to these numbers it is best to refer to Bohr's orbital model of the atom. Although we know that the model is a creation of our imagination and is not a true representation of the reality, it provides an easy graphical representation of those quantum numbers. It helps us visualise the situation and memorise the meaning of those numbers. Let us limit ourselves again to one electron in the neighbourhood of a proton, which is the hydrogen atom. Solution of the wave equation leads to exact expressions for the wave function and to the introduction of a first quantum number, called n, the *principal* quantum number: it is the same as the one introduced arbitrarily by Bohr in his planetary model. It essentially identifies the orbit in which the electron evolves and is connected to the radial coordinate of the electron or its distance from the proton. That number can take any value from 1 to infinity, the orbit increasing in size with larger numbers. When the electron is far removed from the proton, the coulomb interaction is weak. The energy of interaction is zero when the electron is brought to an infinite distance from the proton and is free. We must realize that infinity here means simply a distance large compared to the size of the atom where interactions with other atomic entities are more important than interaction with the original proton to which the electron was attached. This principal quantum number identifies the main energy levels. Thus, up to that point the solution of Schrödinger's equation leads to the same results as Bohr's model, a comforting result since this model was found to explain the bulk of observations.

On the other hand, Schrödinger's equation introduces two new quantum numbers concerned with electron orbits that possess angular momentum and obey special rules. Without going into the details, we may point out that orbits that have angular momentum are quantized, that is to say that the electron angular momentum may not take any arbitrary value.[2] Its associated quantum number is called l and may take values from 0 to (n − 1). Furthermore, as explained in Annex B, section d, the motion of an electron in such orbits looks like a small current in a loop and creates a magnetic moment that take only specific directions in a magnetic field. The associated quantum number is called m and can take values only from −l to +l. These rules all come out of restrictions imposed by Schrödinger's equation; otherwise, a solution does not exist.

These are the rules obtained from the solution of Schrödinger's equation for an electron in the potential of a nucleus with a positive charge. Aside from an additional important property that we will now examine, they provide the basic material for explaining most of the characteristics of atoms and their radiation emission spectra. In particular, the fact that the magnetic moment can take only specific quantized orientation in a magnetic field has the effect of splitting the energy levels and increasing the number of observed spectral lines. This is exactly what is observed in practice. The phenomenon is called the *Zeeman Effect* after the name of the physicist who played a pioneering role in its observation. The interpretation and the conclusions all result from the hypothesis that the electron behaves as a wave and a particle.

Another Fascinating Property of the Electron: The Spin

Consequently, in the study of spectra of atoms placed in a magnetic field, it is observed that the spectral lines split and the phenomenon can be interpreted according to specific rules dictated by the previous considerations regarding the presence of new quantum numbers. It is comforting that many observed atomic spectra can be interpreted within that context. In practice, however, several atoms emit additional spectral lines that cannot be interpreted with such an approach and Schrödinger's equation alone.

To solve that apparent problem, a new electronic property was introduced by Pauli in which the electron had two new degrees of freedom (1924). The idea explained totally the observed spectra. The two new degrees of freedom were associated with the concept that the electron was spinning on itself. The electron spin idea was born and Pauli developed a whole theory around the concept. The spin creates a new angular momentum and magnetic moment within the atom. These are thus new entities characterising the electron and within the atom their value is quantized: they cannot take arbitrary values. If the atom is in a magnetic field, the spin magnetic moment interacts with that field and since it can take two directions, the number of energy levels is doubled. The properties of the electron spin are very much like those of the angular momentum of the electron orbiting the nucleus. However, the value of that new angular momentum is only 1/2 the value of the orbiting angular momentum. This is another fascinating property of the electron. Physicists like to make classical pictures of the atomic world even though they know that

Figure 6.5. The electron may be represented classically as a tiny sphere spinning on itself like a top. Its angular momentum is $1/2\hbar$ and it possesses a magnetic moment $\mathbf{m}$ like a magnet. We do not know the dimension of the electron. Actually, it is believed to be a point in space. It nevertheless carries a charge and has mass. For the purpose of visualization it is naively represented as a small sphere spinning on itself.

those pictures are probably very far from reality. Such a picture of an electron spinning on itself is shown in Figure 6.5.

Since the electron appears to be a point in space with an associated electric charge and a mass it is clear that such a picture is far from representing the real atomic world. The electron spin is just another inner electronic property like mass and electric charge. Where does it come from? In the model in which the electron is seen as a very tiny sphere, almost looking like a point in space, we may assume that the mass and charge are distributed over the sphere. One assumes then that the sphere is rotating and its properties are calculated. The spinning motion creates an angular momentum and a magnetic moment. A property that can be measured exactly in practice is the ratio of the magnetic moment to the angular momentum. When this is done it is found that the ratio calculated on the spinning sphere model is in error, the value found experimentally being a factor of two larger than the calculated value. Consequently, the sphere model for the electron is not right, but it is still used as a tool to describe its behaviour remembering that it is a heuristic representation. The electron appears again as an entity with internal peculiar properties. Then, what is the origin of spin? Schrödinger did not introduce relativity in his wave equation. It was not clear if it was necessary and what it would do in such an equation. Relativity had already been introduced by Sommerfeld in Bohr's model and it was found that some spectral lines would split, an effect that was observed experimentally particularly in the ionised helium atom spectrum. The obvious question was what would happen if relativity was introduced into Schrödinger's equation. This was done by Dirac (1928). It was a major step in the description of the functioning of the atomic world and its conclusions are far reaching. When this is done, the electron is found to have a somewhat more complex wave function than in the simple case where relativity is not included. It is also found to possess an internal property identical to that of the spin assumed earlier by Pauli, with all its characteristics agreeing with experimental data. Spin thus seems to be an effect of relativistic origin and may not represent an actual spinning property. Dirac's finding is another aspect of the influence of relativity in our world, enforcing its universality.

Another finding of Dirac's theory was that his equation had a symmetrical solution and led to the existence of a world of particles with properties like those of the electron, but with a positive charge. They looked like the absence of electrons in a sea that would filled with them. These positive electrons were interpreted later as antiparticles. They look like electrons travelling backwards in time. They are called *positrons*. Such a particle was discovered by C.D. Anderson in 1932. The discovery led to the concept of antimatter. Such entities have been observed in cosmic rays and can also be created in the laboratory by means of particle accelerators. We have introduced the concept earlier. They have rather astonishing properties. The mass of an antiparticle is exactly equal to that of its twin particle and its charge is opposite. However, they have an intrinsic property that differentiates them totally. They cannot exist in the presence of each other. When they collide, a violent reaction follows. Both particles are annihilated and the result is the creation of radiation as gamma rays.

In principle, each elementary particle has its own antiparticle. A positron could orbit an antiproton and form anti-hydrogen. Such an entity was produced recently in small quantity at the large accelerator operated by the CERN consortium. There are no a priori reasons why our Universe should be made of particles and not of antiparticles. It is only a symmetrical change in sign of all the charged particles of the Universe. Could there be some region in our Universe that is made solely of antimatter? This is not impossible. That region of the cosmos would look exactly like a world made of matter. But one should not shake hands with one of the inhabitants of that world, especially if he presents his left hand: a blast of radiation would be the end result. Two colliding stars, one made with matter and the other one made with antimatter, would produce one of the greatest cataclysms, transforming the total mass of the two stars into electromagnetic radiation. A question is raised from the presence of that antimatter: why does it seem that our world is made only of matter? We will come back to that point later.

The prediction of the existence of antimatter in Dirac's analysis and the discovery of the positron by Anderson gave a lot of credibility to the role played by relativity even at the level of atomic physics. Although the electron does not appear to have internal parts, and for that matter have dimension, the introduction of relativity in Schrödinger's equation made explicit a hidden property of the electron, its spin that had been found earlier to be necessary to explain observations. Although it is usual to think of the electron as a spinning object as in Figure 6.5, that property must be interpreted in the same way as mass and electric charge: an intrinsic characteristic. It looks very much that the duality of the electron as a wave and a particle and the influence of relativity are coupled to give it that property.

It should be mentioned that the proton and the neutron, the constituents of the atomic nuclei, also possess a spin. They have the same angular momentum, $1/2\hbar$, as the electron. However, their magnetic moment is much smaller. We will see in the next chapter on the structure of those particles that the introduction of spin in the model is not as straightforward as in the case of the electron.

One of the Most Basic Principles of that Magic Quantum World

In Newton's description of the macroscopic world, some principles were established and the whole theory was based on those principles. The process was rather successful in explaining the behaviour of macroscopic objects. The orbits of the planets became a consequence of universal gravitation and the second principle connecting force, mass and acceleration. The microscopic world, on its side, has its own basic principles embedded in quantum physics. However, the principles are less obvious. Among many of those, we have postulated that matter behaves as if it were a wave and a particle. Another one was an equation, the wave equation, which provides a description of the behaviour of that wave. It was the coherence that developed among those principles and the agreement with observations that validated those principles. Another very important principle introduced by Pauli deserves special attention. It led to a fundamental understanding of the construction of the atomic world: it is called the *exclusion principle.* It was reached in part on the basis of observations on symmetrical properties of the table of elements. It looks very much like evidence that we use all the time in classical physics, the principle that dictates that two objects cannot be situated at the same point in space at the same time. If space is occupied by a solid object it is not possible to place an object at the same point in space. It is *excluded* from that space. This looks very reasonable. However, the principle introduced by Pauli uses a different concept for space. It deals with the state of the electron in the atom. What it says is that no two electrons may occupy the same state within the same atom. We have seen that the interaction of the electron with the nucleus creates energy levels or states and that electrons find stability in those states. Pauli's exclusion principle just says that one state cannot accommodate more than one electron. And this is a very general statement that applies to elementary particles whatever the situation they are in. We will see later that this has important consequences in the evolution of certain stars. From what we have seen above about quantum numbers, the statement amounts to saying that two electrons in the same atom cannot have the same quantum numbers. It should be realized, however, that some of those states may be degenerate. What do we mean by that? Simply that two or more states may have the same energy. Two electrons may have different quantum numbers but be in the same energy level. An example of that situation would be an atom in the absence of a magnetic field. In such a case all levels, identified by the same principal and azimuthal quantum numbers and whose energy depends on the intensity of the magnetic field, have the same energy although they may have a different magnetic quantum number. In complex atoms with many electrons, those electrons distribute themselves among the lowest energy states, one electron per state. We may say that the principle gives diversity to the world. Without it, all electrons would pile up in the lowest energy state. In Bohr's model, all electrons would end up in the same region of space, in the lowest orbit. Without it, the atoms, whatever the number of electrons they contain, would have the same size. It would be a very different world. The principle has important consequences for the actual construction of our modern version of the table of elements.

Fermions and Bosons

Up to this point we have introduced, for describing our world, essentially four types of particles, the neutron, the proton, the electron and the photon. They are either represented by a wave or a particle. Are they of the same essence? It would be nice if they could be treated exactly the same way, say through the same equation and obeying the same law. But we already know that there are important differences between them. Their behaviour is dictated by different laws. The wave associated with the photon obeys the classical wave equation derived from Maxwell's equations while in the case of other particles their behaviour is dictated by Schrödinger's equation. Furthermore, the photon does not have mass while the others do. There is, however, another difference between them, a difference connected to the exclusion principle that we just enunciated. We said that electrons cannot occupy the same region of space. This is a good thing. Actually this is what gives spatial dimension to objects. If that principle did not exist our atoms would invade each other and we would simply go through the floor. But how about photons? They behave very differently: they can occupy the same place. The photons may be interpreted as the excitation state of electromagnetic fields. The more intense the field is the greater the number of photons in the field: it looks as if they pile up on each other to form the wave. Consequently, they are not submitting to Pauli's exclusion principle. They can exist at the same place in space at the same time. In the next chapter we will find other particles that have the same property. This property is considered fundamental in the world of particle physics and the two types of particles have been classified in two different groups. Particles such as photons are given the name *bosons*, while particles similar to the electron are given the name *fermions*.[3] It turns out that fermions, such as electrons, have a spin which is half an integer, such as 1/2, 3/2 . . . , while bosons, such as photons, have a spin given by an integer, 1, 2 . . . It is thus possible to differentiate particles in two groups, bosons and fermions, with the help of their spin properties. All particles with half integer spins are fermions while those with integer spins are bosons. It turns out that the differentiation between them originates from a fundamental property associated with their quantum character. In practice, the wave function that represents them has a characteristic symmetry associated with the spin of the particle. Bosons wave functions are symmetric, while fermions wave functions are antisymmetric. This simply means that when two close-by fermions, electrons for example, are exchanged, the wave function that describes their common behaviour changes sign while for bosons the wave function stays the same under the same operation. That characteristic forces the particles to distribute themselves differently among the states of the energy manifold, the distribution being given by a so-called Bose–Einstein statistic for bosons and a Fermi-Dirac statistic for fermions.[4] We will see in the next chapter that such a differentiation provides a framework that makes possible a somewhat better understanding of the way natural forces are acting.

The Elements and Mendeleev's Table

We have thus seen that quantum mechanics can provide a complete model for the hydrogen atom. The electron evolves in the electric field of the proton and may exist in only definite energy states or levels. These states are characterised by four quantum numbers. However, our world is made of many more complex atoms. How are those atoms built?

To answer that question it is more instructive to jump a step, by going directly to the modern version of Mendeleev's table of elements and by interpreting it in terms of what we have just learned. The resulting arrangement of elements is shown in Table 6.1. It is organised by rows and columns of boxes regrouping 92 elements

Table 6.1. Modern version of Mendeleev's Table of Elements. The upper number is called the Z number and is the number of electrons or protons in the element in question. The table is given essentially as an illustration and for simplicity the full name of the elements have been given only for elements in the first and last columns. The two boxes that include asterisks are followed by a series of elements from Z = 58 (Cerium) to 71 (Lutetium), (they are called lanthanides) and Z = 90 (Thorium) to 103 (Lawrencium), (they are called actinides). The regularity and order can all be explained by means of the quantum rules mentioned in the text and the exclusion principle. Some of the elements, particularly in the high Z numbers, are radioactive. Some do not exist in nature and are produced in the laboratory.

1 H Hydrogen																	2 He Helium
3 Li Lithium	4 Be											5 B	6 C	7 N	8 O	9 F	10 Ne Neon
11 Na Sodium	12 Mg											13 Al	14 Si	15 P	16 S	17 Cl	18 Ar Argon
19 K Potassium	20 Ca	21 Sc	22 Ti	23 V	24 Cr	25 Mn	26 Fe	27 Co	28 Ni	29 Cu	30 Zn	31 Ga	32 Ge	33 As	34 Se	35 Br	36 Kr Krypton
37 Rb Rubidium	38 Sr	39 Y	40 Zr	41 Nb	42 Mo	43 Tc	44 Ru	45 Rh	46 Pd	47 Ag	48 Cd	49 In	50 Sn	51 Sb	52 Te	53 I	54 Xe Xenon
55 Cs Cesium	56 Ba	57 * La	72 Hf	73 Ta	74 W	75 Re	76 Os	77 Ir	78 Pt	79 Au	80 Hg	81 Tl	82 Pb	83 Bi	84 Po	85 At	86 Rn Radon
87 Fr Francium	88 Ra	89** Ac															

* 58 Ce	59 Pr	60 Nd	61 Pm	62 Sm	63 Eu	64 Gd	65 Tb	66 Dy	67 Ho	68 Er	69 Tm	70 Yb	71 Lu
** 90 Th	91 Pa	92 U	93 Np	94 Pu	95 Am	96 Cm	97 Bk	98 Cf	99 Es	100 Fm	101 Md	102 No	103 Lw

found in nature and some others created artificially in the laboratory, which may have a short lifetime. The atoms in the same column in general have somewhat similar chemical properties and a kind of periodicity is observed. That periodicity originates from the electronic configuration of the atoms, from which follows the chemical similarities in each column.

The first atom in the list, in the first row and first column, is of course hydrogen. As we have seen it is made of one proton, its nucleus, and one electron. From the previous description we know fairly well the characteristics of that atom. The next atom should obviously have a nucleus containing an extra proton. However, it is not possible to simply add protons to the nucleus and make more complex and massive atoms. Such nuclei are unstable due to the strong repulsion of the electric charge carried by the protons. Neutrons have been 'invented' by nature in order to stabilise the nuclear structure that holds together by means of the so-called strong nuclear force.

It is recalled that neutrons and protons are known as nucleons and form a special class of particles called *baryons*. For the moment it is sufficient to know that nuclear forces exist and hold nucleons together in some stable arrangement. Looking at the table, the next atom in the first row is helium whose nucleus contains 2 protons and 2 neutrons. Two electrons attach themselves to that nucleus and make the atom neutral. Since the electron has a negligible mass compared to the nucleons if is standard practice to give the weight of an atom by the sum of its nucleons. This number is identified as A. When explicitly given, it appears as a superscript to the symbol representing the element. The number of electrons is called the atomic number Z and is given as a subscript. The representation looks like: $_{Z}(\text{Element})^{A}$. We thus have the following notation for identifying the most abundant isotope of helium in nature: $_2He^4$. This does not mean that other isotopes do not exist. For example, the isotope $_2He^3$, with two protons and one neutron as nucleus, also exists in nature, but in minute quantities, 0.00013%, the rest being $_2He^4$. What is so special about helium? It has two electrons. They attempt to locate themselves in the lowest energy state, but according to the rules we have mentioned above, these two electrons must obey the exclusion principle and thus cannot have the same quantum numbers. What are the possibilities? To accommodate that rule, the two electrons may have the same quantum number n, l, m_l, but must have a different spin orientation. Thus they have $n = 1$, $l = 0$, $m_l = 0$ and a different orientation of their spin. In Bohr's picture they are in the lowest orbit but with their spin oriented in opposite directions. This is illustrated in Figure 6.6(a).

We notice that in that first orbit, $n = 1$, there is no more room for added electrons. All the possibilities for angular momentum and its projection on an axis are filled. This is called a closed shell. In the Schrödinger model, this is a happy world, with all energy states filled, with total spherical symmetry, without handles to grab it. In the quantum mechanical picture, the two electrons envelop the nucleus. The atom is chemically inactive. Helium does not react with any other atoms to form a chemical compound.

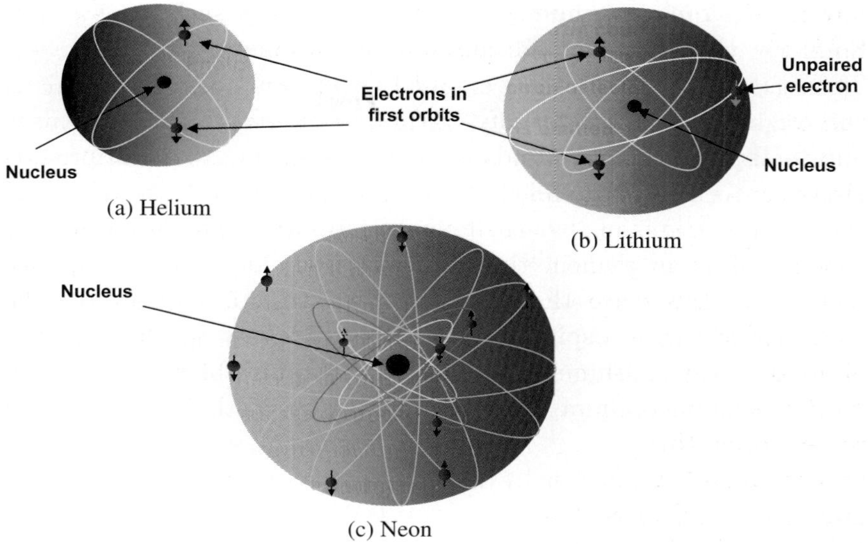

Figure 6.6. (a) The first atomic orbit may contain two electrons with their spin oriented in opposite direction. The nucleus contains two electric charges +e. This is helium. The second orbit with n = 2 may have orbitals with l = 0 and l = 1. (b) When only one electron is added to the first two, we have the case of lithium. (c) That second orbit may contain 8 electrons with their spins paired up in opposite directions. In that case the total number of electrons accommodated is 10. This is the case of neon. The nucleus contains 10 charges +e.

Let us add a proton to the nucleus with sufficient neutrons to stabilise it. The atom is neutral with three electrons orbiting the nucleus. It is Lithium, in the first column, second row, and its most abundant isotope is $_3Li^7$. Its nucleus has four neutrons. In the formation of the elements in the Universe that isotope of lithium appeared more probable than other isotopes of the same element. It exists with an abundance of 92.5%. The rest is the other Li isotope containing three protons and three neutrons. Since the first electron orbit is totally filled, or all the available quantum numbers have been used by the two lower energy electrons, there is no possibility for the third electron to integrate itself to that first orbit: its quantum number would be the same as those of the occupying electrons. The third electron thus installs itself in the second orbit with n = 2 as shown in Figure 6.6(b). In that case, there are many states with different quantum numbers available for that electron. The angular momentum quantum number l may take values such as 0 and 1. Similarly, the magnetic quantum number m_l may be -1, 0 and $+1$. This electron is also totally exposed to the outside world. This is the origin of chemical activity. Lithium is a very reactive element and tends to form compounds with other elements that can provide electrons to fill such unfilled shells. The atom would like to see all the possible states of that shell completely filled. The atom may be thought of as an element that has on its outside shell an *unpaired electron* in the sense that this outer electron is exposed to the outside world with a spin 1/2 not compensated by another electron of opposite spin. Lithium is part of the column of the so-called alkali atoms including sodium, potassium and several other atoms having similar properties.

We can go on like this by adding protons to the nucleus with the appropriate number of neutrons to stabilise it, and adding electrons to make the atom neutral. We can fill all the boxes of the table passing by special elements that find their shells totally filled like helium: they are the rare gases, also called noble gases. The first one to meet on the same row as lithium is neon shown in Figure 6.6(c). Its nucleus has ten protons surrounded by ten electrons. They fill all the states within the quantum number $n = 2$ and they form a complete shell: the atom is chemically neutral as helium is. The next atom after neon is sodium with an added proton and an added electron. The electron is again an unpaired electron and the atom, chemically, looks very much like lithium. This third row ends with argon, also a chemically inactive atom. The fourth row starts with potassium, an element very similar to sodium, and ends with krypton, an atom with totally filled shells and also inactive chemically. And it goes on like this with all the elements, up to element $_{92}U^{238}$, the most abundant of the uranium isotopes.

The rare gases and the alkali atoms are thus reference columns in the table of elements. They have rather interesting structures and are instructive regarding the understanding of the formation of chemical compounds. All alkali atoms are similar to hydrogen. They have one unpaired electron and try to acquire another electron or share their own with other atoms to fill that lower shell. This is the reason why they are so active chemically. For example, chlorine, in the column just before that of the rare gases, is lacking an electron to fill its last shell. It thus reacts chemically with alkali atoms sharing their unpaired electron to form very stable chemical compounds. In the case of sodium this compound is kitchen salt. It is those outer electrons in all atoms that play the active role in chemical reactions. The inner electrons in filled shells are not involved in chemical reactions and are not much affected.

The whole Universe appears to be made of those elements. As we will see in a later chapter the Universe started some 14 billion years ago in an immense, could we say, explosion from essentially 'nothing'. That nothing was 'energy' that transformed itself into elementary particles following some rules about which we keep speculating. Those elementary particles assembled themselves to become atoms that form our world. This was done in many steps at different stages of the evolution of the Universe and we have learned some of the rules that have dictated the relative abundance of the elements presently distributed throughout the world. We will come back to that subject in a future chapter.

The Quantum World Probability Concept

We started by describing an experiment with silver atoms, the Stern–Gerlach experiment. Can we explain it now? Let us look at it in the light of the previous development relative to the personality of the electron. Silver is formed of a nucleus containing 47 protons and 61 neutrons surrounded by a cloud of 47 electrons. Twenty eight of those electrons form so-called closed shells, filling first all available states in the lower orbits with principal quantum number $n = 1$, 2, and 3. The shell $n = 4$ is filled up to $l = 2$, accounting for another 18 electrons. This ensemble formed of 46 electrons envelops the nucleus and shields it. The whole internal package, inner electrons (46) and nucleus (47 protons and 61 neutrons), looks like a single positive charge. At that stage of filling the shells, we would expect the last electron, the 47th, to set itself in the orbit $n = 4$, $l = 3$. However, interaction between all the electrons in the atom perturbs the arrangement and for lowest energy the last electron sets itself in the next orbit, which corresponds to $n = 5$ with $l = 0$. It is an unpaired electron. This electron sees then a shielded nucleus as one single electric charge +e and behaves very much like the electron in the hydrogen atom in its lowest state. It does not have orbital angular momentum since $l = 0$. However, as said above, the electron acts like it were spinning on itself. It has an internal angular momentum like a spinning top and has a magnetic moment looking like a small magnet. This is reflected in the silver atom, which as a whole looks like a tiny magnet. Note that the atom is electrically neutral and the trajectory in the Stern–Gerlach magnet is a function only of a magnetic interaction between the electron spin magnetic moment and the magnetic field. We transpose the laws of our macroscopic world concerning the forces exerted by a magnetic field on small magnets to our atomic world. When an atom enters the gap of a magnet with its magnetic moment oriented against the magnetic field, calculation shows that its energy increases with the strength of the field. Consequently, that atom seeking the lowest possible energy will seek regions of lower field. It will be attracted and deflected towards the pole of the flat magnet where the field is lower. On the other hand, an atom with its magnetic moment oriented in the same direction as the field will see its energy decrease at higher magnetic fields. Consequently, that atom will seek regions of higher magnetic field to lower its energy and it will be deflected towards the sharp edge of the other magnetic pole where the field is stronger. Those atoms with their spins oriented along or against the field, seeking a state of lowest energy, are thus deflected in two different directions. Consequently, if all atoms in the beam were to enter the magnet with their direction either up or down, we would be able to predict the presence of two spots at the screen as was observed by Stern and Gerlach.

However, when we think of the situation in classical terms, atoms leaving the oven at random have their spin oriented with equal probability in all directions. The orientation of the atoms is random when they enter the magnets gap. Those atoms that are not oriented exactly in the direction of the field or opposed to it would thus be deviated less. Thus, the detector should also find atoms in the region between the two spots. However, none are detected in that location. What is happening?

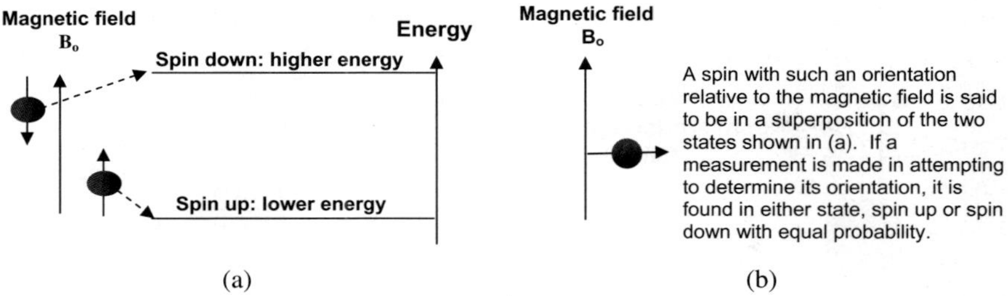

Figure 6.7. Illustration of the states of an electron in a magnetic field. In (a) the electron is in either of the two states, spin up or spin down relative to the field. The state of the electron is determined with certainty. In case (b), however, the electron finds itself in a superposition of the two states and its actual state can only be determined within some probability.

This is the quantum world. As discussed above, an electron spin in a magnetic field is quantized. As illustrated in Figure 6.7(a), it can exist only in two specific energy states corresponding classically to two orientations. These states are identified as 'up' and 'down' for the magnetic moment oriented either along the vertical field or against it. The two states are called eigenstates or proper states of the electron. As in the hydrogen atom, where the electron could not exist between two stable orbits, the electron spin in a magnetic field can only be in one of these two eigenstates. It is a complex way of saying that the spin, when detected, can only be found along two directions in a magnetic field. In such a situation, the spin is oriented either along the field or against the field. But what happens to an atom entering the magnetic field with its magnetic moment or spin at some angle to the field? We say that the spin is in a superposition of the two states. We cannot say in what state it is until we make a measurement, but we still represent it by a single wave function, which is a weighed summation of the wave functions it would have if it were in either of the two proper states. In the case of atoms emitted by an oven we conclude that the eigenstate of individual atoms is not determined until a measurement is made.

A classical picture can also be made of the behaviour of the individual spins in the beam. A spin entering the magnet at a certain angle will start 'precessing' around the field like a top spinning in the earth's gravitational field. This is illustrated in Figure 6.8. All

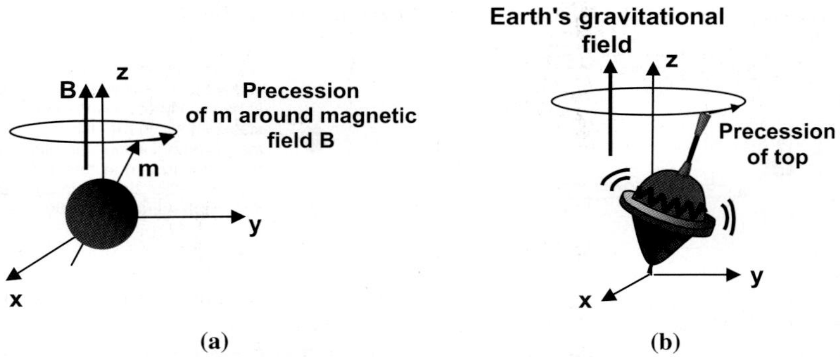

Figure 6.8. In a classical picture, the magnetic moment associated with the spin of the electron (a) may be thought of as 'precessing' around the direction of the applied magnetic field **B** as a top (b) in a gravitational field.

spins will do the same but with different orientations in space around the magnetic field axis. The consequence is that the projection or their magnetic moments in the xy plane or, in other words, their components in that plane will be oriented at random. We say that they have random phases. Their orientation will be random and, due to their large numbers, the components of the magnetic moment of each individual atom in the xy plane, that is perpendicular to the magnetic field **B**, will all cancel out. We will be left only with components of magnetic moments along the magnetic field of the dipole magnet, which is along the z axis. In a quantum mechanics picture we say that owing to our ignorance of the actual states of the individual spins, we have a *statistical ensemble* of atoms in a mixed state, and the atoms are assumed on the average to be distributed equally in the spin up state and in the spin down state. Consequently, as far as we are concerned as observers, not interacting with the atoms until they hit the screen, we can say that the atoms traversing the magnet do so as if their magnetic moments were either up or down and they are deflected accordingly producing two spots on the screen.

That reasoning appears to explain the main result of the Stern–Gerlach experiment. However, it opens the door to a set of most fascinating experiments where the true nature of the quantum world makes itself visible. Let us look at that. To do so we transform

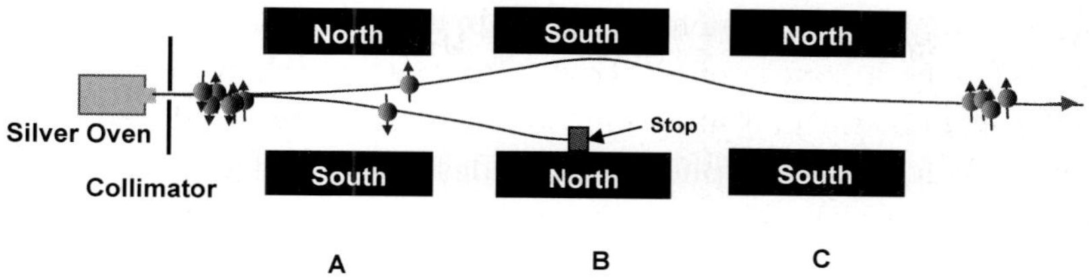

Figure 6.9. Construction of an atomic filter with three Stern–Gerlach dipole magnets.

our Stern–Gerlach device and make a special filter. This is shown in Figure 6.9. The device if made of three of the dipole magnets illustrated in Figure 6.1. The first one, (A), divides the atoms in space according to their magnetic moment. The second one, (B), has its magnetic field oriented in the opposite direction and thus forces the atoms back to the other direction. However, a stop is placed on the path of the atoms having their spin down. Consequently only atoms with spin up continue their path. Then, another dipole magnet (C) follows, which redirects the atoms along their original path. It is clear that we have constructed an efficient atomic filter. Only atoms in the up state go through the apparatus. For simplicity and easy remembering we will call it an 'up' filter.

We may represent that complete device by a so-called black box,[5] with an arrow on it pointing up, as in Figure 6.10, showing the direction of the spin of the atoms coming out of that filter. It is very much like a Polaroid filter that lets light go through with a given polarization, but stops light of a perpendicular polarization. Our device may also be called a polarizer. However, here we deal with atoms, the atomic world. It may be mentioned that we could replace the silver oven by a source of other types of atoms and particles also

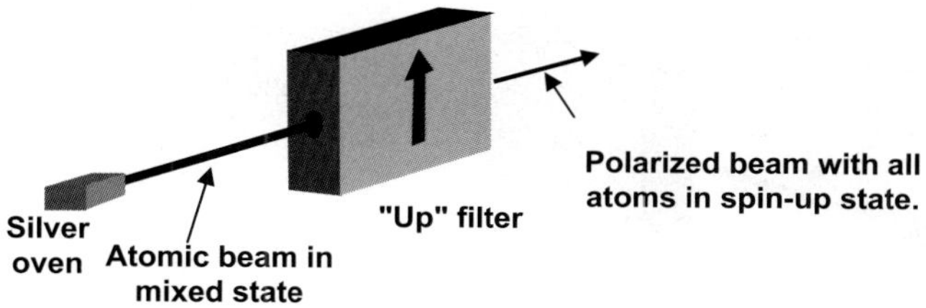

Figure 6.10. Black box representation of the set-up constructed in Figure 6.9, acting as a simple filter of atoms. The arrow indicates that it is an 'up' filter.

having a magnetic moment, such as hydrogen atoms. We may also think of it as a source of electrons, for example. If we neglect their electric charge we may imagine that the electrons would act as the silver atoms since the phenomenon described in the case of silver atoms is created by the atom's unpaired electron. Consequently in the ideal experiments to be described below we may often refer to a source and a beam of spins rather than referring explicitly to a source or a beam of silver atoms with a magnetic moment attached to each atom but really belonging to the electron spin. It must be understood however that when we talk of a beam of spins the particles with which the spin is associated do not have an electric charge, because the electric charge trajectory would be deviated in the magnetic field of the Stern–Gerlach magnet, an added complication that we do not want.

Some Revealing Experiments

We now build several of those filters and use them in series with various orientations. If we follow the 'up' filter shown in Figure 6.10 with another one oriented in the same direction, that is to say, 'up', obviously all spins that emerge from the first filter pass through the second filter without inhibition. However, if we follow the 'up' filter by a filter with its arrow oriented down, which is a 'down' filter, it is clear that no spins will go through, since all atoms reaching the down filter are oriented up. This can be visualised directly from the internal design of the filter shown in Figure 6.9, but with the figure inverted. This is illustrated in Figure 6.11.

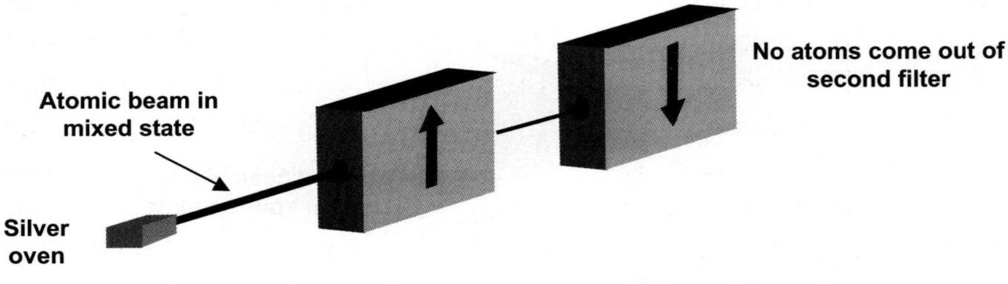

Figure 6.11. A contraption formed of an 'up' filter followed by a 'down' filter stops all atoms. It is like two Polaroid filters set at right angles, an orientation that stops all electromagnetic radiation. The difference is in the relative angle of the two filters.

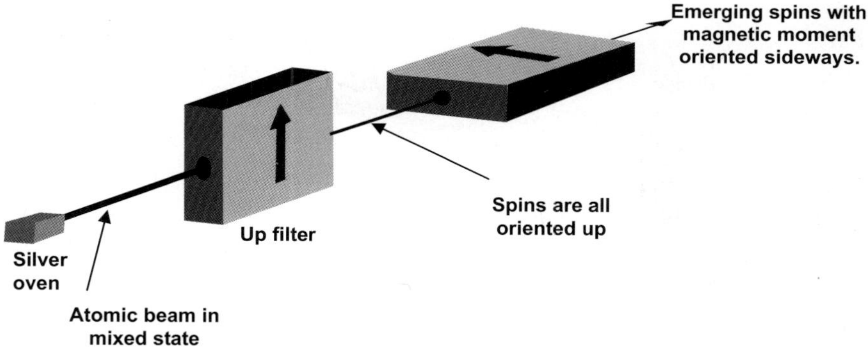

Figure 6.12. Illustration of the case when two filters are oriented at 90 degrees from each other.

All spins are stopped inside the set-up. However, we may wonder what happens if the orientation of the second filter is set at 90° from the orientation of the first one as shown in Figure 6.12. All spins emerging from the up filter have their magnetic moment at 90° from the axis of the second filter set sideways. There are no spins oriented along this second filter axis. The situation corresponds to that shown in Figure 6.7(b) and, in the context of quantum mechanics, the spins are said to be in a superposition of the two allowed states. The motion in space of the individual spins is dictated by a two-component wave function now being a superposition of the two states. The spins may be thought of as having equal probability of being in the up and down state in the reference frame of that filter. Consequently, the spins evolve in the second filter as if they had an equal probability of being in either state. The stopper of the middle dipole magnet interacts with the spins, its effect being finally seen as stopping half of them from going through. The stopper acts in a similar way as a screen and in a sense 'detects' the atoms in that path. We will come back on that subtle 'detection' point below. Consequently, 1/2 of the spins of the original beam emerges from the first filter with their spin orientation up and 1/2 of those spins emerge from the second filter with their spin orientation sideways. The emerging beam thus contains 1/4 of the spins of the original beam. It looks almost as if the filter has changed the orientation of some of the spins emerging from the north filter.

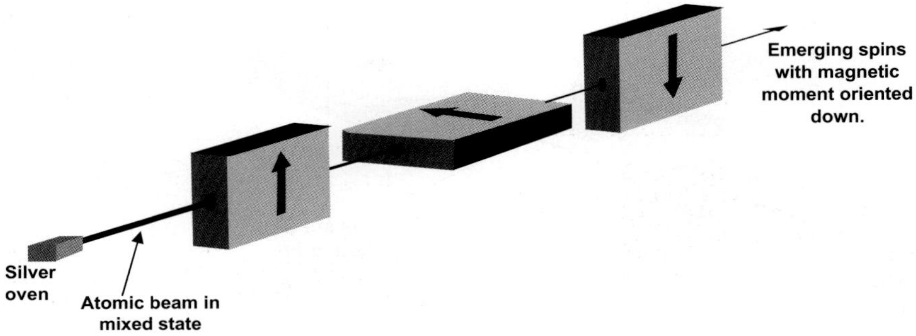

Figure 6.13. Illustration of an assembly of three filters making explicit the quantum mechanical effect of the superposition of states that depends on the frame of reference introduced by the magnetic field within the filters.

We have seen above that if two filters, one oriented up and the other one oriented down, are placed in series all spins are blocked since all spins entering the second filter have a magnetic moment opposite to the filter privileged orientation. However, we may wonder what happens if we put a filter sideways between the two as in Figure 6.13.

Here again a peculiarity of quantum mechanics appears. The spins being in a superposition of states for the third filter, half of them emerge oriented down. When viewed from the angle of classical physics, the situation is not logical: it is as if the middle and third filters were transforming some of the spins oriented north to spins oriented south: the atomic world behaves very differently indeed from the classical world.

In order to understand better that phenomenon, let us examine the trajectory of a single atom or spin through a single Stern–Gerlach dipole-magnet. Let us assume that the atom or spin has already been chosen from an ensemble that has gone through a sideways filter as the second filter in Figure 6.12. Its orientation is known. The situation is illustrated in Figure 6.14.

In entering the Stern–Gerlach dipole-magnet oriented up, the spin finds itself at right angle to the field of the magnet: it is in a *superposition* of the two states as illustrated in Figure 6.8(b) or the inset of Figure 6.14. This is the true world of quantum physics: the electron can exist in a superposition of states in any ratio reflecting its

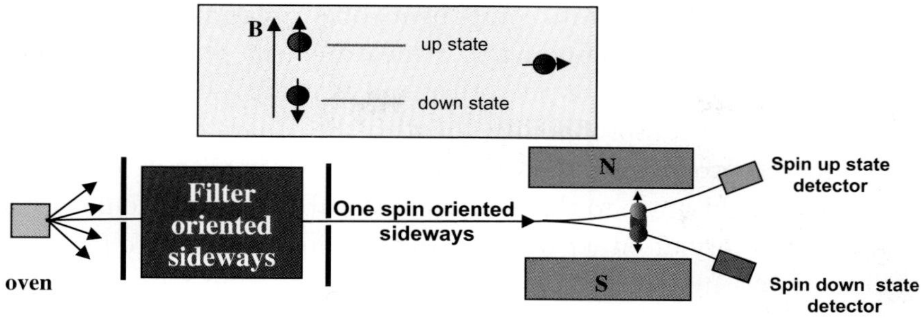

Figure 6.14. In the case shown the spin is in a superposition state and has equal probability of being in either state up or down. In the setup its wave function evolves as a single entity in space but 'collapses' at the detector. The spin is found in either state with a probability equal to the square of the probability coefficient associated with the two states. In the case shown that probability is 50%.

orientation in the classical world. We then associate a number to the proportion of the state that the spin is in, which reflects classically its orientation in space relative to the axis of reference chosen. We call that number its amplitude in a given state. The wave function attached to the spin is then made of two terms representing the two states up and down. The two terms are multiplied by those numbers or coefficients, whose square is the probability of finding the spin in one state or the other when a measurement is made. If the coefficients of the wave function components are equal, as in the case assumed in Figure 6.14, there is equal probability of finding the spin in either state. If a measurement is made on a spin while it is in the magnetic field region, it is found in either state with a probability equal to the square of those numbers. Before the measurement, there is no way of knowing in which state the spin is in. We thus see the concept of probability entering fully into the picture. The world of quantum physics is a world of probability. The spin traversing the field of the Stern–Gerlach dipole magnet evolves according to Schrödinger's equation. Both amplitudes of the wave function components evolve keeping their intrinsic properties. However, in the present case, due to the property of the field and the force difference depending on the two possible states, one part of the wave function evolves in space in one direction while the other part evolves in the other direction. The wave function spreads in space. However, there is a single wave function and its components obey Schrödinger equation. Upon detection, corresponding to a measurement process, the spin is forced into one state or the other and is detected either spin up or spin down with a probability equal to the square of the numbers weighing the two parts of the wave function. A spin with its magnetic moment at right angle to the field, as in the above case, is found with equal probability at either detector but not in between. We may say that the wave function *collapses* upon detection, and reduces to the representation of one state only. This is where the determination of behaviour by means of Schrödinger's equation stops. We do not find one half of a spin in one detector and the other half in the other detector. The spin is detected as a full entity. This is a consequence of the observation process. The collapse of the wave function is a very important property of the detection process: it is the passage from the quantum world to the macroscopic detection world. It is most interesting to observe that if we use only one detector at the screen placed at the position where the down spin could arrive, as in Figure 6.15, that detector may not detect anything, but may still collapse the wave function. We then know, without detecting it, that the spin has gone in another direction. While it looks as if the detection process did not interact with the spin, it did indeed and forced the atom into one of the two states up or down. In that particular case it was the up state.[6]

This brings us to the heart of quantum physics, a world that defies common sense when transposed into our classical world where we deal with objects that, to our eyes, obey Newton's principles. There is a most famous experiment imagined by Schrödinger himself

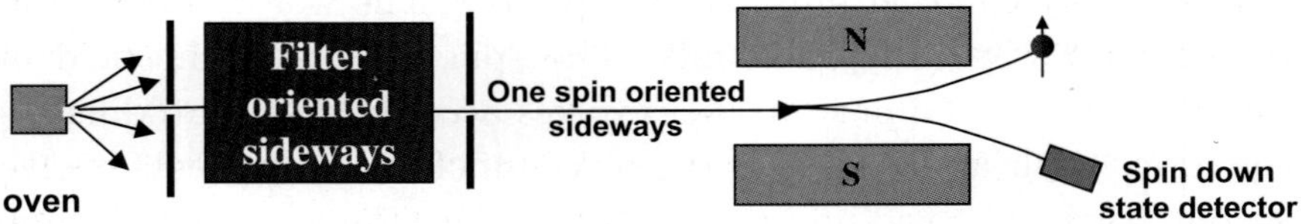

Figure 6.15. A spin in a given state traverses the magnet. Its wave function evolves in space and time. The up state detector has been removed. If the down detector does not count anything, it is still possible that the wave function has collapsed and the atom has gone in the other direction. This looks very much like an action has taken place even if the particle has not been detected.

about a cat in a box that could be in two states, either dead or alive, until we decide to look into the box. It is a fictitious experiment in which a particle emitted by a radioactive material triggers a system that opens a valve letting a deadly gas enter the box in which the cat has been locked. Since the process of radioactivity emission is a quantum mechanical random process, we don't know at a given time if the radioactive element has emitted its particle and we do not know anything about the state of the whole system. We can say that the cat in the box is in a superposition of two states: dead or alive. We will know only when we open the door of the box and measure its actual state, as if by opening the door and looking at the cat had a direct effect on the cat's state, collapsing it into a single state, dead or alive. This defies our common sense and in our world a cat cannot be both dead and alive. This is not the way our macroscopic world works. It is believed that when atoms group together to form macroscopic objects, the coherence that exists at the microscopic level and causes effects such as interferences is lost because of multiple interactions. Repeated interactions such as collisions at the microscopic level destroy the coherence that could exist in the resulting wave function of the macroscopic object. It is that coherence that is at the origin of all those quantum effects that we have just examined. Consequently, at our level, coherence is destroyed and objects behave classically: a cat cannot be both dead and alive.

This kind of thinking brings us to the apparent foundation of our entire microscopic world. It is a world in which the concept of probability plays a major role. We do not know the outcome of events at the atomic scale until we make a measurement. And furthermore, we are basically limited in the precision with which we can make measurements on some coupled variables such as speed and position, for example. This does not mean that we cannot measure one of those variables accurately. It only means that a precise measurement of one prevents precise knowledge of the other. This concept was introduced by Heisenberg on deep thinking about our involvement in the measurement process. We are used to classical physics where the position and speed of a billiard ball hitting another ball can be calculated as accurately as we wish. The laws of conservation of energy and of momentum are strict: no objects deviate from them unless disturbed by an outside force that, if we know its strength and direction, we are able to include in the calculation. Our observation of the billiard balls does not disturb them. The photons that are falling on them, to make them visible, have so little energy compared to the kinetic energy and mass of the ball that they do not alter its position and speed. The atomic world is different. Let us examine briefly an ideal measurement suggested by Heisenberg. Suppose you use a powerful microscope and you attempt to measure at the same time the position and the speed of an electron. You need to send a photon on the electron to see it and determine its position under the microscope. Since you want to localise it accurately you use a short wavelength photon, that is to say a photon with a lot of energy. We know well that the resolution power of the microscope increases with shorter wavelengths. What is the effect of the photon on the electron when it is scattered? The photon bounces back in a given direction but the electron is kicked off by the energetic photon and acquires some momentum. You have thus localised the electron accurately with the short wavelength photon, but you have changed its speed or momentum. If now you wish to measure its velocity you need to make another measurement of its position a moment later. However, the photon has altered its speed. Consequently, you cannot know accurately the speed it had since your measurement process has altered it. You could use a less energetic photon, one with a longer wavelength. However, you will no longer be able to locate the electron accurately since due to diffraction, the resolution of the microscope reduces with longer wavelengths. This is somewhat like the speed radar used by traffic controllers. They have to use short wavelength devices in order to pinpoint a particular car. Thus, a simultaneous measurement of position and velocity is limited in precision.

The microscope experiment is a fine example where the observer is involved in the measurement process itself. However, uncertainty is an inherent property that characterises the dynamics of the electron and, for that matter, of all atomic particles. Atomic particles live in a world where their position and speed are uncertain to some extent. The product of the uncertainty in position times the uncertainty in momentum may not be less than the reduced Planck's constant, $\hbar$. The microscopic world lives with this restriction and there is nothing one can do about it. If the electron finds a spot in space like in a box, for example a potential well that fixes its position within certain borders, it acquires at the same time some momentum that to some extent alters that position and it enters into oscillation or vibration in the well like a pendulum in a gravitational field. The electron dances around and its position and momentum influence each other to the extent that the product of their fluctuations is of the order of $\hbar$.

We have introduced previously the Rutherford–Bohr model of the hydrogen atom. We have said that the electron revolves around the proton in a kind of orbit. We have introduced the duality wave-corpuscle for the electron and the wave function describing the electron behaviour was assumed to give the probability for its presence at a distance from the proton. However, we have not really answered the question as to why the electron simply is not in the proton. The proton and the electron have opposite charges and attract each other very strongly. Since the electron looks very much like a point particle, why does it not simply plunge into the nucleus? That situation should be the most stable and have the least energy. As outlined by Feynman[7] the situation may be described by means of the uncertainty principle just introduced. One observes that the proton is very small, something like 10^{-15} m. If the electron in the hydrogen atom were inside the proton it would be extremely well localised within that volume. Consequently the uncertainty in its position would be extremely small. The uncertainty principle would then require that its momentum be undetermined to a level that makes the product of the uncertainties on position and momentum equal to at least $\hbar$. To satisfy that condition the electron would then jump around and would acquire a momentum with a large uncertainty on its value. It could even escape. In practice, we conclude that the electron, on the average, sets itself at a certain distance and dances around satisfying the principle. On the average it appears as a cloud with a diffuse spatial distribution. Knowing the minimum momentum as required by the uncertainty relation we can calculate the kinetic energy of the electron and its potential energy.

The electron seeks minimum energy in the proton field and we can calculate its energy and the average distance from the proton where its density is maximum. The result is most fascinating: the distance found is $a_o = 0.053\,\text{nm}$, the atomic radius calculated in Bohr and wave mechanics approaches. The minimum energy found is $-13.6\,\text{eV}$ and is also the same as the one calculated in the wave mechanics model. This is the energy of the electron in its lowest state, very much like the potential energy of a mass in a gravitational field that we encountered earlier. It is the energy required to extract the electron from the proton field and bring it far enough that the interaction with the nucleus becomes negligible. The origin of the minus sign is the same as in the case of gravity: we have set the electron energy zero at an infinite distance from the proton. That number, $-13.6\,\text{eV}$, is just the energy we must spend to extract the electron from the atom. It is also known as its ionization energy. It is interesting to note that the discussion just made and the principles invoked do not forbid the electron to have probability of being inside the nucleus. As a matter of fact, in its lowest state, the electron wave function given by Schrödinger's equation is a maximum inside the nucleus and decreases exponentially with distance from the nucleus.[8] Actually when we calculate the interaction of the magnetic moment of the electron with that of the proton in the hydrogen atom we find that we have to include the penetration of the electron within the proton to obtain a result compatible with experimental data. This is also compatible with the fact that the electron is insensitive to the strong force acting within the nucleus. It is thus the uncertainty principle and the wave nature of the electron obeying Schrödinger's equation that forces the spherical symmetry of the wave function enveloping the nucleus.

The concept just explained and known as the Heisenberg uncertainty principle is inherent to the atomic world. It is possible to prove it using some reasoning on the basic principles involved in the foundation of quantum physics. In classical physics, position and momentum may be thought of as concepts that make possible the mathematical writing of laws of motion as introduced by Newton. When we transpose these concepts in quantum physics they become operators acting on the wave function telling us how that wave function behaves in specific circumstances. They take special forms that are dictated by the rules of quantum physics, and they are called conjugate operators. In the classical world they commute in the sense that the product (position $\times$ momentum) is equal to the product (momentum $\times$ position), like numbers. This is not the case in quantum physics. When these variables are replaced by their quantum mechanical operators, the difference between the two products is $i\hbar$, yes, that mysterious imaginary number again multiplied by Planck's constant. This is what makes quantum physics so confusing for us: we are used to the motion of macroscopic objects in a visible, deterministic world.

An important point is that the uncertainty principle applies to other conjugate operators. We have the same rule for the combination of time and energy, also conjugate operators. For example, due to the finite short lifetime of an electron in a given state, as we have seen above in the case of hydrogen, for example, it is concluded that the energy of the electron in that state cannot be known to better than a certain value. As for the position-momentum combination, we have now the energy-time combination and the product of their uncertainty must be equal or larger than the reduced Planck constant, $\hbar$. A most visible effect is the broadening of spectral lines emitted by atoms since the energy level from which the transition starts has an uncertain value and, consequently, is broadened.

It thus appears that the Heisenberg uncertainty principle plays a most important role in the dynamics of the microscopic world. The validity of the principle has been questioned by several well known scientists, amongst whom was Einstein who did not accept it as such. Einstein, in fact, appeared to be sceptical about the validity of many concepts that were introduced in the new quantum mechanics. This is reflected in his famous statement that God would not play with dice in the determination of the properties of the Universe. He claimed that the theory of quantum physics must be incomplete. Indeed it is a very provocative theory that transforms completely our vision of classical concepts. We do not understand why it works that way, but we have found how it works. Furthermore, we have found that the theory works extremely well and is in full agreement with observations.

Photons and Electrons Interact

We have assumed above, either in Bohr's model or in the wave mechanics model, that electromagnetic waves or photons may be emitted by atoms or more precisely by the electrons within the atom. How can that be? We said that it was a transformation of energy: the electron falls from a higher orbit to a lower one and the difference in energy is transformed into electromagnetic energy. The process is rather intriguing. We have in the macroscopic world similar transformations between different types of energy. For example, potential gravitational energy of water at a certain height is transformed into mechanical energy at hydroelectric dams and the resulting flow of water through turbines is transformed into electrical energy by electric generators. Similarly, the reverse process takes place in our house in all electric motors, refrigerator, fans, washing machine and many others. We know the rules involved in those processes. They are derived from experimental observations: currents are induced in wires when they are forced to move in magnetic fields; wires carrying currents start moving when placed in a magnetic field. Maxwell used those observations to derive his famous four equations that form the basis of the theory of electromagnetism.

In the case of the atom a direct process coupling an electron to an electromagnetic wave takes place. Can we explain this with the knowledge we have developed up to now in the previous paragraphs? Let us look at it. We said that electromagnetic waves consist of electric and magnetic fields oscillating with time at the frequency that we associate with the waves travelling through space at speed 'c'. Let us look at an atom with an unpaired electron, one of those alkali atoms, say sodium. The unpaired electron may be raised to an excited state by some means, say a collision with other free electrons in an electric discharge, as in city street lights. When it falls back into its lower stable state that we call the ground state, it emits the characteristic yellow light to which our eyes are most sensitive. We call that yellow light resonance radiation: its frequency corresponds to the energy difference between the first excited state and the ground state of the unpaired electron in the sodium atom divided by Planck's constant. We may think of the electron in either of those two states as oscillating with characteristic frequencies about the nucleus. The frequency of the wave emitted in that context is then simply the difference of oscillation frequency of the electron in either state. Now, suppose that instead of exciting the electron by means of an electric discharge we send on the atom a wave at the same frequency as the one that was emitted. What will happen? We have seen in a previous chapter that we may interpret the force exerted on an electron by another charged particle as the effect of the electric field created by that other charged particle. Since the wave is made at least partly of an electric field it thus interacts with the electric charge of the electron. Since that electric field is oscillating, it drags the electron into oscillation at the frequency of the wave electric field. If the wave frequency is that corresponding to the difference of oscillation of the electron in either state, which is the frequency of the resonance yellow light emitted when the electron falls from a higher state to the ground state, it will cause the electron to oscillate strongly: it is like a swing receiving a push in the right direction and at the right time. The wave enters in resonance with the natural frequency of the electron. If the electric field of the wave is sufficiently strong it will not be long before the electron oscillates strongly, gaining energy and being raised to the higher excited state. In the process, energy is conserved and the wave energy is absorbed by the atom. We have just described in simple words the process of absorption of radiation or light by atoms when they are exposed to electromagnetic radiation of the right frequency. If the light wave is turned off, the atom, seeking lowest energy, falls back in a short while to its ground state emitting yellow radiation of the same frequency. We call that spontaneous emission and we will see shortly why the electron does that. These processes are taking place all the time in the Universe. For example a star emits radiation that, on its way to us on Earth, is absorbed by an agglomeration of gas on its path somewhere in space. The gas atoms absorb radiation and are excited to any of their upper states. If the gas is hydrogen, as is very likely, the atoms re-emit radiation at many wavelengths corresponding to possible transitions from upper states to lower states of hydrogen. One of those transitions is reddish and corresponds to Balmer α radiation that we mentioned earlier. This is the origin of the light emitted by those nebulae that we see in the sky with modest telescopes. The hydrogen atoms in the nebulae are excited to higher states and in decaying to their lower state re-emit the energy absorbed as radiation in the red.

The explanation can also be used for the case where the atom or the electron is characterised as the seat of a magnetic moment. The electron in hydrogen has a magnetic moment due to its spin. As we mentioned, the proton also has a spin and a magnetic moment. It produces at the site of the electron a magnetic field that is rather strong due to the close proximity and even overlapping of the two particles. The spin of the electron can be parallel to the field of the proton or opposed to it, creating two energy levels. The difference in energy between these two levels corresponds to a frequency of 1420.4 MHz, in the so-called microwave range of electromagnetic radiation.[9] Is it possible to couple an electromagnetic wave to the electron to excite transitions between these two levels? Yes, but in that case the interaction will take place through the magnetic field component of the electromagnetic wave coupling to the spin magnetic moment of the electron. Otherwise the whole process is the same as with visible radiation.

What we just described is an approach called 'semi classical' because the atom is treated as a quantized object and the electromagnetic wave as a classical entity with its oscillating electric and magnetic fields. The explanation seems to work well in the case of absorption of radiation. It also works well in the case of emission if the emission is induced by an electromagnetic radiation intentionally sent on the atom. In this case, the same type of interaction takes place. However, when the atom or electron is in an excited state without any radiation incident on it, we may wonder what is the force or process that induces it to fall back to its ground state. We said that the atom seeks lower energy, but there must be a mechanism that induces that electron to make the step. After all, any allowed atomic state derived from Schrödinger's equation appears to be stable. This is where all the force of quantum physics gets into play. When quantum mechanics is coupled to Maxwell's electromagnetic theory, it is found that the corpuscular nature of electromagnetic radiation is made explicit, and the photon appears in a most natural way. It has become one of the most successful theories of the whole of physics. It has led to the field of *Quantum electrodynamics* which, through a somewhat complex but elegant mathematical framework, has been able to provide, at the microscopic level, a full description of the interaction of electromagnetic radiation with matter. The new theory makes possible a fully quantized calculation of phenomena such as absorption and emission of radiation by atoms exposed to electromagnetic radiation with the same results as those obtained in the semi-classical approach described above. But it also makes explicit spontaneous emission of radiation by atoms prepared in excited states. For example, the lifetime of the electron, when excited to the first upper level in the hydrogen atom, is calculated as 1.6 ns, in total agreement with measurements. On average an electron excited to that level stays in that level for that length of time. Why does it not stay there forever? The theory provides the answer.

A basic step in the formulation of that theory is the so-called 'quantization' of the electromagnetic field. What does that mean? We have seen above that Maxwell had arrived at the formulation of a full theory of electricity and magnetism predicting the existence of electromagnetic radiation based on empirical observations and experimental laws. The whole theory was classical in the sense that nothing from the microscopic or quantum world was involved. However, it is possible to represent that oscillating radiation field as an entity by itself responding to its own rules as if it were an oscillator such as a pendulum. In classical mechanics, such an oscillator is called a harmonic oscillator, has a definite frequency of oscillation, and can oscillate at various amplitudes. A best example is the pendulum of the old grandfather clock. The pendulum oscillates at a given frequency and given amplitude. However, in quantum mechanics the amplitude of oscillation of such an oscillator is not arbitrary: the amplitude giving the energy content is quantized and cannot take arbitrary values. For a classical pendulum the mass is so large and the energy states so close to each other that the spectrum of allowed amplitudes of oscillation looks continuous. The quantization of electromagnetic waves is like the case of the electron in the atom, which cannot exist in arbitrary orbits: it can only exist in certain quantized states. In the case of electromagnetic waves, the energy in the wave must be a multiple of '$h\nu$', the quantum of energy associated with the wave of frequency ν. We say that the field amplitude is quantized and we arrive at the concept of the photon. The hypothesis that light is formed of particles, photons, finds a mathematical basis totally coherent with quantum mechanics. The number of photons in a radiation field is essentially a measure of the oscillation amplitude of the wave. The number of photons is the level of excitation of the fields forming the wave.

When the concept of photon is obtained in the process of *quantizing* the electromagnetic field, a most fascinating result is found from the calculation: the field cannot be made equal to zero by reducing its amplitude. This is illustrated in Figure 6.16. The energy of the field is a measure of the number of photons n contained in the field. In this quantized representation the field cannot be totally extinguished: even by setting the number of photons equal to zero there is a quantity of energy, $1/2\,h\nu$, that remains. It is called *zero-point field fluctuations.*[10] This phenomenon is common to all oscillators: a pendulum cannot be totally at rest. We have seen that classically the minimum temperature that could be reached by any process is zero Kelvin, a temperature at which we said that all motion stops. We have to correct that statement. At zero K, there is a minimum energy remaining in any system and we cannot extract that residual energy. In the quantum world, there is no such thing as total rest. That residual energy is interpreted as vacuum energy. Vacuum

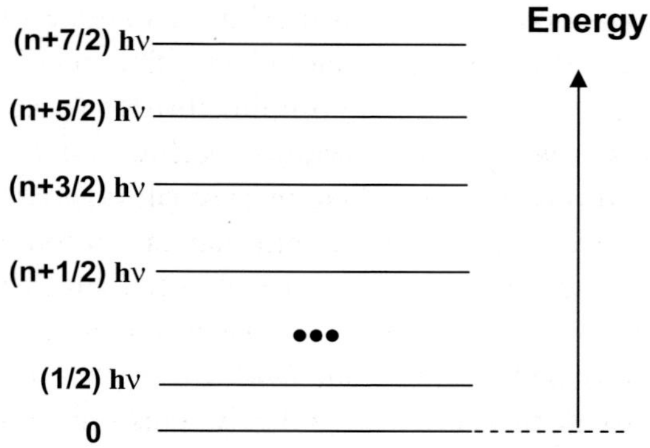

Figure 6.16. Representation of the energy contained in an electromagnetic field of frequency ν. The electromagnetic field is quantized, that is to say, its intensity increases by step of one unit of $h\nu$.

is filled with photons and in our scale they constitute an infinite number since they spread to all frequencies. Consequently, following that reasoning each cubic centimetre of space is filled with an infinite number of photons. This leads to infinite energy, a situation that must be avoided at all costs, otherwise we cannot do meaningful calculations. It is possible to surmount that difficulty by assuming in the calculations a cut-off frequency since in any case in practice our instruments have a finite bandwidth.[11] Furthermore, in our experiments involving electronic transitions between atomic energy levels, we deal with difference among the states of the electromagnetic field and it is standard practice to renormalize the zero of the whole energy manifold.

Nevertheless, at the microscopic level those zero-point field fluctuations interact with matter on a continuous basis creating phenomena and causing effects that are part of our daily lives. We raised the question above about the lifetime of an electron in an excited state within an atom. We were wondering why it did not stay there. Its decay to a lower state within a short time is caused by those zero-point field fluctuations. Those random photons interact with the electron in the excited state and stimulate it to fall to the lower state: a photon is emitted by the atom. The theory of quantum electrodynamics in fact goes much further in its predictions. As we will discuss later in more detail the uncertainty principle allows the presence of electromagnetic field fluctuations in vacuum. These fluctuations may be created out of nothing as long as they exist for a very short time obeying the uncertainty principle. The zero-point fluctuations mentioned above may be interpreted as a manifestation of those photons created from nowhere, from vacuum.

The presence of those so-called virtual photons may also cause measurable effects other than spontaneous emission mentioned above. For example, they may cause small shifts in the energy levels of atoms. The effect has been measured for the first excited state of hydrogen by W. Lamb in 1956 and is in total agreement with the theoretical predictions. Similarly, it is calculated that such random fields cause attraction between two metallic plates separated by a very small distance: the space between the plates, being limited, certain modes of the vacuum fluctuations are inhibited. The equilibrium is disturbed and a net unbalanced force per unit area appears on the plates. The phenomenon has been demonstrated experimentally and is called the *Casimir Effect*.[12] Another effect a little more subtle, but real, is concerned with the electron spin itself. Those zero-field fluctuations interact with the electron. This perturbation is reflected by a small shift of the precession frequency of the spin in the magnetic field, or in quantum language, the zero field fluctuations alter the energy levels of the spin in the magnetic field and changes its resonance frequency. This effect is calculated to a great precision and agrees with experimental data to more than ten digits.

The concept of zero-point field fluctuations has been re-addressed more recently and given a broad and fundamental role in the quantum world. It is assumed that it gives rise throughout space to a random field and the effects of that field are studied by means of the techniques of so-called stochastic electrodynamics, which simply means random fluctuations of electromagnetic fields. A rather interesting analysis has been made.[13] For example, the spectrum of black body radiation has been derived directly from that field without the need for quantizing the electromagnetic field. Similarly, it was shown that the lowest energy level in the hydrogen atom could be calculated by assuming that the electron in its lowest Bohr orbit exchanges energy with that field: the electron absorbs energy from the field, but also emits energy due to its circular motion around the proton. The system is in equilibrium and when these two energies are made equal, the results of Bohr's planetary model of the atom for the first orbit are obtained.[14]

There is no doubt about it: the vacuum virtual photons created in vacuum are there. They cannot be used as such, but their effect is observed and measurable. Those observations and calculations have prompted some to view the vacuum fields as responsible for something deeper than rational physics is ready to fully accept at this stage of our understanding.[15] However, it is legitimate to question the true nature of vacuum, normally considered as the total absence of any physical phenomena or objects. We find theoretically that it is the locus of tremendous agitation with measurable effects. We will come back to that point.

Where is the Electromagnetic Force Coming From?

The theory of electricity starts with Coulomb's law that describes the force between two electrically charged particle, an action at distance something like the concept of gravity introduced by Newton. In that case, the action at distance was interpreted by Einstein as originating from a curvature of space-time caused by the presence of mass.

That question of force at distance in the field of electromagnetism has also been one that remained for some time a kind of mystery and was brushed under the carpet. As in the case of gravitation, the concept of field was introduced and it appeared to address the question rather well; we can make calculations with it and solve essentially all problems. However, one may ask what is exchanged between the two charged particles interacting through their electric field? The theory of quantum electrodynamics[16] provides an answer. As just mentioned the quantum world is a world of fluctuations. For example, an electron travels around surrounded by its own electric and magnetic fields. These fields, as per the uncertainty principle, are in continuous agitation. It very much appears that photons are being created by the electron, and are reabsorbed by the same electron. During that process, however, energy is not conserved since a photon is created. From our macroscopic point of view, this is not acceptable, but the probabilistic aspect of the quantum world provides an escape from that rigid rule. The presence of uncertainties in the microscopic world allows the borrowing of a certain quantity of energy ΔE as long as it is repaid in a short time Δt. This loan period is allowed by the uncertainty principle and is of the order of $\Delta t = \hbar/\Delta E$. This means that a photon of very high energy must be reabsorbed rather rapidly while a low energy one may live for a long time. These are called virtual photons.

How can we apply that reasoning to the interaction between two electrons at a certain distance from each other? The interaction is interpreted as an exchange of those virtual photons. The interesting point is that if the photon energy is very low, corresponding to low frequency or long wavelength radiation, there is essentially no energy transported by the photons. Since photons travel at the speed of light the interaction of the two electrons can take place on a very large distance because Δt may be very long without violating the condition $\Delta t = \hbar/\Delta E$, the energy ΔE being very small. We thus have an interaction that propagates at the speed of light as expected. It may happen that the emitted photon is reabsorbed by the same electron. In that case, the electron interacts with itself through its own field. It gives rise to the concept of self energy. It is readily seen that the vacuum in the neighbourhood of an electron is rather active or effervescent with photons of various energies and lifetimes.

The process of photon emission or absorption of a photon by an electron, has found an elegant representation in so-called Feynman diagrams. Two processes, absorption and emission are illustrated in Figure 6.17. The particle is represented by a solid line while the photon is represented by a dotted line. Time is running upward. In Figure 6.17(a), a photon is absorbed by the particle while in Figure 6.17(b) a photon is emitted. The point of junction of the paths where the photon is absorbed or emitted is called the 'vertex'. This is the point of interaction of the electron and the photon and this is where quantum electrodynamics plays its main role: it provides a model for calculating the strength of the interaction, which is proportional to the charge. As mentioned previously, when compared to the strength of gravity it is very large, but when compared to nuclear interactions it is very small. This is a rather subtle explanation of the electrical effect over distance through space, but it appears to be an excellent model that has not yet been falsified.

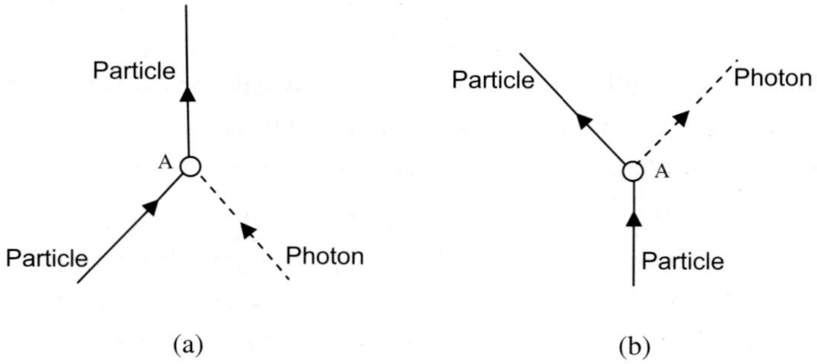

Figure 6.17. Feynman diagrams representing the interaction of photons and charged particles. The interaction takes place at point A, which is called the vertex. In (a) the photon is absorbed while in (b) the photon is emitted.

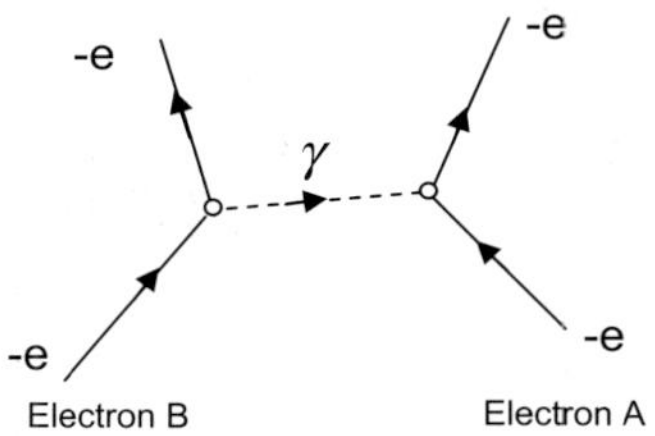

Figure 6.18. Feynman diagrams for the electromagnetic interaction of two electrons A and B through the exchange of a photon. The photon γ may have a very small energy. It may then exist for a relatively long time not violating the relation $\Delta E \times \Delta t = \hbar$. Since it travels at the speed of light it may be felt at very large distances.

On the other hand, the interaction between two electrons may be represented as an exchange of photons as in Figure 6.18.

In the approach just described, it is observed that the electromagnetic field can either be a wave or a particle while the electron is a particle. In the interaction process, there is destruction or creation of photons while there is no creation or destruction of electrons. Thus the total number of particles changes. Schrödinger's equation by itself does not lead to the creation of destruction of particles. It only provides a study of the dynamic behaviour of a particle such as an electron, for example. There are processes in nature, such as the so-called radioactive beta-decay, in which an electron is created by a nucleus. Furthermore, it would appear reasonable to include in the interaction processes the findings of special relativity, such as the finite speed of light and equivalence of mass and energy. This last equivalence provides an avenue for the creation and annihilation of particles.

This kind of reflection has led to the development of a more general theory in which all particles are thought of as ripples in a field. As the photon is interpreted as a level of excitation of the electromagnetic field, the electron may be thought of as being the excitation level of a field, the electron field. We will introduce later many other fundamental particles and they are interpreted according to this way of thinking. This approach brings in an explanation concerning the identity of families of particles, such as electrons. All electrons in the Universe are identical. Why is that so? Why would an electron here have the same property as one in the atmosphere of Jupiter? This is understood as the consequence that electrons are 'ripples' of the same kind of fundamental substance, the electron field. We may say that the electron field is the reality while the electron is an excitation of that field. Photons are also understood as excitation of the electromagnetic field. The interaction system electron–photon becomes then somewhat symmetrical. In that kind of formalism, however, we must remember that photons are bosons and can pile up while electrons are fermions and cannot occupy the same state. The consequence is that they obey different statistics in distributing themselves in energy levels. This approach is called *Quantum field theory.*[17] Its development started in the late 1920s in the study of the quantization of the electromagnetic field. That early work led to the development of quantum electrodynamics, which provided a basis for generalising the concept of fields and associated particles. The development has continued up to this day in applying the approach to all interactions, for example, at the level of nuclear forces. It provides the framework for the interpretation of those forces as an exchange of particles that are called mediators of the force in question. Gravity should not escape that refinement and the question is the subject of major efforts. We will come back to it in later chapters.

A basic conclusion of quantum electrodynamics is that our perception of the world originates from atom- or electron-photon interactions. This statement is valid for inanimate objects as well as living beings. The chemicals, which all objects and living beings are made of, are an ensemble of atoms formed of nuclei and electrons interacting with each other through the exchange of photons. And the vacuum that we believed empty of matter is simply boiling with photons zipping in all directions. Vacuum is effervescent with photons, which are present even if all accessible electromagnetic fields have been removed. The quantity of energy contained is enormous.

The Quantum World Contains Some Real Mysteries

Let us do a simple experiment. Suppose for example that you have split a coin exactly into two thin slices. You place these two thin discs in a bag and you ask a friend to choose one disc without looking. He does so and goes away with his slice without looking at it. You pick up the other part that is left. You look at it and you know immediately what part your friend has in his pocket. There is nothing strange in this. Our macroscopic world is a deterministic world, it is causal and actions take place locally: the nature of the slice your friend took with him was decided at the instant he removed it from the bag. He does not even have to look at his slice: you can inform him about the situation by sending him a message. There is no mystery in this. This is physical reality: this reality is independent of any observation that can be made on each slice of the coin.

The same experiment transposed at the level of the atomic world leads to results that defy common sense. Let us look at a *Gedankenexperiment* that was proposed by Bohm and Aharonov in the late 1950s.[18] Suppose you have a situation in which a molecule disintegrates and emits two particles in opposite directions. We assume that the molecule does not have angular momentum to start with. We may say that the molecule is spherically symmetric and its orientation in space is totally arbitrary. Assume then that the particles emitted are similar to silver atoms as in the Stern–Gerlach experiment described above: they both have angular momentum due to the unpaired electron in each atom. Since the molecule did not have angular momentum to start with, the sum of the angular momentum of the two emitted atoms must be zero in order to satisfy the principle of conservation of angular momentum. Thus the two electrons, attached to each atom, must have spins oriented in opposite directions in space or using the interpretation of a spinning sphere as in Figure 6.5, one must be rotating clockwise and the other one counter-clockwise. It is an intrinsic property of the ensemble, forced by the absence of angular momentum in the molecule from which the two atoms originate. We say that the two electron spins form an *entangled* pair. In what follows, since the angular momentum of the atom is really a property attached to the spin of the electron, we will talk indifferently of the electron spin or of the atom angular momentum as we have done previously. The orientation of the pair, however, is not determined in the emission process although the spins are opposite. Quantum mechanics does not dictate which direction the spins will take after emission. In fact, we may say that the ensemble of the two atoms does not have a precise orientation in space. There is no net angular momentum, and the direction relative to the individual spin is not determined by any known rule. However, whatever that orientation, the two electrons must always have opposite spins. They started from the same molecule and their state was characterised by a single wave function representing their entangled state. Now they are flying apart to large distances, in principle no longer interacting, and logically in our mind we conclude that they should act as two independent particles.

We can thus say that the spin of either electron has no determined orientation. However, the angular momentum associated to a spin is a quantized entity and it can take only two directions along a chosen axis. There is no such thing as a fraction of spin angular momentum along an axis. A measurement necessarily results with the spin in one of the two directions: along the direction of the axis or opposed to it. In other words, the spin is either rotating clockwise or anticlockwise relative to the measurement axis. Thus, when a measurement is made, the spin is 'forced' necessarily along the measurement axis and the electron is found in either of those two situations. It is said that its wave function *collapses*. From an undefined direction it takes a direction determined absolutely. Furthermore there is no way we can go back and attempt to determine what its state was before the measurement. All information about its orientation along any other axis is lost in the measurement process. This is in a sense an application of the uncertainty principle saying that the determination of a spin orientation along a given axis forbids the retrieval of information along the other axes.

With this background let us go back to our experiment that can be represented as in Figure 6.19 where we use our colleagues Joe and Mary to do the observation. They both have a filter such as the one we have described previously when we examined the Stern–Gerlach experiment shown in Figure 6.9. We assume that their filter contains a detector. They can orient their filter in any direction they wish. Both Joe and Mary do not initially know the orientation of the spin emitted by the source between them. The direction is not given by any rules of quantum mechanics aside from being opposite to each other as dictated by the principle of conservation of angular

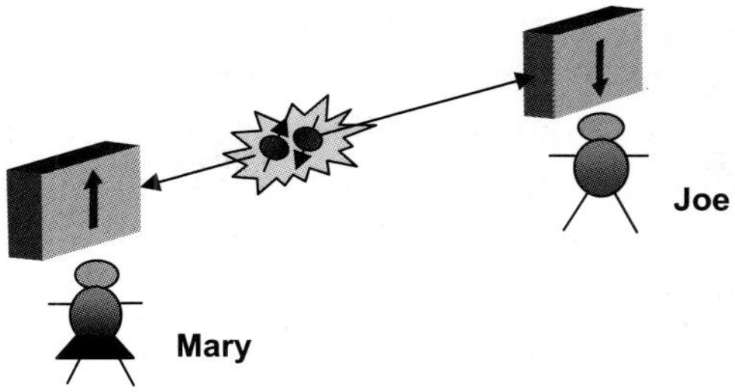

Figure 6.19. An ideal experiment in which a molecule having zero angular momentum disintegrates, sending in opposite directions two atoms with spin angular momentum. The atoms must have opposite spins because the original molecule had no angular momentum. We say that they are entangled. Joe and Mary are set to measure the direction of the angular momentum of the atom or of the spin reaching their detector.

momentum. Mary orients her filter as in Figure 6.19, that is in the up direction. She gets a click on her detector. The detection process collapses the wave function. Consequently, she knows that the spin she receives in her filter is oriented up. Since the two particles emitted by the source are entangled, their spins are opposite, she is certain that Joe, whatever his distance, finds the spin coming towards him with an orientation opposite to the one she detected. Joe does not even have to check with his detector. There is total correlation between the two results and the spin going towards Joe has an orientation totally determined. It looks as if the two spins were connected somehow, having a common wave function that collapsed when Mary detected the spin with her filter. Or it looks as if the result was written beforehand in the process: either the particles emitted were programmed somehow to pass information to the detectors or that information was transmitted from one detector to the other.

Did Mary's detector communicate its orientation to Joe's detector telling Joe's spin which orientation to take? It should be pointed out that Joe and Mary could be on different planets and the phenomenon would still be observed. The effect observed is instantaneous and the information would have to travel faster than the speed of light. This is not possible because of the principle of relativity that says that no information can be transmitted faster than the speed of light. We will see below that physicists, studying the phenomenon, have been very careful in their experiments using sophisticated switching techniques in order to avoid exactly that problem of communication. It looks as if a totally unconnected measurement on one system affects instantaneously the result obtained in a distant system without the possibility of interaction between them. The action is said to be non-local. It has been qualified as spooky action at distance.

In the early 1930s, Einstein objected to this element of quantum mechanics and believed in the existence of a physical reality independent of the observer. Einstein's point of view was made public in a famous article with collaborators Podolsky and Rosen.[19] They concluded on the basis of a thought experiment similar to the one just illustrated, but related to speed and position instead of spin orientation, that if effects like those arose contradicting the presence of a physical reality, it must mean that quantum theory is incomplete. Within that context this simply means that if Joe's spin has a direction determined by an action made by Mary, without any connection between them, it must have had that direction to start with. Or, although not said directly in their article, the conclusion reached may be interpreted as a requirement of hidden variables or instructions that have not been included in the theory. In other words, since quantum mechanics, through Schrödinger's equation for example, does not predict specifically such an effect, it must be incomplete. The objection gave rise to long discussions and exchanges between Einstein and Bohr, Einstein insisting on the existence of a reality independent of the observer while Bohr, using the complementarity principle, particularly in connection to the determination of position and momentum, insisted on the concept that the observing apparatus and the objects observed are inseparable and cannot be analysed separately. In such a context, they are assumed to form a whole and that some of the questions asked, such as how correlations appear in those experiments, have no meaning.[20] The spin has a random direction not because we do not know it but because nature has not determined it. It will take a direction when we interact with it with a measuring apparatus. The positions taken led to lasting philosophical arguments giving rise to unrealistic questioning such as: 'If we do not look at the moon is it still there?' The discussion among physicists lasted for decades. One of the problems is connected to the collapse of the wave function as we have seen in the case of the atom detection when the atom is in a superposition state as shown in Figure 6.7. There were many contributions offering diverse points of view, like the presence of a potential or a field. However, it seemed that the problem was insoluble since there was no experimental way of testing the validity of either position. This lasted until the early 1960s when John Bell published an analysis that, when applied to such an experiment as the one described above, led to the possibility of verifying the presence or not of hidden variables or instructions embedded in the particles' properties.[21] The main idea in Bell's analysis is the use of a two-state system, such as an electron spin up or spin down, and three conditions under which these two states can be observed like three orientations in space of the filter equipped with a detector. He provided a proof that if a programming is present in the system it will be possible through statistics to detect its effect on the experimental results obtained from a large number of measurements. It should then be possible either to confirm or belie the assumption of the presence of a program.

In order to illustrate Bell's analysis, which leads to a criterion to decide on the existence of those actual instructions or programming, we will use the following heuristic approach inspired from a paper published by Mermin.[22] It is based on simple logic using classical concepts. We construct a device and use again the help of our colleagues Joe, Mary and Bob to actuate it. Assume that we have two identical detector boxes and a source that emits two particles in opposite directions situated between the two detectors as illustrated in Figure 6.20. When Bob presses the knob on box B, two particles, which contain the same instructions or program, are emitted back to back. When one of those particles enter either detector it triggers an internal mechanism that lights up a bulb, either blue or red, depending on the switch position, 1, 2, 3, as set by Mary and Joe on their respective detector M and J. We do not need to know the actual mechanism connecting those switches to the properties of the arriving particles. There are no connections between the detectors and the emitting box B, as well as between themselves. There is no communication either through light or radio waves between any

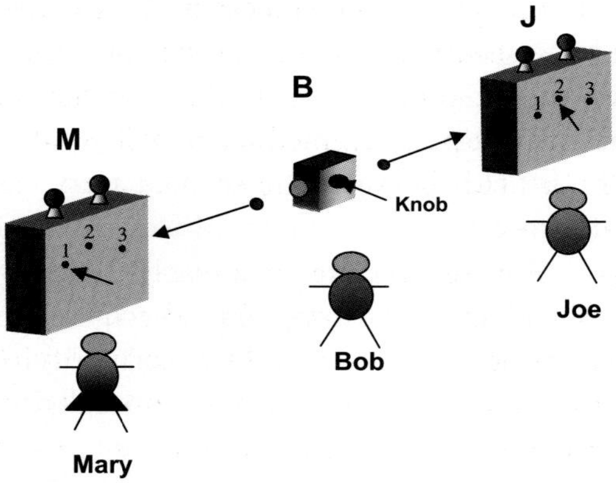

Figure 6.20. Mary and Joe with Bob's help attempt to observe non-local effects or spooky action at distance in a classical world with programmed particles.

of them. They can be placed at different distances from the source B and arrive at the detectors at different times. Consequently, if for example, Joe is farther away from the source and he knows that Mary has already detected a particle he can change the position of the switch on his detector in order to confuse the situation. The particles may be programmed in pairs in any of the eight combinations: RRR, RRB, RBR, RBB, BRR, BRB, BBR, and BBB, but each member of a pair has the same program. This programming simply means that in the case of RBB for example, if Mary sets her switch to 1 and Joe to 2, the red light will turn on in Mary's detectors and the blue light in Joe's detector. If Mary sets her switch to 2 and Joe to 3 both Mary and Joe will see the blue light turn on. On the other hand the number of combinations of the switches amount to nine: 11, 12, 13, 21, 22, 23, 31, 32, 33, the first number applying to Mary and the second to Joe. If you think a bit about it and do the exercise, in the case, for example, of a RBB programming of the particles in one pair, Joe's and Mary's detectors will light up with the same colour in 5 out of the 9 combinations of their switches: these are 11, 22, 23, 32, 33. This is also true for whatever colour program is associated with the particles. We note that in the case the particles are programmed RRR or BBB, Joe's and Mary's detectors obviously will always turn on the same colour. Thus, in any case, we would expect lights to be the same colour in more that half the cases. Thus, when instructions are present, the probability of getting the same colour is greater than 50% by a measurable amount (>55%). Thus, experiments on a large number of particles, pre-programmed as described above, would give results in which the two detectors show the same colour for more than 50% of the cases.

This simple exercise illustrates essentially Bell's finding in connection to the paradox raised by Einstein and colleagues, regarding spooky action at distance. Bell arrived at the conclusion that if entangled particles contained a kind a program which determined their behaviour, a detection of their statistical properties on a large number of measurements would provide information on the existence of a program or not. His reasoning led to the imbalance or inequality shown above and it can be verified experimentally. It is called *Bell's inequality.* The whole problem is transformed from a philosophical argument to a practical measurement that can be made in the laboratory.

The experiment concocted above in Figure 6.20 for Mary, Joe and Bob can be transposed to the atomic world as follows. On may think for example that the source B emits two entangled electrons with their spin having opposite directions. When Bob presses the knob on box B two electrons are emitted back-to-back, with their spins opposite but oriented at random in space. They are thus entangled. The detectors can be Stern–Gerlach dipole filters with a particle counter at their exit. The arrangement is thus similar to that of Figure 6.19 in which the source of entangled spins is controlled by the experimenters. The detectors can be oriented at three angles that we chose as either 0 degree, +120 degrees, and −120 degrees corresponding to the position of the three switches on Mary's and Joe's detectors. The orientations thus provide detection along different axes. We thus have exactly the situation that we have described above with switches and lights and the experiment with spins can be done by our friendly group as in Figure 6.20. Now, we can calculate what quantum mechanics predicts for such an experiment. Such a calculation shows that the detectors should not agree in more than half of the cases.[23] Thus the conclusion of the two scenarios, classical reality with an internal program and quantum uncertainty determined by probability alone, leads to two different results verifiable by experimental measurements.

Experiments were carried out in the early 1980s by Alain Aspect and his colleagues. They did the kind of experiments proposed using entangled photons spins.[24] Why use photons instead of electronic spins? It was easier, using a calcium source, to create entangled photons with left and right polarization corresponding to opposite orientations of their spin. Measurements were also easier through the use of optical polarizers as detectors rather than bulky Stern–Gerlach dipole magnets. In their first experiments the polarizers could be moved to 6.5 m from the source. Furthermore, in later experiments they used time-varying polarizers that change orientation in a time short compared to photon transit time avoiding correlation through the possible existence of a transmitted unknown signal between the polarizers and detectors.[25] The results were found to violate Bell's inequality; the predictions of quantum mechanics were verified showing the absence of local predetermined instructions.

Those results were confirmed in many other experiments. Experiments have been done over distances as large as 10 kilometres between observers and Bell's inequality was still violated.[26] Einstein and colleagues were wrong. There is no programming present in the entangled photons. In simple language, spooky action at distance is present and it is concluded that the direction of the spins in Mary's and Joe's experiments were not predetermined. It appears that a pair of entangled electrons forms a single object and cannot be split into separate physical entities until those are detected. In the quantum world, the notion of locality is lost and it is not possible to attach independent properties to the members of the pair. The members of the pairs, electrons or photons, 'stay in touch' over space. It looks as if the two members of the entangled pair are described by a single wave function whatever their distance. We have seen something similar above in the case of a spin, prepared in a given state, going through a Stern–Gerlach dipole magnet (Figures 6.14 and 6.17). In that case, the spin evolves in the magnet according to the rules of quantum mechanics; it obeys Schrödinger's equation, its wave function spreading over the two possible paths to finally collapse at the detectors with only one direction surviving. No hidden instructions need to be invoked and it does not appear that the detectors need to communicate to provide information to each other. The finding is a shock: the effect opens the door to the non-locality of phenomena in the quantum world. A measurement here dictates instantaneously the result of a measurement there without any physical connection between the two objects by means of any forces or interactions known. This points to the difference between the quantum mechanical world and the classical world, the transition taking place at the collapse of the wave function. However, the collapse is not described by either quantum mechanics or classical mechanics.

The phenomenon, however, cannot be used to transmit information at speeds greater than that of light. The effect is not in violation of the principle of relativity as the speed of light is a true limit for information transfer. We can say that information transfer takes place at the macroscopic level while the effect of entanglement takes place at the atomic level. Nevertheless, it can be used as an encoding mechanism in secure communication systems to transmit a key. The effect is then used as a deterrent to eavesdropping since the observation by a third party of transmitted information with entangled photons would obviously collapse the wave function of the elements forming the transmitted key, thus making the intrusion evident to the partners.[27]

This behaviour of the quantum world does not make sense in our classical world that operates locally. We generally tend to think like Einstein about the reality of the existing world with causes and effects being closely connected. Spooky action at distance is not part of our normal way of thinking, but we have to accept it since we observe it. Its true origin is not known and understanding it in terms of our acquired awareness of the world surrounding us is a real challenge to the mind.

Usefulness of Quantum Physics: Understanding of the World and Its Use in Our Day-to-Day Activities

Quantum physics was born of a desire to understand the way the atomic world works. And it does so very well. If this were the only result, it would still be a great achievement. We have been looking at galaxies for nearly a century, we know that they are millions of light years away and we also know that they do not have much influence on our daily activities. But we are still looking at them, we are learning about their dynamics, and we look forward to understanding a lot more about the rules that govern all objects of our Universe.

Although we do not know why the microscopic world needs to be quantized, this is the way it works and we must accept it as such. As mentioned above, our knowledge of the components of that microscopic world is rather good and we have found that with that knowledge we can make predictions that are verified experimentally. Our work in quantum physics, however, has been driven for other reasons than just knowledge: its understanding has led to applications that have improved our way of life and every day our understanding is getting better, resulting in more applications. There are many such applications and it would be impossible to cover in a few pages all those that have followed from that understanding. These extend from the field of medical diagnostics and surgery, to communication, navigation, entertainment, and even in better implementation of standards of physical measurements that have a direct effect on our daily activities. Let us look at a few examples.

The electron microscope

The wave property of the electron provides a new tool in high resolution microscopy. It is well known that the resolution of a microscope is inversely proportional to the wavelength of the radiation used. This is due to diffraction phenomena taking place in the instrument since the magnifying lens has a limited aperture and light, passing through a small hole, spreads in space. Shorter wavelengths allow a greater resolution since the diffraction spot caused by the limited aperture is smaller. Electrons with their shorter wavelength could thus in principle provide a tool for reducing that effect. The wavelength of electrons accelerated under a voltage of, say, 10 kilovolts is of the order of 0.01 nm, which is very short as compared to around 500 nm for visible light. The resolution power of a microscope using an electron beam would thus be several thousand times greater than a microscope using a visible light beam. The magnification power of the electron microscope is of the order of a few millions while that of a visible light microscope if of the order of a few thousands. The advantage of the electron beam over short wavelength electromagnetic radiation, such as X rays, is that electron beams can be focused more easily by means of electric and magnetic fields while the use of glass lenses is limited to longer wavelengths. The electron microscope has become a very useful tool in many scientific sectors such as biology, material physics, as well as in forensic activities.

Magnetic resonance imaging

In medicine, the magnetic resonance image (MRI) apparatus has become a preferred tool for making diagnosis of disorders in our organism. What does it do? It detects the presence of water. Water is a major constituent of our body and its molecule contains two atoms of hydrogen and one of oxygen. As we have seen, the nucleus of hydrogen is a proton. Protons possess a spin angular momentum in a similar way as electrons. That spin gives rise to a magnetic moment that settles in two states in an applied magnetic field, with its magnetic moment along the field or opposite to the field. In the instrument a very strong magnetic field is applied to the body under investigation. Two proton energy levels are created by the magnetic field as in the case of the electron examined above. The spacing between the levels is proportional to the magnetic field applied. A radio frequency field (RF) is also applied, that corresponds in energy, $h\nu$, to the exact spacing between the two energy levels in question. That oscillating field forces the protons into oscillation like a swing is forced into back and forth motion by means of a push from our hand at the right time. This is called a resonance effect. The RF field is applied as a strong pulse. After the pulse the hydrogen nuclei continue their oscillation, like the swing in the example above, and they emit energy at their resonance frequency. The radiated energy creates a signal in a small coil or antenna nearby. The intensity of the signal is proportional to the density hydrogen or water stimulated by the RF field. If the applied constant magnetic field were perfectly homogeneous, all the protons in the whole body under examination would be resonating at the same frequency and the signal would originate from the whole body. No detailed image of the various parts would result. However, if the magnetic field is made to vary in space, the spacing between the magnetic moments energy levels vary also in space and resonance is observed with frequencies different from one part of the body to another part in a known fashion. Consequently, if the frequency of the applied radio frequency field is fixed, and the magnetic field is made inhomogeneous, a scanning of the magnetic field (change of value with time) will make different parts of the body resonant with the applied radio frequency field at different times and will give information on the hydrogen (water) concentration in those various parts providing an image of the distribution. The noise that is heard by the patient under examination in one of those instruments is due to those abrupt changes or scanning of the magnetic field. A very rough sketch of such an arrangement showing the principles of operation is shown in Figure 6.21. In practice the constant field is produced by means of a solenoid providing the bore into which the patient is introduced and coils are added to create field gradients. The signal induced in the coil originating from various parts of the body is processed by means of a computer and form the basis of the desired image. The use of a superconducting coil magnet is dictated by the need of obtaining a large magnetic field and a large spacing of the proton energy states, to raise clarity and resolution of the final image.

I remember a period of my graduate studies when I was investigating magnetic resonance phenomena of a similar nature in

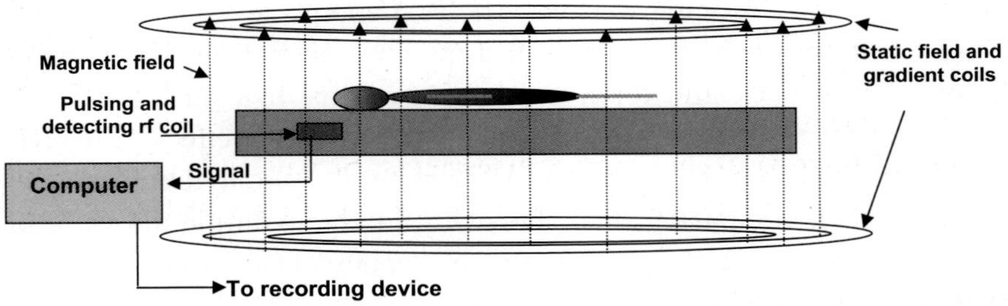

Figure 6.21. Simplified illustration of the principle at the basis of magnetic resonance imaging used in health diagnostics. In a practical device the uniform static magnetic field is created by means of a horizontal solenoid in which the body is entered. Coils are added to create the necessary field gradients.

solids. The signals were weak and the study and analysis were rather painful. It is an amazing feat of engineering to be able to make clear images with such a device and we must admire those that have achieved such an exploit over a limited number of years of research and development.

What a beautiful application of quantum theory! It uses the spin concept, which is a quantum as well as a relativistic effect, and a superconducting magnet, the functioning of which is explained only by means of a special coherent property of the electron within quantum theory. This instrument is used daily for making diagnostics and it does it better than any other instruments.

The transistor

Another application that deserves mention is the transistor that is embedded in millions in those integrated circuits that provide the working heart of our computers, calculators, cellular phones, iPods, compact disks players, DVD players, radio receivers, television sets, GPS receivers, to name a few. The functioning of the transistor, invented some 60 years ago, is based on a deep understanding of the behaviour of electrons in solids relying heavily on quantum theory. The subject is called Solid state physics and is still one of the most studied fields today. As another example, one most useful device that was born out of that field is the so-called Charged Coupled Device (CCD) which makes possible the fast recording of images in a digital form that can then be processed by means of computers. The device is made of arrays of microscopic semiconductor islands, acting as capacitors, which act also as light sensors through the photovoltaic effect for which Einstein received the Nobel Prize in 1921. It is used as a very efficient image collecting device in digital cameras as well as in advanced terrestrial and spatial telescopes. Scientists W. Boyle and G. Smith received the 2009 Nobel Prize in physics for that invention.

Superconductivity

Elements such as metals conduct electricity at room temperature. When a voltage is applied to a wire made of copper, for example, electrons within the metal start moving in the same direction under the effect of the electric field created. An electric current starts circulating in the wire. However, due to the presence of the copper atoms interacting slightly with the electrons, there is a small impediment to the flow of those electrons and we say that there is resistance to the flow of electric current. This is good in some applications because energy is lost for the same reason and the wire carrying the current starts heating up. We can use the effect to make toast in the morning, boil water for the coffee, light up our houses at night and heat them up in winter. In copper and silver, that resistance is very small and we say that they are good electrical conductors. If we lower the temperature we find that the resistance decreases. Can it get to zero when the temperature is lowered to zero kelvin? Not quite. There is always some resistance left in those metals. However, some metals such as lead and niobium, at a critical low temperature (7.19 K for lead and 9.26 K for niobium), become perfect conductors. Their resistance goes completely to zero. This is called superconductivity. There is no energy loss in the circulation of the electrons in the metal and the electric current may last indefinitely, as long as the temperature is kept below the critical point. This is a most remarkable effect and its existence did not find an explanation for decades after its discovery by Onnes in 1911. It took the genius of the same scientist who had invented the transistor to find an appropriate physical explanation.[28] It is based on the quantum properties of electrons that develop coherence in pairs, the metal atoms acting as a catalyst through their vibrations. The electrons are so well correlated that they do not find any impediment in their motion. The resistance to current is totally absent. However, not all metals react that way. In fact, the best conductors at room temperature, such as copper and silver, do not become superconductors at low temperatures. Recently some ceramic composites have been found to exhibit such a property at a temperature as high as 138 kelvin. Superconductivity is a most extraordinary property allowing the passage of enormous electric currents without losses. They can thus be used to create very strong magnetic fields such as those needed, for example, in the functioning of MRI machines as described. These instruments require a strong magnetic field in order to obtain intense resonance signals to create a clear image. The high temperature superconductors also raise hopes in the field of electrical power transportation, which is plagued with loss of energy with ordinary transmission lines over long distances. Superconductivity is also responsible for the Josephson effect, an electron tunnelling phenomenon that takes place between two superconductors separated by a thin insulating layer. As we will see, this effect is used to implement a measurement standard for the volt with great precision and accuracy. Superconductivity is also used extensively in basic research in several laboratories studying fundamental properties of matter.

The atomic clock

I mentioned several times the atomic clock, particularly in connection with relativity experiments and time dilation. It is one of the most fascinating applications of quantum physics. A clock is simply an oscillator that is kept running continuously. We are very familiar with the common clock constructed with mechanical parts. The heart of the oscillator can be a pendulum oscillating back and forth under the influence of gravity and a spring supplying the energy required to compensate for friction losses. The period of oscillation provides the unit of time that may be made exactly one second if the length of the pendulum is properly chosen. The internal mechanism of the clock, consisting of a regulating escapement and driving gears, adds the seconds to drive needles indicating time of day after a proper origin has been agreed upon. In such a setup, the period depends on many factors such as length of pendulum, which is a function of temperature, tension on springs, friction and several other perturbations such as gravity fluctuations and effects of magnetic fields on the material used in its construction. All those are called environmental effects that influence the functioning of the clock. Clocks are so important to navigation that in England, in the 18th century, a large prize (£ 20,000) was offered to the person that could design a clock that could be transportable on a ship and had a stability such as to keep time better than two minutes for a trip from England to the Caribbean. Instability of the order of three seconds per day after a trip of forty days would cause such an error. In navigation, this results in an error of about half a degree in longitudes or 30 miles at the equator. A clock maker named Harrison spent forty years in creating such a clock and won the prize at the end of his life.[29]

It had been suggested at the end of the 19th century that if internal properties of atoms could be used, a better clock could be built since its operation would be dependent mainly on the internal parts of the atom, whose functioning should be independent of environmental effects. Although the physics behind the construction of such a clock was well known in the early parts of the 20th century, it took nearly half a century to see its practical realisation. The reason of this lapse of time is mainly a question of available technology.

How is such a clock built? The key point is the use of an internal property of the atom, a property that is inherent to its existence. And that property must be cyclic in order to obtain *second ticks* or in other words a *tick-tock*. This is not simple because the gears within the atom are not directly accessible. The atom must be prepared first to make one of its internal properties easily accessible. Then, that property must be investigated (interrogated!) in order to use it as a pendulum. Finally, we have to connect to it and count the cycles. This is called the detection process. This whole idea is illustrated in Figure 6.22.[30]

Several atoms can be used to implement such a clock. For example, hydrogen, the simplest element provides the pendulum for one of the most stable clocks. Its medium term frequency stability makes it ideal, for example, in applications such as studies of far celestial objects by means of radio telescopes using interferometry techniques. It is also used in navigation systems and basic research

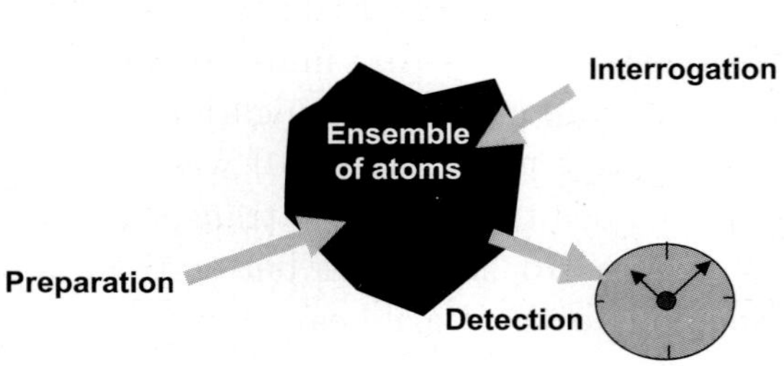

Figure 6.22. Illustration of the processes involved in creating an atomic clock.

as well as time keeping. Rubidium is also used for implementing clocks for providing accurate timing in navigation satellites and in communication systems. On the other hand, caesium is used to implement the most accurate and stable clock. It is also used in navigation and communication systems, but has acquired its primacy mainly through its use as the clock in essentially all National Institutes in charge of time keeping for their country. In fact, the caesium atom is used as the basis for the definition of the second in the International System of units: 'The second is the duration of 9,192,631,770 periods of the radiation corresponding to the transition between the two hyperfine levels of the ground state of the caesium 133 atom'.

A device implementing that definition is thus an essential item in a National Institute in charge of basic standards. There are several ways of constructing such a clock, but the most common and most reliable one is shown in Figure 6.23. It works in the following way. First a vapour of Cs atoms is formed by heating a small reservoir containing Cs metal. It is called the caesium source. A small hole in the source provides an exit for the atoms and a narrow beam of those atoms is formed by appropriate collimation. The beam is then

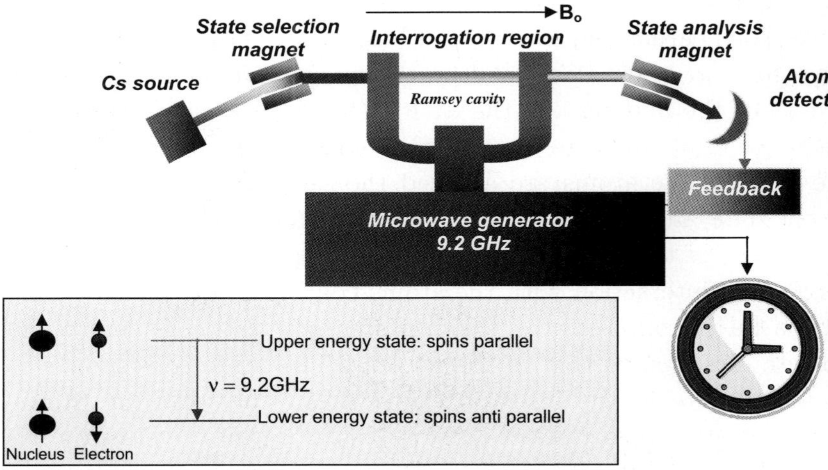

Figure 6.23. A Cs atomic clock using Stern–Gerlach's state selection approach. The inset shows the two possible energy states of the unpaired electron in the ground state of the caesium atom.

made to pass through a so-called state selection magnet, which is none other than a Stern–Gerlach dipole magnet that we have shown in Figure 6.1. As explained in that section, in such a magnet, the beam is split into two parts according to the state the atoms are in. Those states are shown in the inset of Figure 6.23. This is the state preparation action shown in Figure 6.1. Atoms in the upper state with nucleus and electron magnetic moments oriented in the same direction are selected and made to pass in a so-called Ramsey cavity,[31] which is fed at the resonance frequency of Cs by means of the microwave generator at 9.2 GHz. This part of the instrument is identified as the interrogation region. The last section includes a Stern–Gerlach magnet similar to the first one, now used as an analyser, followed by a detector or counter of atoms, and a feedback system to lock the frequency of the microwave generator to the resonance of the caesium atoms. This last section plays the role of the detection part identified in Figure 6.22. The properties of the atom are transferred that way to the microwave generator that drives the clock needles. The Ramsey cavity has a special feature. It is a doubly symmetrical structure. Atoms in their upper state, selected by the dipole magnet and passing in the first part of the cavity, are excited by the injected microwave radiation from the generator. Their internal state is altered. They are set into a superposition state as described earlier and as shown in Figure 6.8, the magnetic field in that figure now being provided by the nucleus. In simple words, at the exit of the first part of the cavity, the electron spin orientation is now at right angle to that of the nuclear spin orientation. We note that all the atoms are affected the same way: they evolve in unison. The atoms in the beam are then allowed to drift to the second part of the Ramsey cavity. If the frequency of the generator is right, the interaction with the atoms continues and their internal state is further altered: the electron spin orientation is sent to an opposite direction to that of the nucleus. This state corresponds to the lower state in the inset of Figure 6.23. The following Stern–Gerlach dipole magnet select those atoms in which the electron has made a full transition, thus oriented opposite to the nucleus magnetic moment direction. Those atoms are counted by means of a process of ionization on a hot wire. We can see readily that if the frequency applied to the Ramsey cavities is not that of the Cs transition an interference will take place in the excitation of the atoms by the two sections of the cavity and the number of atoms detected as having made a full transition will be reduced. Thus the atoms are used as a discriminator and the signal detected contains that information for the locking of the generator frequency to that of the atoms by a feedback loop. In that implementation the electron spin interacting with the Cs nucleus is thus effectively driving the needles of the resulting clock. It is known as an atomic clock.[32]

We thus have a clock the functioning of which depends on the internal properties of an atom and should be quite independent of environmental conditions. Other clocks using atoms such as rubidium and hydrogen are implemented in a somewhat different configuration. The basic principle of operation, however, is the same: atoms are forced to make transitions in unison between energy levels or states and their frequency of resonance is detected to obtain 'second' ticks. The best of those clocks keeps time so accurately that it will not lose one second in a million years. This is far better than the pendulum clocks which keep time to maybe one second per day or still the best quartz wrist watches that may keep time to one second per week.

The Cs clock is made commercially and is the most commonly used clock in navigation, communication and time-keeping applications. It can be made in various sizes depending on the application, frequency stability required, and accuracy desired. For greatest accuracy the Ramsey cavity must be made as long as possible, since the finesse of the resonance is dictated by the spacing of its arms. A typical very large size laboratory system is shown in Figure 6.24. The U shape Ramsey cavity is visible on top of the several-metres-long cylindrical vacuum vessel. It is the primary atomic clock of the National Research Council of Canada. It was the first Cs atomic frequency standard operating continuously as a clock and was for a long period a most important contributor to time-keeping at the *Bureau International des Poids et Mesures* (BIPM). Those clocks are now used throughout the world to keep time. An atomic time scale is formed by the BIPM out of the individual time scales maintained by each National Institution. We have also seen a most beautiful application of an atomic clock, the hydrogen Maser,

Figure 6.24. The primary clock of the National Research Council of Canada. It is a Cs Clock about three metres long, and has an accuracy of about one part in 10^{13}. In the future it will be replaced by a laser-cooled clock with an accuracy improved by a factor of at least 100.

for verifying a prediction of general relativity regarding the rate of clocks in different gravitational fields. The Hydrogen Maser, although using the same basic principles as those used in the Cs clock just described, operates somewhat differently in that it is a maser.[33] This means that instead of having a microwave generator to interrogate the atoms, the system oscillates by itself and emits radiation at the frequency of the atom, 1.42 GHz in the case of hydrogen. The signal obtained is then used as the driver of the clock needles.

One of the greatest achievements has been the use of atomic clocks in implementing a navigation system on the surface of the Earth and its neighbourhood. The system is called the Global Positioning System (GPS), implemented by the USA Department of Defence. A similar system has been implemented by Russia and is called Glonass. The European community is also planning such a system in the near future. It is called Galileo. The development of so many navigation systems appear to be a question of states' independence relative to security, but certainly serves the purpose of redundancy. The application of clocks in navigation is based on the fact that the speed of electromagnetic radiation is a constant.

Consequently, it is sufficient to determine the time a given pulse of radiation takes to travel from an observer to another one to evaluate the distance between them. This can be done by determining exactly the time of departure of the pulse of radiation from the first observer and by a measure of its time of arrival at the second observer. Electromagnetic radiation travels about 30 cm in 1 nanosecond. It is thus required to have a clock stable enough to make the measurements to 1 nanosecond to obtain a precision of 30 cm in the measurement of distance. In practice, for navigation on the totality of the Earth's surface and its close surrounding, atomic clocks are placed onboard satellites that transmit coded information on the satellites orbit and the time onboard. The relative time of arrival of the radiation pulses from those satellites is measured at the location where position is to be determined using small sensitive receivers. Using several satellites at the same time, the system allows the determination of position in three dimensions, longitude, latitude, and altitude, as well as providing exact time. In the GPS, a constellation of more than 24 satellites is placed in orbit around the Earth at an altitude of about 20,200 km as illustrated in Figure 6.25.

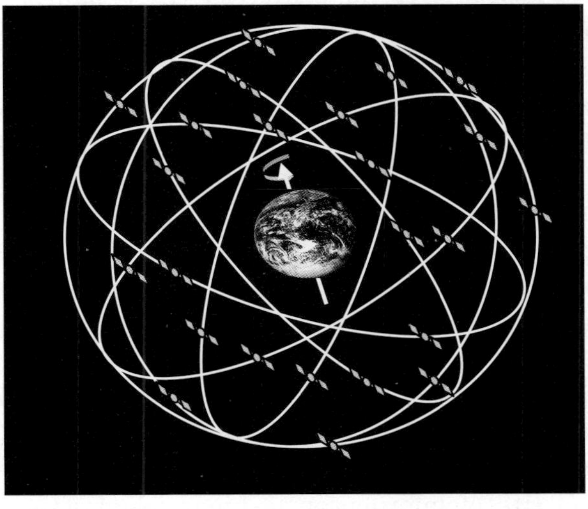

Figure 6.25. GPS satellites configuration consisting of 24 units. In principle, at least four satellites can be detected simultaneously over the whole surface of the Earth. The system provides determination of longitude, latitude, altitude and time.

This number ensures that the totality of the Earth's surface is covered and that at least four satellites are visible at all times. For redundancy each satellite in equipped with at least three atomic clocks using either Cs or Rb atoms. Time synchronization is kept between all satellites and day to day time stability is maintained to better than 20 nanoseconds, by continuous monitoring and updating the satellites from special stations on precise location on the Earth's surface. In practice with the present system, accuracy of the order of a few metres is possible. The concept has known tremendous success in many kinds of application, such as those concerned with military needs, marine and airborne navigation, rescue, surveying, car navigation on road networks, sports like hunting, fishing and hiking, and scientific experiments of various sorts.

Entanglement, computers and secure communications

Another application is the use of the previously mentioned phenomenon of entanglement for coding communication messages. It is possible to exchange quite securely information in communication networks by using complex codes or keys based on random numbers. These keys are secure, in the sense that they cannot be decoded unless a powerful computer is available, but there is always a possibility for a code to be broken. However, if the key exchanged between the partners were coded with entangled photons, it would not be possible to break it without the communicating persons becoming aware of it.[27] It does not prevent the key from being attacked, but the communicators are informed of such an action. This is because the wave function of entangled particles collapses when a measurement is made on them. Consequently, after detection, the quantum particles are no longer entangled and the partners exchanging information become aware of it.

It may be mentioned that a similar kind of thinking can be applied to the possible realisation of extremely powerful computers using quantum properties. In computer science the 'bit' represents the minimum size of information, a binary digit either 0 or 1. In quantum computing, the basic information is contained in a 'qubit' for quantum bit representing a two-state system. The qubit can be a superposition of states as studied above. This approach leads in principle to the possibility of having a far greater capacity of information storage and handling and consequently higher speed. The approach is at present the subject of intense research and it appears that important results could be obtained in the near future, although the achievement of a practical computer remains a challenge at this time.

The laser

The word laser stands for Light Amplification by Stimulated Emission of Radiation. It works on the same principle as the maser mentioned above but delivers light at its output instead of microwave radiation. In view of the much shorter wavelength involved, the structure is different. In practice, mirrors are used to form the resonant cavity rather than a box, which is appropriate mainly at microwave frequencies. The light wave is reflected on the surface of two mirrors and is amplified by the medium in between, as shown in Figure 6.26. One mirror is made semi-transparent and a small amount of light is transmitted outside the structure. If the amplifying medium has sufficient gain, the system maintains itself in oscillation. The energy is supplied by the high-voltage source that creates in the amplifying medium, by means of electrodes, an electrical discharge called plasma. The two mirrors form a so-called Fabry–Pérot cavity. Any light ray or photons that does not travel parallel to the longitudinal axis is reflected out of the system after a few bounces on the mirrors, For this reason the system amplifies only rays nearly parallel to the axis and the result is a very narrow

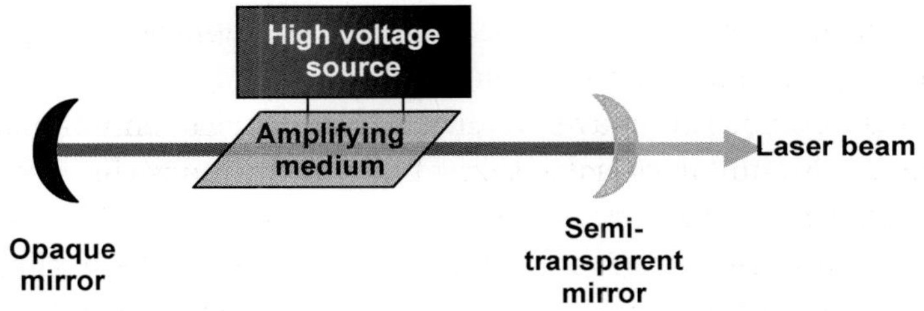

Figure 6.26. Basic components of a laser.

beam of radiation that keeps its parallelism for large distances. A widely used system is that using Neon as its amplifying medium. How does quantum physics come in the picture for the functioning of such a device?[34] First, as usual, light comes from electrons making a transition between two energy levels within the Neon atom. Thus the light coming out is characteristic of Neon, reddish at a wavelength of 633 nanometres. In the H maser, we were talking of radiation at 21 cm. This is a large difference in wavelength and this is the main difference between lasers and masers. But how do we get the system to amplify? To start with, we must get the energy into it. This is done by means of a high voltage source. With the help of a bit of helium that is mixed with neon at the time of filling the glass enclosure, neon electrons lodge themselves in upper states. We have what we call an inversion of populations and the Neon atoms natural action is to fall into the lower state emitting a photon. That photon interacts with another neon atom and induces it to fall to a lower state and emit another photon in phase with the original one. A chain reaction follows with the help of the mirrors that reflect most of the photons back and forth inducing more atoms to emit. The excited atoms emit photons that add and produce a coherent light wave that propagates within the structure and exits through the semi-transparent mirror. In practice it is possible to obtain a narrow beam with a power of the order of a few milliwatts. The beam is narrow because any photon that does not travel parallel to the axis is reflected out of the apparatus after a few reflections. Only those photons travelling parallel to the axis contribute to the beam.

What are the properties of such a light beam? How is this light beam different from light emitted by an incandescent light bulb in the ceiling or neon tubes used in electrical signs? That difference is coherence. In a neon tube, the photons are emitted one by one through a process called spontaneous emission (a direct effect of the vacuum photons), they have a range of frequencies, and their phase is random. A similar process takes place in an incandescent lamp. In that case when light is made to pass through a prism it is found to separate into various colours like a rainbow showing the many frequencies or wavelengths contained in the light emitted by such a device. The photons are independent of each other; they travel in space like rain droplets falling to the ground, each with their own energy. The droplets arrive one by one independent of their neighbours.

In the laser, all the photons emitted have the same wavelength (or frequency) and the same phase. They are working in unison. The wave travelling out of the semi-transparent mirror is coherent: there is a single wave front like a wave on the surface of water with all molecules at the forefront going up and down together. Such a beam of radiation may be very intense because it is emitted as a very narrow beam. Furthermore, the radiation emitted has a single wavelength because it originates from transitions between a single pair of energy levels within the neon atom. If the laser radiation is made to pass through a prism it is found that it does not decompose into other colours. It stays red. That type of laser may be made with several other atoms, each of course, emitting at their own specific wavelength. For example carbon dioxide emits at a wavelength of 10 micrometres, not visible. Because of its efficiency, that laser can produce power in the kilowatt range.

Very small lasers can also be implemented using the same type of material as that used to make transistors (semiconductors). In that case the amplifying medium is a small single crystal block of semiconductor (a cubic mm) of which two opposing surfaces serve as mirrors (cleavage surfaces). The light originates again from electronic transitions within the semiconductor atomic energy levels manifold. The energy is supplied by passing a current in the specially designed structure, a current that forces electrons into an upper energy level causing a population inversion. Otherwise the system works very much like the Neon laser. The radiation emitted may have various wavelengths depending on the type of impurity atoms that have been integrated to the semiconductor material during its growth as a crystal. These lasers are used at large in communication systems taking advantage of the purity of the radiation emitted (single frequency) to transmit signals by means of optical fibres over long distances. They are also used in entertainment systems, such as in video and music reproducing systems (DVD and CD), their coherent property and the narrowness of their beam allowing concentration of light to a very tiny dot by means of a lens. Due to the very small size of the dot it is then possible to store large quantities of information on very small surfaces. Lasers are also used in medicine for treating specific problems with minimum invasion. For example the retina may be repaired by focusing a laser beam to the problem point in the eye, fusing tissues together over a very small spot without affecting the surrounding. Treatment of minor myopia may also be done by reforming the curvature of the surface of the eye. In all those cases, the main advantage of the laser is its focusing to a very small spot, limiting invasion to a small region and leaving intact surrounding tissues.

Cooling of atoms

In a gas, atoms are in motion. They travel at great speeds in straight line like a rifle bullet until they hit an obstacle, another atom or the wall of the container in which they are located. That motion is random and is perceived macroscopically as temperature or pressure that the gas exerts on the container walls. The temperature is a measure of the average energy contained as molecular motion in the gas. The average energy per atom is given by an expression such as kT, where k is the famous Boltzmann constant that plays a central role in thermodynamics and T is the temperature in the Kelvin scale. We said that such a temperature scale was characterised particularly by the existence of a minimum temperature called absolute zero where all motion stops. Can we reach that point? Helium is a gas that liquefies at 4.2 K. If we lower the pressure in the space where it is contained, forcing it to evaporate a little faster than under a pressure of one atmosphere, it is possible to obtain a much lower temperature. With special arrangements it is possible to reach a few hundreds of millidegrees above absolute zero. It does not appear that we can go much lower by simply cooling with macroscopic approaches such as forced evaporation. The technique can remove a portion of the heat contained in the substance studied but not all.

To cool atoms in a gas what is needed is a technique that alters mechanically the motion of the atoms. This is possible by taking advantage of Doppler Effect. We have seen earlier that Doppler

Effect alters the frequency of the radiation emitted by a moving atom. An approach for cooling atoms uses that phenomenon to our advantage. A photon is sent towards an incoming atom at a somewhat lower frequency than the actual frequency that the atom is tuned to intrinsically when it is at rest. Because the atom is moving towards the radiation, the atom, due to Doppler Effect, sees the incoming radiation shifted up becoming equal to its own resonance frequency and it absorbs the photon. The atom has thus absorbed a photon with lower energy than it would have absorbed if it were at rest. The atom, after a short time of the order of several tens of nanoseconds re-emits a photon at its resonance frequency, which is higher than the frequency of the photon it has absorbed. The energy re-emitted is larger than the energy absorbed. This energy must come from somewhere: it comes from the atom's kinetic energy, and the momentum lost in the cooling is not recovered in the same direction. On the average, after absorption of many incident photons in succession, the atom loses its kinetic energy. Its speed is thus reduced by a small amount every time it absorbs and re-emits a photon. Every time it absorbs a photon and re-emits it, the atom is cooled by a small amount, approximately 0.05 degree. The cooling is done with laser radiation that can be tuned exactly to the frequency desired and can be intense enough to provide the quantity of photons required. In that process it is possible to stop an atom almost completely with about 100,000 photon-atom interactions. A slow atom is a cold atom and an atom at rest in a given frame of reference is at zero Kelvin in that frame. Cooling of atoms and its complete explanation in terms of quantum effects was realised by C. Cohen-Tannoudji, W. Phillips, and S. Chu.[35]

What can we do with slow atoms? This was one of the first questions that were raised after a first success in the early 1980s. It did not take long for applications to follow. One of the first applications was the implementation of a super stable atomic clock. Yes! A clock again. It consisted in cooling a vapour of caesium atoms in three directions. This is done with six laser fields propagating in pairs in opposite directions and cooling all motion along the three axes of the reference frame attached to the system. A small ball of atoms, about one cm in diameter, is formed containing over a million caesium atoms almost at rest within the ball. Their residual velocity after the cooling cycle has taken place is of the order of 1 cm per second, which means that the ball does not expand in space by more than one cm in one second. By slightly detuning the vertical cooling lasers, the whole ball is made to move vertically at a speed of the order of one metre per second. When all the cooling lasers are turned off, the ball keeps travelling up in the gravitational field of the Earth to a height of about one metre. The ball follows Newton's gravitational law and at that height, like a macroscopic object sent in the air above ground level, the ball reverses direction and falls back to its point of origin in about one second. The atoms in the ball, due to their residual velocity, act like water droplets in a fountain and move transversely by a small amount: the apparatus is called an atomic fountain. During their path up and down the caesium atoms are made to pass through a microwave cavity and resonate with the radiation present in that cavity as in the Cs clock described earlier. Due to their very low speed, the atoms spent a long time between the passage up and the passage down in the cavity. It looks as if the distance between the two arms of the cavity shown in Figure 6.23 is greatly increased and the atomic resonance frequency is better determined than in the clocks using so-called thermal atoms that travel at a speed of 300 metres per second. While in standard clocks working at room temperature the spread of the resonance is of several hundred Hz, in the Cs Fountain the spread is of the order of 1 Hz allowing a better determination of the resonance frequency. Its stability and accuracy are improved by orders of magnitude. Fifteen years after its implementation, that type of clock is now the primary clock providing the most accurate *mise en pratique* of the basic unit of time, the second, in most National Institutes responsible for time keeping in their country. It has replaced the room temperature Cs beam standard described earlier.

It may be mentioned that an atomic clock using laser cooling called PHARAO (*Projet d'Horloge Atomique par Refroidissement d'Atomes en Orbite*), will be combined with another atomic clock, a Space Hydrogen Maser (SHM), to form ESA's Atomic Clock Ensemble in Space (ACES) on the International Space Station. The launch should take place in the period 2010–2020. The two clocks will provide an accuracy of 1×10^{-16}, corresponding to a time error of about one second over 300 million years. They will be used to study General Relativity and among other applications, such as remote sensing and geodesy, provide an accurate reference for global time scales.

Macroscopic quantum effects

Another accomplishment is concerned with a curiosity predicted by Albert Einstein and Satyendra Nath Bose early in the previous century. It is concerned with a phenomenon embedded in quantum mechanics and associated with the wave property of objects in the microscopic world. Those scientists predicted that if the temperature of a gas were lowered, the de Broglie waves that are attached to the atoms of the gas would become longer, start to overlap, collect together coherently and form a single unit. The ensemble of atoms would act as a single super quantum mechanical object. It is called a Bose–Einstein condensate. What temperature is required? A few billionths of a degree at most. How is that done? In two steps. First, the atoms are cooled by means of lasers as in the example of the fountain clock described above. They reach a temperature of the order of a few hundred nanokelvin. Then they are left in a trap formed by a magnetic field acting as a kind of containing wall for the atoms. The field is adjusted to just barely keep the slowest atom in the trap. In that situation the fast atoms are left to evaporate like vapour from a cup of coffee. The faster atoms that leave the cup bring energy with them and the coffee left in the cup is formed of colder atoms. The same process is used in the magnetic trap, but in that case the height of the trap walls can be adjusted at will by lowering the magnetic field. Those faster atoms escape the trap and only atoms with low speed are left. Very interesting! Because now a million very slow atoms are left with their wave function overlapping and all together form a single quantum ensemble. We have a macroscopic quantum entity in our hands. The Bose–Einstein condensate fulfils a dream that had been left unfulfilled for decades because of the lack of tools to realize it.[36] It is a most wonderful tool to study fundamental properties of matter passing from a microscopic quantum world to an almost macroscopic classical world with quantum properties.

How to make our measurements really reliable in our macroscopic world

When we fill up the fuel reservoir of our car at the pump, how are we sure that we get the volume that is written on the dials? When we buy a kilogram of sugar at the market, how are we sure that the scale the merchant is using is the same as his competitor's scale, and is accurate? When a technician in a manufacture orders parts from another manufacture situated thousands of km away, parts that are to be assembled with components ordered from another company in another country, how is he/she sure that everything will fit together correctly? When scientists from various laboratories report results of their measurements, how are they sure that the quantities they obtain relate uniquely to the quantities reported by their competing or collaborating laboratories? When a patient in a hospital receives radiotherapy how is he/she sure that the dose received is accurate, according to the prescription and not threatening his/her health?

These activities and many others rely on an international consensus agreed upon by all members: they use the same system of units of physical measurements called *Système International d'Unités*, SI. Standards are maintained in National Institutions that are used to guarantee the coherence of measurements within a country. The accuracy of national standards is insured either by comparison with standards maintained by the International Bureau of Weights and Measures (BIPM) in Sèvres, France, or by means of protocols on their implementation. The integrity of the world system of units is maintained by means of an international treaty to which most industrialised countries participate. In such a world it is then possible to trace back a measurement to a standard either maintained at the BIPM or agreed upon on paper. Each country is responsible for maintaining the integrity of its own system internally by verification of the measuring instruments used within its borders.

But what is a system of units? And how many quantities and units do we need to operate in our daily activities and to quantify the Universe we live in? We have been talking about many quantities such as mass, dimensions, light intensities, magnetic field, electric charge and so on. Are there some of these units more fundamental than others? Could some of the units be obtained by combination of other units? For example there is no need for a standard of area since the area is the product of two lengths. It is sufficient to have a single standard of length.

Decades ago, it was concluded that seven base quantities defined independently were sufficient to cover all measurements that are involved in our daily activities, commercial, industrial and scientific. These are mass, length, time, electric current, light intensity, temperature, and amount of substance. Instead of those we could have used others quantities, such as electric charge, voltage or power, for example, but it was found that the ones chosen were practical and covered a large spectrum of common usage. What remained, however, was the question of expressing those quantities by units and represent the units by practical standards that could be reproduced, distributed to partners and then easily compared to verify their integrity. The units agreed upon to represent the chosen quantities are: kilogram, metre, second, ampere, candela, kelvin, and mole.

These were then defined and implemented in practice as physical standards, representing the definition as well as possible. With those units as the base of the system it is possible to define derived units that are combination of those base units. For example the unit of volume is just the metre raised to the cubic power. Some derived units, however, have special names. For example force is expressed as Newton and corresponds to the combination, kilogram $\times$ metre/second^2; pressure being a force exerted on a given surface is expressed Newton/metre^2, called Pascal. There are nineteen of these derived units with special names.

But what should we use as standards representing the base units? This is where surprises are waiting for us. In the beginning, those standards were defined on the basis of the knowledge of the time. For a period, the mass of a cubic decimetre of water at 4 degrees Celsius was identified as a kilogram. However, for various reasons it was believed that a solid piece of metal would be more appropriate and practical, and the kilogram was defined as the mass of a cylinder of platinum containing 10% iridium, made as equal as possible to that of the mass of a cubic decimetre of water. That kilogram is kept in a vault at the BIPM in Sèvres, France. However, it was no longer easily available as a litre of water was, and consequently copies were made and distributed to countries for their use as their own primary standards. Such a copy was shown in Figure 3.4. On the other hand, the metre was defined as the fraction 1/10,000,000 of a terrestrial meridian. As is obvious from the definition, it was not readily available for day-to-day use. It was thus defined by a more practical artefact as the length of a platinum bar maintained also in the BIPM vault with copies distributed throughout the world. On the other hand, the second was simply accepted as the fraction 1/86,400 of the average day without formal definition.

The basic electrical unit, the ampere was defined as the current, which carried in two infinitely long wires, separated by a distance of one metre, produces a force 2×10^{-7} newton between those wires. The unit of temperature, the kelvin, was defined in terms of basic thermodynamic properties and setting on that scale the freezing point of water as 273.16 K. The light intensity standard was defined in terms of the human eye response and the amount of substance, the mole, was itself defined in terms of the so-called Avogadro constant well known from earlier work. The bureaucratic establishment behind the maintenance of such a system of units ensured a rather conservative approach to the management of those definitions and standards, and the system stayed unchanged for a long period.

However, there were no guarantees of the constancy of those standards, particularly those connected to the kg, the metre, and the second, while the ampere through its definition remained slippery, extremely difficult to implement and providing very little accuracy when implemented. In fact, it was soon found that the standards fabricated as artefacts tended to vary with time when copies were compared to the primary standards or to each other. It was also found that the Earth rotation was fluctuating and slowing down, due to energy lost, presumably caused by the effect of tides, and easily detected with stable quartz clocks. The length of the second was changing with time.

During the second part of the 20th century, developments in technology and scientific discoveries relating some physical phenomena to fundamental constants came to the rescue. It is now reasonable to believe that very soon all units of physical measurement will be defined in terms of fundamental physical constants, many of them related to phenomena discovered within the last few decades. The most surprising evolution is the use of quantum phenomena at the atomic scale level to implement standards of physical measurement at the macroscopic level. I do not wish to review this whole field, which could cover a whole volume, but let us mention some of those quantum phenomena that are now commonly used in standard laboratories to fulfil their task on maintaining integrity in the measurement system of their countries.[37] One of the first actions was to transfer the basis of the definition of the second from an astronomical unit to one using an atomic clock. This was made necessary because of the fluctuations of the period of rotation of the Earth on itself, as mentioned above. That period was found to change randomly by milliseconds per day and was a function of several other uncontrolled factors such as, for example, the motion of the poles. In the 1950s, the definition of the second was changed using as formal basis, the revolution of the Earth around the Sun. But that definition made its implementation impossible to most laboratories. It was soon found that the atomic clock based on the ground state resonant frequency of caesium was actually much more stable and more accessible than most astronomical phenomena. In 1967, the second was thus re-defined in terms of that frequency as we have mentioned above. We may mention that the hydrogen Maser described above could have been used for the definition, but the caesium clock appeared to be more promising regarding stability and accuracy. This was the start of the use of actual physical phenomena at the atomic level to implement standards of physical measurements. Soon after, realising the constancy of the speed of light and the practicality of using laser coherent radiation to make length measurements by interferometry, the metre was re-defined as the length of the path travelled by light in vacuum during a time interval of 1/299,792,458 of a second. The speed of light thus became a constant having an exact value, no longer measured experimentally. This effectively made the metre a secondary unit defined in terms of the second. It was then the turn of electrical standards to be implemented in practice by means of quantum mechanical phenomena such as the *Josephson Effect*, related to electron behaviour in a special superconducting structure, and quantum Hall effect, another phenomenon taking place in semiconductor materials at very low temperatures and large magnetic fields. The Josephson effect provides a standard for the unit of the volt in term of Planck's constant, the charge of the electron, and the second. Quantum Hall effect provides a standard for the unit of electrical resistance based on a different combination of the electron charge and Planck's constant.[38] Using the connection between voltage, resistance, and current, the ampere becomes essentially defined in terms of fundamental constants. It is then possible to design an instrument that measures the force between wires carrying a known current in terms of the kilogram in the Earth's gravitational field; this instrument is called an ampere balance, since it measures the force developed by a current in terms of the force developed by the gravitational field of the Earth on a mass of one kilogram. The whole measurement system SI becomes entangled around a single unit, the second, and the kilogram becomes merely a unit defined in terms of fundamental constants. In such an approach, other units become derived units, including the unit of light intensity that becomes simply a definition of power in terms of electrical units and the kelvin, which also becomes a measure of power connected to the random motion of electrons in a solid, with Boltzmann's constant acting as the proportional factor. The quantity of matter is already defined in terms of the Avogadro number, a fundamental constant. That constant has also been proposed as the basis for implementing the unit kilogram through the counting of atoms in a pure crystal of silicon.

It is readily observed that the SI, from its original implementation by means of artefacts, has evolved considerably in recent years. It is no longer as vulnerable to the loss of those artefacts as in the early days of its implementation. In fact, all the kilogram standards of the world could be destroyed with very little effect on the SI. The standard could be recovered with an accuracy of about one part in 10 million. The operation would simply consist of calibrating a small piece of material in terms of fundamental constants by means of a few special measurements routinely done in some of the National Institutes. The SI is now the system used throughout the world and is at the basis of the integrity of international exchanges. It is most interesting to realise that the whole international network of trade and commerce using weights and measure is now based essentially on phenomena of quantum origins through fundamental constants.

Of course, other systems of units have been developed over time. There is for example the CGS system that stands for Centimetre, Gram, Second, with very special units for electricity. It was used in science until the advent of the MKS, standing for Metre, Kilogram, Second, a predecessor of the SI and very similar to it. There is also the British system using pound, foot and second, which is still used in a few countries, but is convertible to the SI directly by means of exact definitions such as 1 inch (1/12 foot) = 2.54 cm, or 1 pound = 0.453,592,37 kg.

Natural units

The international system of units as described above is of course an invention of the human mind. The original definition of the metre as a part of an Earth meridian was quite arbitrary and has resulted in the measured speed of light in terms of that metre as 299,792,458 metres per second. If we had known more about physics at the time we would probably have used a different metre bar, which would have made the speed of light a round number and simplified calculation involving that constant. It is the same for mass. We chose water as the substance connecting a volume to mass: 1 cubic decimetre or a litre of water with a mass of 1 kilogram. A litre of water is a very manageable volume and a mass of 1 kg sounded attractive, but the choice was arbitrary.

In the case of electricity we chose the ampere as the base unit for current, which looked very practical at the time. In countries using voltages of the order of 100 volts as distributed by electrical companies, the ampere is a current going through a 100-watt light bulb. However, in the SI, its definition is quite elaborate: 'The ampere is the constant current which, if maintained in two straight parallel conductors of infinite length, of negligible circular cross section, and placed 1 metre apart in vacuum, would produce between these conductors a force equal to 2×10^{-7} Newton per metre of length'.

Just trying to remember it is difficult. Implementing it in practice is a real challenge. It is possible that if the properties of the electron had been well known another definition would have resulted. Consequently, we have a system whose units were implemented well before a deep knowledge of physics was acquired to the present level.

We have seen that in essentially all laws of physics there is a fundamental constant that shows up. It is characteristic of the law that describes a given phenomenon and its value depends on the system of units used. Then, why not use fundamental constants as units and set them equal to unity? They would become standards at the same time. The difficulty would be in their implementation as practical devices and in the performing of measurements. We could say that the unit of charge is the charge on the electron. We could use the mass of a given atom as the unit of mass. We could multiply the number of atoms by an arbitrary number to make that mass practical for day-to-day use. This would give us standards of mass and charge. How about time and length? We know that they are entangled through the speed of light. We think we know what length is. It is part of our three-dimensional space and we can visualise it. We have a certain control on it. But time is a different concept. Its control escapes us totally. We 'travel' through it and we cannot do anything about it. In our mind it is connected to a flow of something or a cycling of a physical phenomenon. We used to determine its flow from the rotation of the Earth on itself. We found that this was unstable. Now, to define the second, we used a cycling property of an atom, a resonance in the ground state of the Cs atom, because this was the best clock we could build experimentally at the time. It may have been better to use a similar resonance in hydrogen, the simplest atom. However, at the time the definition was introduced, we could not make an instrument using hydrogen with the same accuracy as the one that was implemented with Cs. In any case, the definition was made coherent with the previous definition in terms of astronomical phenomena and the number chosen for the resonance frequency of Cs was nothing simple. Since we live by the time of day as given by the Earth's rotation, the number chosen could not be arbitrary and had to be compatible with that requirement. We are living with a system of units whose implementation was driven by practicality of measurements on the human scale.

It is interesting to note that theorists often sidestep the concept of practicality in the system. Often, they set various physical constants, such as the speed of light, Planck's constant, the charge of the electron, equal to unity. However that has no effect on the final result. At the end of the calculation the constants have to be introduced in the results in order to have meaningful numbers in a given system of units.

The following question, however, may be asked: is there not a possibility of combining fundamental constants in such a way as to obtain units reflecting the fundamental properties of the Universe? The answer is yes, but the entities obtained in terms of our present units are totally impractical for day-to-day use. They are called Planck's Units or sometime Natural Units. The definition of these units is an interesting exercise in what is called *dimensional analysis*. The exercise couples the fundamental quantum constant ($h/2\pi$) to macroscopic constants such as the gravitational constant, G, the speed of light, c, the permittivity of free space, ε_o, and Boltzmann's constant, k. We note that these definitions do not involve entities such as the electron mass and charge or the proton properties. They involved only so-called 'vacuum constants'. The expressions for such units are given in Annex B. Their values in term of SI quantities are:

Length: $l_p = 1.62 \times 10^{-35}$ metre.

Time: $t_p = 5.39 \times 10^{-44}$ second.

Mass: $m_p = 2.18 \times 10^{-8}$ kilogram.

Electric charge: $q_p = 1.87 \times 10^{-18}$ coulomb.

Temperature: $T_p = 1.42 \times 10^{32}$ kelvin.

As is readily seen, these units are totally uncorrelated to our normal activities. Their size is out of proportion with the dimensions we deal with every day. They are either extremely small or extremely large. They all involve the reduced quantum fundamental constant, $h/2\pi$. In quantum physics, Planck constant h introduces granularity of radiation energy or the photon. The numbers do not appear to be related to anything we have described up to now. However, through the use of Planck's constant they appear to set a frontier between the quantum world and the macroscopic world that we live in. They are more fundamental then it appears at first sight. In fact, we will see in a later chapter that those natural units of length, time and temperature seem to play a fundamental role in the evolution of the Universe in its original phase. We will use them and will show that they are quite appropriate numbers to describe the Universe state at its beginning, at the time of the so-called Big Bang.

VII

Elementary Particles and the Atomic Nucleus

We have been talking about the electron and described its characteristics particularly when it is within an atom attached to a nucleus. We have seen that it is responsible for the presence of light that makes us see objects around us as our eyes are sensitive to a range of wavelengths. The electron is responsible for the existence of chemical compounds that are important in our lives. The electron is responsible for the electricity that is at the basis of operation of our modern society. As living beings our activities rely on a nervous system that uses the flow of electrons for transmitting signals from the brain to the cells of our body and activate some reaction. As an ensemble of chemicals, our body's functions are ruled by the electron. It is also at the heart of operation of all the devices that have improved our well being, in day-to-day living, in communication, in transportation, in health care. The electron is what makes the world the way we see it, from objects under the microscope to the galaxies millions of light years away.

We said that the electron appears as a small, shy particle that does not like to be observed and likes to attach itself to a nucleus. In hydrogen, in the lowest energy state, it appears in its motion to envelop totally the proton as a three-dimensional crown. It has the greatest probability of being found on a spherical annulus at an average distance of about 5×10^{-11} m from the proton, which itself is extremely small, of the order of 10^{-15} m in diameter. The relative dimension of these particles is better realised when the whole structure of the atom is magnified and imagined as a hollow geodesic dome 100 metres in diameter. The nucleus then looks like a fly a few mm in diameter in the middle of the sphere with the electron, looking like a small speck, a dust particle, not too well localised, wandering and jumping around on the surface of that dome. The space in between appears to be totally empty. Is it really? Or is that space a kind of substance whose nature is not accessible to us? Strangely enough, that vacuum appears to be a non-substance characterized by well-defined properties. It would appear to be more logical to talk about the properties of something, not of a non-existing substance. However, we have to accept that it is something that we call 'vacuum', characterized by constants introduced earlier: speed of light c, dielectric constant ε_o, magnetic permeability μ_o, gravitational constant G, Plank's constant h and Boltzmann constant k.

We know that the electron acts in some circumstances as a particle. It makes a definite click when detected in a photodetector. A single electron can be stored in vacuum in a trap made from a combination of electric and magnetic fields (called a Penning trap) and studied for long periods of time.[1] The particle appears to be extremely small: we have not been able to measure its radius down to the 10^{-19} level. However, we have given it a dimension through the equivalence of the energy associated with its mass to the electromagnetic energy content of a charged particle of the same radius. This dimension, called the classical radius of the electron, is of the order of 2.8×10^{-15} m. Its electric charge e is 1.602×10^{-19} coulomb and its mass, m_e, is 9.109×10^{-31} kg. We are allowed to ask ourselves the following question: if the electron had real dimensions what would be the nature of the substance forming it? What would be the support of its electric charge, of its mass? It seems to have no internal components and to be truly elementary. However, it also acts like a wave since it causes interferences. It is a particle or a wave depending on the circumstances of its observation, a personality proper to the atomic world embedded in the concept of complementarity developed by Bohr. The electron is described by both a wave and a particle and we have to live with that.

On the other hand, what is that small nucleus to which the electron likes to attach itself and that serves somewhat as an anchor? The nucleus has dimensions as shown by scattering experiments with electrically charged particles. Its measured radius varies from about 1.5×10^{-15} m for the hydrogen nucleus to about 15×10^{-15} m for the heaviest elements We have mentioned earlier that the nucleus is made of two types of particles, neutral particles called neutrons and positive particles called protons, and that they contribute to the differentiation of elements and isotopes among themselves. These particles are called *nucleons*. Their identification is a great step in understanding the nature of the nucleus, but it is just a step. Are they elementary particles? If not, what are they made of? Let us recall their main properties.

The proton

The existence of the proton was inferred early in the 20th century following scattering experiments made by Rutherford. It is found to be positively charged (+e) and assumed to be the core of the hydrogen atom. It was the particle giving mass to the hydrogen atom. Its measured mass is 1.672×10^{-27} kg, thus 1836 times heavier than the electron. From the results of scattering experiments it was found to have a dimension like that given above for the hydrogen nucleus. The proton acts as if it was spinning having a spin angular momentum of $1/2\,\hbar$ like the electron. In a magnetic field it acts as a small magnet and has a magnetic moment measured as 2.79 nuclear magneton.[2] Hydrogen is a stable non-radioactive element and the proton is a stable particle.

The neutron

The neutron is the other constituent of the nucleus. How was it found experimentally? It goes back to the work of Joliot, Curie, Rutherford and Chadwick in the early part of the 20th century. It was found that when beryllium is bombarded with alpha particles some mysterious invisible neutral radiation is emitted. The first conclusion was that the radiation was X rays. However, when that radiation was allowed to enter a sample of paraffin it was found that protons, the nuclei of hydrogen atoms that are one of the chemical constituents of paraffin, were ejected. It was not possible that the assumed X rays, with their limited energy, could eject protons from the paraffin. The conclusion was that such a phenomenon was the effect of energetic particles colliding with the hydrogen nuclei within the paraffin and that those particles were neutrons, assumed earlier to exist in order to explain the atomic mass of elements. They were indeed the invisible particles looked for. The neutron has no electric charge. Its measured mass is 1.675×10^{-27} kg and its dimensions are the same as that of the proton. Like the proton, it has a spin angular momentum of $1/2\,\hbar$ and has a magnetic moment of 1.91 nuclear magneton. As mentioned previously, the neutron, when free, is unstable. It transforms itself into a proton, an electron, and an antineutrino in about 15 minutes.

It has been found convenient to normalize nucleons masses to a reference. Carbon was chosen with its 12 nucleons and 6 electrons as a reference and is said to have a mass of 12 'atomic mass units' (amu). On that scale the proton mass is then 1.0073 amu while the neutron mass is 1.0087 amu. Another convenient measure is by means of the mass-energy equivalence relation $E = mc^2$ expressed in electron-volts. The proton mass corresponds then to 938.3 MeV and the neutron mass corresponds to 939.6 MeV. On that scale the electron mass is 0.511 MeV. The small difference in mass between the proton and the neutron leads to a neutron behaviour that is fundamental in the formation of elements.

Are protons and neutrons elementary or do they contains parts? Why are they sticking to each other to form nuclei? As we will see from their behaviour they appear to be complex structures or have parts. On the other hand, it appears that a strong force exists, holding them together. Following the general lines of our introduction on the subject in Chapter II, we will see that physicists have arrived at a rather stunning model for the nucleus itself and the internal components of those nucleons, a model that provides a very satisfying explanation for the phenomena observed experimentally.

Radioactivity: A Source of Information

When we hear the word nucleus, we generally make an association with nuclear energy, power plants, the atomic bomb, Hiroshima, the Cold War. Early studies of the nucleus properties led to the field of nuclear physics, concerned with the behaviour of the nucleus, the understanding of its properties and the use of those properties in diverse applications, particularly, to be on the positive side, in the production of cheap energy. During the early part of the 20th century extensive work was done in connection with that aspect of the nucleus and models were made on its overall nature and bulk properties while, in the last part of the century, interest was concentrated mainly on the constituents themselves, the nucleons, and their internal parts.

This field was started at the end of the 19th century and beginning of the 20th, when a tremendous amount of work was done on the nature of the radiation emitted by some of the heavy elements found in nature (Curie, Joliot, Becquerel, Rutherford to name a few). Nuclear radiation appeared to be emitted spontaneously without a need for an external agent acting on the substance and was given the name *radioactivity*. In other words, some elements were found to be active, emitting radiation by themselves, without being induced to do so by external means. After a large number of studies, analysis and reflection, it appeared that nuclei emitted three different kinds of radiation. They were simply called alpha (α), beta (β) and gamma (γ) radiation. These radiations were found to carry a rather large amount of energy and could not originate from the electronic cloud enveloping the nucleus. For example, jumps from internal electronic orbits in heavy atoms to orbits close to the nucleus can create X rays of relatively large energy, that is to say hundreds of eV. However the kind of energy we are talking about when we refer to γ rays is in the thousands and millions of eV. The origin of those rays had to be in the nucleus. On the one hand, the alpha radiation is actually a particle and was found to have two positive electronic charges with a mass equal to that of four hydrogen atoms or amu. It was found to be identical to the nucleus of the helium atom. It is thus made of two protons and two neutrons. This is the particle that Rutherford used to study the internal structure of atoms by directing such particles on a gold film and studying their deflection. On the other hand, the beta radiation was also found to be a particle. It has all the properties of an electron. It was concluded that Beta radiation is an electron emitted by the nucleus, a rather surprising phenomenon, since we said that the electron's normal place is in a space around the nucleus. The gamma radiation

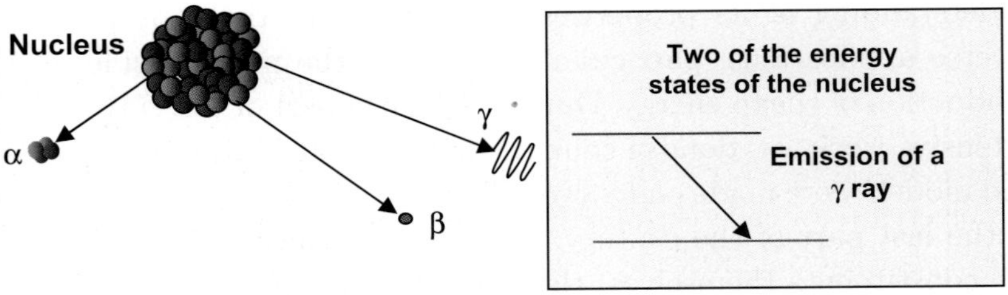

Figure 7.1. A heavy nucleus can emit alpha particles (helium nuclei), beta particles (electrons) and gamma radiation (energetic photons). In emitting alpha and beta particles the nucleus transforms itself into another element, while in emitting a photon of high energy, gamma radiation, it falls from and excited state to a lower energy state.

was found to be electromagnetic radiation of very short wavelength or high frequency, through the relation $E = h\nu$, thus photons of very high energy. The three types of radioactivity are illustrated in Figure 7.1

These findings showed that the nucleus itself was a complex structure and was the seat of energetic interactions. The emission of gamma radiation, that is photons, showed that the nucleus can exist in an excited state with high energy and can make transitions to a state of lower energy. This is very much like the emission of a photon by an electron when it falls from an excited atomic state to a lower state. This situation can be realised only in the case of a complex structure. The emission of alpha particles showed that the nucleus was indeed an object made of such particles as those constituting the alpha particle itself. The emission of such a particle as a whole was a sign of its own cohesion and stability, even within the nucleus. The emission of an electron, on the other hand, showed that one of the constituents was transforming itself through some internal unknown mechanism. The emission of an alpha particle results in a drastic change of the nucleus losing two electric charges and becoming another element. Similarly, the emission of an electron by the nucleus means, through the principle of conservation of electric charge, the appearance of one positive charge within the nucleus: the element is transformed into another element. In both cases the transformation fulfils the dream of the alchemist of the possibility of transforming some elements in gold.

Nuclear Forces

Let us now look at the elements existing in the Universe with an eye on the nucleus. The first one is hydrogen. It is difficult to have something simpler. Both particles forming it, the proton and the electron, are very stable. They interact through electromagnetic forces and hydrogen is a stable element in nature. The next nucleus that exists is the deuteron, formed of a proton and a neutron, slightly more complex than hydrogen. It is the heavy isotope of hydrogen that is found to exist in sea water in minute quantities ($\sim 0.0015\%$). There is no electrical repulsion between the two particles since the neutron has no electric charge and since they stick to each other, it appears that a very strong short-range force holds the structure together. It is the nuclear force, an internal property of all nucleons, which acts independently of electric charges. It also acts only at very short range, when those particles are less than a distance of the order of 10^{-15} m. On the other hand, there is no such stable nuclear structure as one containing two protons alone. Why not? It turns out that the Coulomb repulsion of the two protons at such close range is so large that those protons, having the same positive electric charge, would simply fly apart. However, if neutrons are added, they act as a kind of catalyst through the strong nuclear force and stabilise the structure. Put simply, the attraction of the neutrons to the protons through the nuclear forces is so large that it compensates for the Coulomb repulsion of the protons and stabilises the whole structure. The helium nucleus, formed of two protons and two neutrons, is stable. We have shown earlier a simple way to represent such a structure by the chemical symbol of the element with indices attached to it. Let us recall that notation: ${}_2He^4$. The symbol 'He' identifies the chemical element, in this case helium, whereas the superscript 4, determines the number of nucleons or the atomic mass A. The subscript 2 determines the charge of the nucleus. It is the so-called Z-number (Z = number of protons) that characterises the element in question in Mendeleev's table of elements. Interestingly enough, a structure of 2 protons and 1 neutron is also stable. It is called ${}_2He^3$. It is chemically inert as is ${}_2He^4$, but has slightly different physical properties. Is a structure formed of one proton and two neutrons possible? Yes, it is an isotope of hydrogen, called tritium, ${}_1H^3$.

However it is unstable. It exists in nature only in traces produced by cosmic rays interacting with the atmosphere. It may be produced artificially in nuclear reactors. It is radioactive with a half life of 12.3 years, emitting a β article and thus becoming another element, ${}_2He^3$. This means that a certain quantity of tritium just created would be reduced to 1/2 of its original quantity after 12.3 years by the emission of beta particles.

The table of elements is built of atoms with nuclei having an increasing number of protons and neutrons. The lower weight natural elements are found with nuclei formed of a number of neutrons about equal to or somewhat larger than the number of protons. However, in the case of heavier elements, the number of neutrons may exceed the number of protons by 50%. Some of those elements are stable and others are radioactive showing, in part, a strong sensitivity to the equilibrium between Coulomb repulsion of the protons and the nuclear forces.

What is that nuclear force? We have seen earlier that the Coulomb force between charged particles could be interpreted as an exchange of photons. This was a conclusion from the very successful theory of quantum electrodynamics. Could a similar approach be used in the case of nuclear forces? As we have seen, quantum field theory opens the door to such an interpretation. This is the avenue that was taken by Yukawa in the early 1930s. His hypothesis was that a particle must be the mediator of the strong nuclear force as the photon is for the electromagnetic force. The particle responsible for the force was called a *meson*. However, since the nuclear force has an extremely short range, existing only within the nucleus (10^{-15} m), the meson should not live for a long time in free space. Consequently, according to the uncertainty principle it should carry a lot of energy. In order to fulfil that condition it was concluded that the meson would have mass preventing it from travelling at the speed of light. A calculation made by Yukawa showed that the mass of the particle should be about 140 MeV or 200 m_e.[3] A particle looking very much like the one predicted was soon found in the study of cosmic rays by Anderson and Neddermeyer in 1936 and Yukawa's work was recognized for such a clever prediction. Unfortunately, the particle found was not the right one. Having approximately the right mass, it was believed to be the particle predicted by Yukawa, but it was not interacting with nuclear particles through the strong force. It was a new kind of particle that had not been observed yet. It was actually a heavy electron and was later given the name *muon*. The particle predicted by Yukawa was found later at high altitudes (Powell, Lattes, Occhialini, 1947) and was also produced in accelerators around that time. It is called a *Pion*. It is interesting to note that Yukawa was recognized first for a correct prediction of a massive particle, but on a wrong identification. He and Powell were recognized later with the Nobel Prize.

That kind of interpretation of the nuclear force thus provides a model based on previous experience relative to electromagnetic forces. A diagram similar to the one made for Coulomb interaction in the quantum electrodynamics context can be made. Nucleons are then said to exchange a Pion identified by the symbol π that mediates the force of attraction between them. Pions are spinless. Three types exist, electrically charged, positive and negative and neutral.

A Word about Nucleus Stability and the Neutrino

A most interesting property of the neutron is that when it is free, not attached to a nucleus, its lifetime is about 15 minutes, as mentioned previously. What happens to it? In its own frame of reference, it appears to eject an electron and transforms itself into a proton. However, as mentioned, it is found that the energy and the momentum of the two particles produced are not equal to the energy and momentum content of the original neutron, in violation of the conservation laws that we have looked at in an earlier chapter. After disintegration of the neutron, the sum of the energy of the resulting proton and electron, taking into account the equivalence between mass and energy ($E = mc^2$) and the kinetic energy of the constituents, does not correspond to the original energy of the neutron. Furthermore, the sum of the electron and neutron momentum is not the same as the momentum of the neutron before disintegration. This observation can be explained by the existence of another neutral particle, with the required momentum and energy, created at the same time as the disintegration of the neutron.

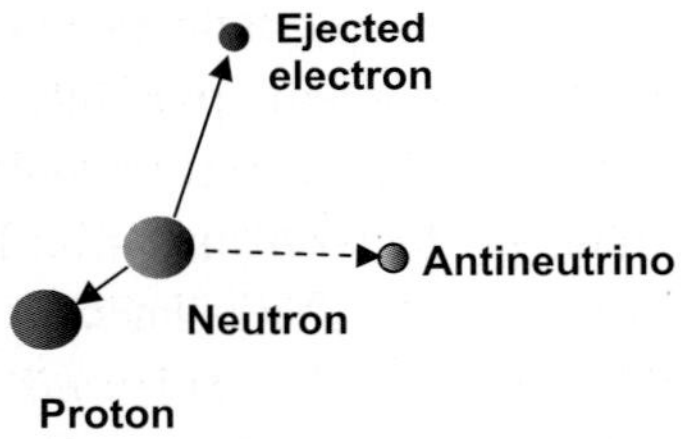

Figure 7.2. Disintegration of a neutron. The residues are a proton, an electron and an antineutrino whose existence is required on the basis of conservation of energy and momentum.

It was imagined by Pauli in the early 1930s. It was said to be essentially undetectable since it had no mass, no charge and did not interact through the electromagnetic and strong forces. That particle was later given the name *neutrino* by Fermi who developed a theory describing the phenomenon. The process of emission may be represented somewhat naively as in Figure 7.2.

It is assumed that the neutrino having negligible mass or no mass at all travels essentially at the speed of light. It was detected in a reactor where large quantities were produced and made to interact with protons, producing neutrons and positrons (Cowan, Reines, 1956; Nobel Prize, 1995). Thus, the process of β particles emission by a nucleus is the transformation of a neutron into a proton, with the emission of an electron and an antineutrino.

But why does the neutron disintegrate when free? As mentioned previously, the mass of the neutron is slightly larger than that of the proton. From the properties of those particles given above, we observe that the extra mass of the neutron relative to the proton corresponds to an energy difference of 1.3 MeV. As we mentioned, nature always attempts to find a state of lowest energy. Apples fall to the ground, a state of energy lower than when the apple is up in the air. If there is a path available that can reduce the energy of a system, nature will search for that path and use it. Consequently, if the neutron can find a way to transform itself into a proton, which has a lower mass, thus less energy, there is a probability it will do so: it does it by emitting an electron and an antineutrino. Why does it choose that path? There are certainly many other paths, but it appears to be the path of least effort. We will see later, in view of the actual complex nature of protons and neutrons, that such a path makes a lot of sense. The neutron ends up as a proton with lower mass, thus lower energy, and two ejected particles of rather low energy content. This is an internal transformation and it shows us that the neutron is not an elementary particle: if it can go through such a complicated process it must have internal components.

Why do we say that beta decay results in an antineutrino rather than a neutrino? This choice is forced by the observation that in nature baryon and lepton numbers are conserved as inferred from gauge symmetry that we introduced earlier. The conservation rule is based on the assumption that baryons and leptons have respectively positive baryon and lepton numbers while their antiparticles have negative baryon and lepton numbers. Beta decay ends up with a proton thus maintaining the same baryon number of unity. However, there were no leptons to start with and we end up with two leptons. The way to solve that problem is to conclude that the mysterious particle required for conserving energy and momentum is an antineutrino with a lepton number -1. The electron having a lepton number $+1$, the total lepton number stays zero, and the conservation law is obeyed. Furthermore the electric charge is conserved, a proton and an electron of opposite charges being created.

On the other hand, atoms with heavy nuclei exist and a great number of them appear to be rather stable. They do not disintegrate. Since the neutron disintegrates when free, then why is it stable in most nuclei composed of many nucleons? To answer that question, let us look, for example, at the deuterium nucleus. The neutron attaches itself to the proton and is glued to it by the strong nuclear force mediated by the exchange of Pions. It is readily seen that if the neutron transforms itself through the process explained above, it leaves behind a proton in very close proximity of another proton, a situation with higher energy because of the strong Coulomb interaction between the two protons. The neutron thus becomes stable in the deuterium configuration because the whole assembly has lower energy than it would have if the neutron transformed itself into a proton by the emission of an electron. Alternatively, is it possible that two neutrons group together to form a single entity? It appears that such a system is not stable. The extra interaction, resulting from the sticking of a neutron to another one, corresponds to an additional energy that is larger than the energy that would result by sticking to a proton instead. As can be seen, a very delicate equilibrium exists in the selection of combinations just because of the difference in mass between the proton and the neutron and the interplay of nuclear and Coulomb forces. The stability is created by the difference between the reduction of energy by neutron decay and the increase of energy caused by electromagnetic interactions when the proton is added as a result of that decay. That interplay is the origin of the radioactivity of certain elements and their isotopes.

This equilibrium exists at various degrees with heavier atoms. In general it is found that there is great stability when the number of neutrons is approximately equal to the number of protons or somewhat larger. A model was developed for the behaviour of the neutrons and protons within the nucleus, using concepts borrowed from the quantum theory of the electron in interaction with the nucleus. It is called the *shell* model and provides some insight into the question of stability of some specific nuclei (Wigner, Goeppert-Mayer, Jensen, 1949). In the model, a potential well created by the strong nuclear forces is assumed to exist in which the nucleons evolve in discrete levels or shells, obeying quantum mechanical rules and Pauli's exclusion principle in a similar way as the electron in an atom does. It is found that certain 'magic' numbers for the assembly of nucleons, as in the case of electrons in orbits, are privileged. Those with nucleon numbers such as $2, 8, 20, 28, \ldots$ appear to have stronger binding energy than their neighbours. This rule can be applied to the individual number of protons and neutrons and the number can be doubly magic through a combination of those numbers. Nuclei of atoms such as He^4, and O^{16}, both doubly magic, $2p + 2n$ and $8p + 8n$, are examples of such a rule and are stable elements. At the other end of the table of elements, heavy elements tend to be unstable. The best-known radioactive elements are probably the heavy atoms, uranium ($Z = 92$), and radium ($Z = 88$), a decay product of uranium, both having numerous radioactive isotopes. These atoms have played a crucial role in the development of nuclear physics from the discovery of radioactivity by Curie, Joliot and Becquerel at the end of the 19th century to the implementation of the first nuclear reactor by Fermi in the late 1930s.

A Global Model of the Nucleus, Fission and Fusion

In the thirties, a model proposed by Gamow was developed by Bohr and Wheeler in which the nucleons, due to their close packing and their fast motion within, were forming a kind of amorphous ensemble held together by the strong force. The nuclear force is very short range and exists essentially only within a radius of the size of the nucleus itself. Those nucleons near the centre of the structure experience interactions from all their neighbours while those at the surface experience a cohesive force towards the inside of the ensemble. This is very similar to a drop of water whose atoms stick together and are submitted to surface tension: the model was called the liquid-drop model. This droplet may be thought of also as a potential well, trapping the nucleons. The emission of α particles could then be interpreted as a quantum mechanical escape from the nucleus or a quantum tunnelling through the barrier formed by its surface (Gamow, 1928).

What happens then when that nucleus is bombarded with other particles? First can we get some of those particles inside the ensemble, inside the nuclear droplet? The process is difficult with protons, due to the strong Coulomb repulsion when a proton gets very close to the nucleus. The process is much easier with a neutron. Since it is neutral, a neutron can penetrate the nucleus and distribute its energy to the ensemble. This process may set the nucleus in vibration, in resonance, create instability, and even split it in smaller pieces: this is the process of *fission* (Hahn-Strassmann 1938) with the release of an enormous amount of energy. Let us look at that phenomenon in more detail.

When a nucleon is added to a nucleus, some energy is released. This is the same thing as when an object falls to the surface of the Earth from a given height: energy is released. If we want to remove the object from the surface of the Earth and place it at a certain height we have to do some work and it is that work that is stored in the object and released when it is set free to fall back on the ground. It is the same kind of process that takes place with the nucleus. If we wish to take away a nucleon from a given nucleus we need to do work against the attraction due to nuclear forces. The opposition to that action is called the *binding nuclear energy*, looking very much like the process involved in gravitational energy or still the binding chemical energy of atoms in chemical compounds. The energy, however, in chemical compounds is due to electromagnetic forces and is of the order of electron-volts, while in the case of nuclear forces we are talking in terms of millions of electron-volts. The difference is enormous.

The nuclear binding energy may also be expressed in terms of mass through the Einstein relation $m = E/c^2$. This means that we will find the mass of a given nucleus to be different from the simple addition of the mass of the constituents, neutrons and protons. We recall that A is the measured mass in so-called atomic mass units, amu. The nucleus of the carbon atom, with its 6 protons and 6 neutrons, is set as a reference equal to 12 atomic mass units (amu). One amu is equal to 1.660×10^{-27} kg. If we calculate the difference between the measured mass value for a given isotope nucleus and its number of nucleons, and divide it by the number of nucleons, we then have the extra mass (positive or negative) per nucleon. This can be done for all elements and what is found is most interesting. It is found that there is a minimum of mass per nucleons for elements with about $A \sim 100 - 150$. For A larger or smaller, the mass per nucleon increases. Thus, a nucleus with a number of nucleons of 200 has more mass than two nuclei of mass 100. This is illustrated qualitatively by a graph as shown in Figure 7.3.

An important consequence follows. If a heavy nucleus can be split into smaller constituents a certain amount of mass disappears. It is energy and since energy is conserved it appears as radiation, other particles and kinetic energy (heat) of the smaller constituents. This is the essence of nuclear fission. If a neutron is sent into an uranium 235 nucleus, that nucleus, which is not very stable to start with, is profoundly perturbed and splits into two parts, for example radioactive isotopes of barium and krypton, much lighter elements, with the release of three neutrons. In their turn the isotopes of Ba and Kr spit out beta particles to become stable elements. A rough

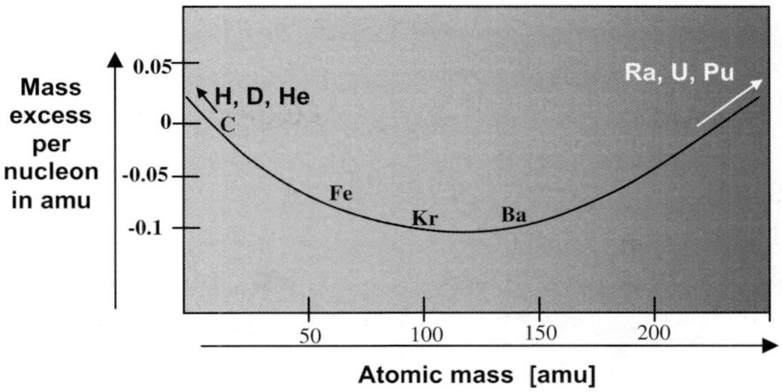

Figure 7.3. Illustration of the fractional mass excess of various atoms over the sum of the masses of the individual particles. Elements with atomic mass A around 100 have the lowest mass excess per nucleon. We note that carbon is defined as having 12 nucleons with A exactly equal to 12. It is a reference for mass value A of other elements.

calculation, using the numbers shown in Figure 7.3, shows that the fission of an uranium nucleus results in less mass by a quantity of about 0.2 amu. The equivalent energy released is obtained by simply transforming that mass by means of the relation $E = mc^2$. It corresponds to a total energy released of the order of 200 MeV.

We see immediately the enormous amount of energy available in a single fission as compared to that involved in a chemical reaction, which involves energy of the order of a few eV. If the reaction can be maintained on a continuous basis it becomes a source of tremendous energy. It can be done. The extra neutrons that are released can, in turn, cause the splitting of surrounding uranium atoms: this is called a chain reaction and provides the heart of the nuclear reactor used to produce electricity. And if the whole process is let go the number of neutrons produced triples at each splitting and an explosive chain reaction results. The result is the so-called nuclear fission bomb, commonly called the atomic bomb. The processes are illustrated in Figure 7.4.

From Figure 7.3, it is also seen that a symmetrical process can be imagined, but this time with the smaller elements. For example, if hydrogen, deuterium or helium can be fused to create heavier elements, extra mass is freed and transformed into energy. The difference of mass between the original particles and the resulting one is much larger than in the case of the fission of heavier elements.

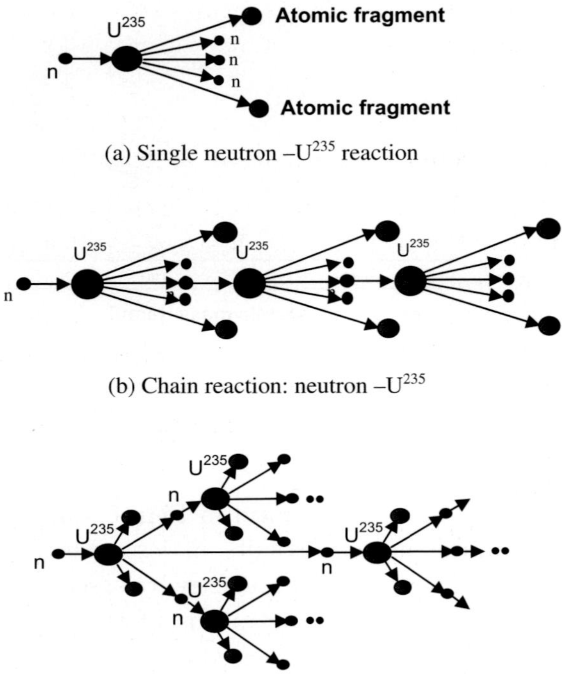

(a) Single neutron $-U^{235}$ reaction

(b) Chain reaction: neutron $-U^{235}$

(c) Explosive reaction: neutron $-U^{235}$

Figure 7.4. Possible results of the reaction of a neutron with a uranium nucleus.

This is the basis of the claim of a much larger possible energy production in fusion reactors than in fission reactors. The higher efficiency of fusion over fission for the same mass of material has been demonstrated in explosive thermonuclear devices. The fusion requires the violent collision of the lower mass nuclei. This means very high energy collisions. This can be done by heating the sample to a very high temperature. In thermonuclear bombs, the required high temperature is obtained by means of an explosive fission reaction. This is a one-shot result. Up to now, the difficulty in obtaining fusion on a continuous controlled basis as in a fission reactor has been the difficulty in reaching the required high temperature in a constrained environment. As we will see later, stars are enormous reactors supporting nuclear fusion in their core, fusing neutrons and protons to form deuterons and helium nuclei. It is that fusion process, transforming mass into energy that heats them up and that gives them their tremendous luminous intensity allowing them to shine for billions of years.

Elementary Particles: A Persistent Illusion

There was a time when the atom was believed to be the indivisible corpuscle of matter that gave characteristic properties to a particular element and thus was considered to be an elementary particle. Then it was found to have internal constituents, a nucleus and surrounding electrons. It was also found that the nucleus was constituted of internal components with mass and called protons and neutrons. They were thought to be indivisible entities with the proton having a charge equal to that of the electron but positive and the neutron without electric charge, adding mass and acting as a catalyst for the strong force. Although the mass of those particles was about 1,800 times that of the electron they were found to occupy a very small volume as demonstrated by means of various scattering experiments. The proton and the neutron were common to all known elements and gave mass to matter in the Universe. In that Universe, those particles appeared to be the ultimate step in reduction thus a limit to the divisibility of matter. After all, it is natural to think that there must be a limit somewhere. However, it was an illusion.

A Zoo of Elementary Particles

In the early 1930s the number of so-called elementary particles known to exist was very limited: it consisted of the proton, the neutron, the electron and the ghost neutrino 'invented' to explain beta decay. Physics was living well with that simplicity and physicists were relatively happy to have to deal with that limited number of constituents. However, at about that time, studies were initiated on the nature of radiation coming from the sky, so-called cosmic rays. That radiation was observed either at ground level or, to avoid atmospheric effects, at the top of mountains and with the help of balloons sent to high altitudes. Particles were detected by various means such as photographic emulsions sensitive to the traversal of electrically charged particles or chambers containing a saturated vapour where charged particles left a trace of condensation. The sign of the charge, their mass and their lifetime could be determined from studies of their path in magnetic fields applied within the instrument, fields that altered their path in a direction that was function of their mass and the size and sign of their charge. One of the first particles to be discovered was negative, with a mass equal to $\sim 200\,m_e$ (Anderson, Neddermeyer, 1936). It was mentioned above and is called the *muon*. This led a physicist like I.I. Rabi to question the reason for the existence of such an entity that caused confusion in the peaceful landscape of particle physics known at that time. The muon was thought first to be a meson. Later it was properly identified as a heavy electron, a lepton.

In the following decades a rather large number of new particles were discovered either in the laboratory or in cosmic rays. The advent of particle accelerators also provided a means of getting data in laboratory-controlled experiments. For example, protons or deuterons could be accelerated to very high speeds and made to collide with target particles. In those experiments the target is demolished and particles emerge with various masses, charges and energy. Mass can be transformed into radiation and vice versa. The number of 'elementary' particles observed grew at a rate that was not expected and rapidly constituted a zoo of more than 100 entities. It did not make sense that so many elementary components were needed to constitute our Universe. It caused uneasiness. It looked like the situation before Mendeleev proposed his table: it could not be that all those atoms identified as the smallest part of a given element were elementary particles. Could the same approach be used, could those particles be organised in a table of some kind in an attempt to discover some common properties that would provide insight into their internal constituents, if any? The approach used by Mendeleev with the elements had been very successful. In that case, some common physical and chemical properties were found among the elements and Mendeleev had the intuition that if they were put in a kind of arrangement, a table, symmetry would become evident. And so it was. By listing them successively by atomic weight and taking into consideration similar chemical properties, the elements appeared to group among themselves somewhat symmetrically. But one of the greatest finding in doing that exercise was the existence of vacancies in the table, as if other elements should have existed in nature and had not yet been found. Symmetry provided a mean of extrapolating and predicting new results. It was not long before the existence of the predicted elements was confirmed by experimental observations. Nowadays the table of elements appears to be complete and we can describe in great detail the nature of all elements. However, an important lesson could be learned from the early results. If symmetry existed among the elements and if repetitions were observed, it appeared that there was something else common to all elements that provided the basis for the symmetry and the repetition. The elements had to be made of internal parts. It was inferred that the same internal constituents were common to all and were forcing symmetry due to some unknown rules. We now know that these common constituents are the electron and the proton: going in the table from one box to the next consists of adding one proton and one electron, obeying the exclusion principle. What it shows forcefully is that the atoms of an element are constituted of particles that appear to be 'more elementary' than the atoms themselves.

True Elementary Particles

Can a similar approach be used with the so-called 'elementary' particles forming the zoo observed in the study of cosmic rays and of the by-products of particle collisions in the laboratory? What do we have as information that can provide some insight into finding symmetries and correlations among those particles? First, it is found that some particles have properties that associate them to the strong interactions and some have a nature that makes them look more like being connected to the electronic properties. Then, there are those that are the mediators of the forces. The particles are thus divided first in three large classes: (1) *hadrons* that include mesons and baryons introduced earlier, that in turn include the protons and neutrons, particles being subjects to the strong force; (2) the *leptons*, particles that are outside the nucleus and not subject to the strong force; (3) the *force mediators*, those particles that carry the force between particles. The names chosen by physicists for those particles have generally Greek origins, with a meaning somewhat related to their main property. For example, baryon is from a Greek word meaning *heavy.* The word hadron means *thick* while the word lepton means *thin.* Meson on the other hand means *intermediate.* Remembering them all requires some effort. We have to live with it.

Leptons

The lepton group includes of course the electron. It includes also the muon and the neutrino. In beta decay we have said that an electron and a neutrino were produced. Are they not particles existing inside the neutron to start with since they are its siblings? It turns out that they are the product of a transformation and do not exist as such within the neutron before the decay takes place. We will have a more coherent picture of the process later. We mention that the muon was a by-product of the reaction of cosmic rays with nuclei of some molecules of the atmosphere at high altitudes. It appears to be an elementary particle, but is unstable. Its observation at the surface of the Earth originating from cosmic rays requires the effect of a relativistic time dilation because its lifetime would not allow such a long travel from the upper atmosphere at non relativistic speeds. It decays in a few microseconds transforming into an electron and two neutrinos. In the period of 1974–77 another lepton similar to the electron, called "tau", was found in the by-products of collisions between very energetic electrons and positrons in particles accelerators. It has a negative charge $-e$ similar to the electron, but has a very large mass, about 3,500 times the mass of the electron, or 1,777 MeV. It exists only for a very short time, of the order of 10^{-13} s. Of course, these three particles, electron, muon, and tau, have their own antiparticles and also have their associated neutrinos and antineutrinos. This is best represented in an arrangement such as that shown in Table 7.1.

Leptons do not react to the strong nuclear forces that we have introduced above. The electron, muon, and tau leptons do respond to the weaker electromagnetic force through their electric charge. They

Table 7.1. Listing of some properties of the known leptons. Leptons do not react to the strong force.

Leptons				
Flavour Electric charge = −1e Spin = 1/2	Lifetime	Mass (In terms of m_e)	Mass (MeV)	Corresponding Neutrino (charge = 0, Spin = 1/2) (very small mass)
Electron	Stable	1	0.511	Electron neutrino
Muon	2.2×10^{-6} s	$m_\mu = 207\, m_e$	106	Muon neutrino
Tau	2.9×10^{-13} s	$m_\tau = 3{,}500\, m_e$	1,780	Tau neutrino

also respond to the so-called 'weak interaction' that takes place in beta decay, for example, and that we will again examine below.

As we have mentioned, the energy emitted by stars originates from a fusion reaction that takes place in their core and an enormous amount of neutrinos are produced. There are many other sources of neutrinos such as in the case of the interaction of cosmic rays with the nuclei of the atmosphere molecules. However, the main source of those reaching the Earth's surface is the Sun. It is evaluated that 2×10^{38} electron neutrinos are produced per second in the Sun's core. Of those, 6.5×10^{10} per sq cm reach the surface of the earth. Our body is continuously traversed by an enormous number of neutrinos. Their study has drawn interest as their detection adds information on the fusion process that takes place in the core of stars. Detection of neutrinos is extremely difficult, however, since their interaction with matter is very weak. To detect them one must then count on their large number and the use of extremely large detectors, essentially a large tank containing pure water. Such an experiment relies on the statistical probability that a certain number of collisions may take place in the tank due to the large number of neutrinos involved. As well, one also relies on the large number of hydrogen nuclei in the tank, protons with which the incident neutrinos may interact. Without going into a detailed description of the complex detection techniques that needed to be developed to observe their presence,[4] the results obtained may be summarised as follows. It was found that the number of electron neutrinos detected as originating from the Sun disagreed with the expected number by a factor of about three. One way the results could be explained was to assume that in their journey to Earth the electron neutrinos change nature or flavour, that is to say they changed from electron neutrinos to mu or tau neutrinos. This is called neutrino oscillation. This phenomenon was confirmed by means of another type of detector that was sensitive to all three flavours. This was done through the use of a heavy water detector, where deuterons allow the detection of mu and tau neutrinos as well as electron neutrinos. A detector, six metres in radius, containing 1,000 tons of D_2O, and 9,600 light detectors sensitive to the radiation produced by the interaction, was constructed in a mine 2 km underground in Sudbury, Canada. The site was chosen to avoid contamination of results from unwanted interactions, cosmic rays and others. The results obtained with that detector confirmed the expectation in connection with the reaction taking place in the Sun's core, in agreement with the Solar Standard Model. Furthermore, the electron neutrino number was found to be one third of the total, thus demonstrating that some of them changed flavour on their route to earth. A theory, based on quantum mechanical principles, shows that such a change may take place only if the neutrinos have a mass that depends on their flavour. The exact mass of neutrinos is not known but is expected to be very small, of the order of 10^{-5} that of the electron. They travel nearly at the speed of light.

Those are the leptons of which, at present, we know the existence. Mu and tau leptons are elementary particles having an electric charge equal to that of the electron. Their lifetime is so short that one is allowed to question their role. Unlike the electron, they do not appear to have anything to do with the visible elements. However, they exist and we have to deal with their presence. Those charged particles are matched by neutral particles, neutrinos, one for each flavour. This is the world of leptons and it is legitimate to ask, like I.I. Rabi in learning about the discovery of the muon, 'Who ordered that'? The electron is entirely sufficient and explains well our visible world. It looks that we don't need those short-lived muons and taus aside from a question of symmetry with the quarks that we are going to describe.

How about the hadrons?

Hadrons are particles that respond to the strong nuclear forces. The situation is much more complex than with leptons: they are not elementary particles. More than 100 of those particles were identified with specific characteristics such as mass, charge, lifetime, intrinsic angular momentum, magnetic moment and some other characteristics that help in differentiating them and making possible their classification in groups, categories or families. A few of those particles are listed in Table 7.2 along with some of their properties. One notices the short lifetime of some of those particles. Some others, not listed, have lifetimes of the order of 10^{-23} s, very ephemeral indeed. In the history of their study it became clear that they could not all be elementary particles. Nature could not be that complex. There had to be something common to all those particles, something that was hidden to the experimenters and theorists, something that was their essence, their soul, waiting to be discovered.

Unscrambling such a list was a demanding task requiring a lot of imagination. What was required was a classification by means of

Table 7.2. Partial list of particles identified as hadrons that are reacting to strong nuclear forces, with some of their properties.

Hadrons				
Name	Symbol	Electric Charge (units of e)	Mass (amu)	Lifetime
Proton	p	+1	1	N/A
Neutron	n	0	1	~ 15 minutes when free
Pion	π^+	+1	1/7	10^{-8} s
	π^-	−1	1/7	10^{-8} s
	π°	0	1/7	10^{-16} s
Kaon	K^+	+1	1/2	10^{-8} s
	K^-	−1	1/2	10^{-8} s
	K°	0	1/2	10^{-8} s
Sigma	Σ^+	+1	1.2	10^{-10} s
	Σ^-	−1	1.2	10^{-10} s
	Σ°	0	1.2	10^{-20} s
Xi	Ξ^-	−1	1.3	10^{-10} s
	Ξ°	0	1.3	10^{-10} s
Lambda	Λ°	0	1.1	10^{-10} s
...				

properties that can be given a number and that appear to reproduce from particle to particle. When such properties are found, one can then form, with some ingenuity, geometrical arrangements, like new listings, connecting those particles with similar properties. A possibility was to regroup particles with similar masses thus forming small families. For example the neutron and the proton have almost identical masses, ~1 amu or 938 MeV. They should be made of very similar stuff. There were other particles found being 20% to 30% heavier than the proton. They were called Sigma and Xi particles, with the Sigmas forming a family of three and having either +e, 0, −e charges, and the Xi forming a family of two having 0 and −e charges. Similarly, there were those π mesons or Pions, the Yukawa particles acting as the mediator of the strong force between the nucleons. They also formed a family having similar masses of about 1/7 of the proton mass and existed in three varieties with electrical charges +e, 0 and −e. Another type of meson, called the Kaon or K meson, was also found with a mass of about 1/2 that of the proton. It also exists with charge +e, 0, −e.

Another readily identifiable property was the spin angular momentum. It turns out that two distinct groups could be identified, one group with half integer spin, $1/2, 3/2, \ldots$ and another group with integral spin $0, 1, 2, \ldots$ That information is rather important in providing a clear partition between the two groups: all mesons have integer spin and all baryons have half integer spins. As we have seen earlier, mesons are thus bosons like photons and obey Bose–Einstein statistics. On the other hand, baryons, such as the neutron and the proton are fermions with half integer spins and obey Pauli's exclusion principle and Fermi–Dirac statistics.

It was also realized that all hadrons except the proton and the neutron were rather ephemeral beings, some living for a mere 10^{-23} s. Strong interaction appears to oblige some particles to have short lifetimes. However, a certain number of particles had a lifetime longer than the others, by something like ten orders of magnitude. Because those particles did not fall in the norm, they were called *strange* particles. *Strangeness* became a new property of particles (Gell-Mann, Nakano, Nishijima, 1953), with the same importance as those mentioned above. In an attempt to quantify that property, an empirical rule can be developed giving the strangeness as twice the average charge of the particles in a given group minus the so-called baryon number.[5] We have introduced the baryon number concept before in describing beta decay. It is a quantum number introduced as follows: protons and neutrons have a baryon number equal to 1; leptons have a baryon number equal to zero; particles that decay into a proton have a baryon number of 1. In such a reaction, the baryon number is conserved. Following that procedure, it is possible to give a 'strangeness number' to all particles identified in cosmic rays or in by-products of particles collisions in accelerators, a number that is conserved in strong interactions. As we will see later, the strangeness number acquires a simpler and more understandable meaning when the secrets of the basic constituents of all those particles are finally revealed and correlations between the properties of those particles are established by means of their basic constituents. The introduction of those properties and 'numbers' may be somewhat confusing at this stage, but their interpretation will be simplified shortly.

As can be immediately realized through this short description, the kind of exercise that physicists had to go through to make a model of our microscopic world at the hadron level, based on presumably true elementary constituents, and consistent with observations, was looking like acrobatics. From the simple model of the Universe provided by Newton's and Galileo's principles, to the more complex relativistic world of Einstein and finally to the dual quantum-wave world of de Broglie and Schrödinger, we were arriving at a world of complexity never imagined before. The properties of those particles were leading to some basic information on their 'inner self', but nature was really secretive, and mean in providing clear-cut information. It was not obvious at all what created the observed resemblance between the particles and rationalisation had to be created first in their complex listing. That rationalisation was done in the early 1960s by Gell-Mann and Nee'man. The technique can best be illustrated by an example using, for instance, the two types of mesons, Pions and Kaons, with abbreviated names π and K. First, the strangeness and the electric charge of those particles are established. With some imagination, using strangeness on the vertical axis and charge on an oblique axis, the particles can be organized in

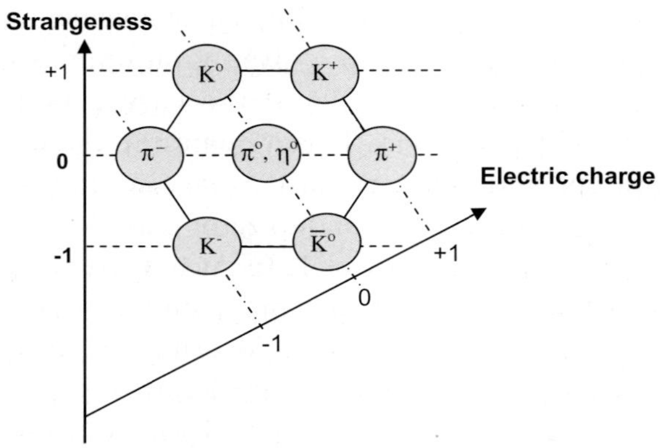

Figure 7.5. When Kaons and Pions are grouped together according to their properties relative to charge and strangeness, they organise themselves in a hexagonal pattern. The bar over K^o identifies it as its antiparticle.

an hexagonal pattern. This is shown in Figure 7.5. A particle similar to π°, called η°, but with a somewhat larger mass was also found to fit in the picture and is included in the diagram.

A similar kind of pattern is obtained with the baryon, neutron, proton, Sigma and Xi. The arrangements were known as the 'eightfold way'. But the story does not end there. At the time when that rationalization was made, other types of baryons existed forming a family with larger masses than the ones mentioned. They are called Delta (Δ), Sigma star (Σ^*) and Xi star (Ξ^*) particles. Their properties, strangeness and charge were such as to allow also a grouping of them into a symmetrical figure, actually, a triangle. However, the arrangement with the known particles left a vacant hole at one of the apexes of the figure. The symmetry was such that it was concluded that a particle should exist to fill the empty hole, an approach similar to that used by Mendeleev in the case of his table of elements. Extrapolating from the properties of the other particles it was concluded that the missing particle should have a strangeness of -3, a charge of -1e and a mass of about 1,680 Mev. Very soon such a particle was found in two laboratories in New-York (Brookhaven) and Geneva (CERN). The result appeared as a confirmation of the validity of the direction taken by Gell-Mann and Nee'man.

The patterns found provided some degree of order and symmetry in the list of particles, which an arrangement in tables does not make evident. It was concluded that there must be something common to all those particles that lead to symmetrical geometrical figures. The organisation of elements in Mendeleev's table is based on the presence of electrons and protons providing chemical properties and an atomic number, situating precisely the elements in the rows and columns of the table. What was then that common substance to all those particles that made possible such regular geometrical arrangements? The answer came from Gell-Mann and Zweig independently (1963). Put simply, it was realised that if one assumes the existence of three types of elementary particles, assuming at the same time the existence of their antiparticles, one could construct the whole set of geometrical figures mentioned above by grouping them in various combinations. Zweig called the particles 'aces' while Gell-Mann called them *quarks*, a word that was inspired from a fiction story. The name quark survived. The three quarks needed to be identified by a name. They were called 'up', 'down' and 'strange', identified as u, d and s giving their flavour. Why not? We don't know what they are anyway and they sure don't look like anything we know of that exists. Our vocabulary is adapted to our macroscopic world and is very limited. A thing to remember is that the names chosen have nothing to do with what they mean in our daily activities. What were the properties of those three quarks? They needed to have electric charge. Since there were three flavours, the charge of each had to be a fraction of the unit of elementary charge in order to accommodate the integral charge of the hadrons. They also needed to accommodate the strangeness number of the particles. Since they formed hadrons, they needed to have mass. Or did they really? Since mass is energy, could hadrons not have sufficient motion within their own enclosures to appear to have mass? We will come back to that point. Table 7.3 shows the choice that was made.

Interestingly enough, it is possible with these three quarks and their antiparticles to construct the meson hexagonal pattern we have shown above. As well, we can construct the baryon pattern that includes protons and neutrons. After some thinking (and deep affection for symmetry properties) one arrives at the

Table 7.3. Basic quarks required to create protons, neutrons and mesons and make a model for visible matter in our Universe. A similar table can be made for their antiparticles with a change in sign of the electric charge.

Quarks forming matter in our world			
Flavour	Symbol	Electrical charge	Spin
Up:	u	2/3 e	1/2
Down	d	−1/3 e	1/2
Strange	s	−1/3 e	1/2

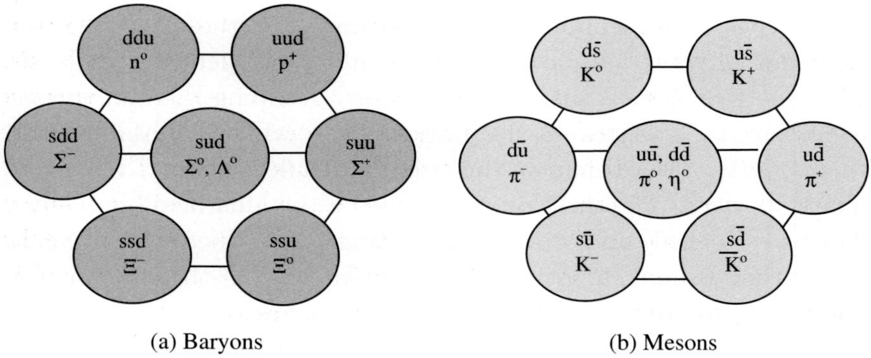

Figure 7.6. Combination of the assumed quarks that duplicate the symmetry found in the representation of a family of baryons (a) and of mesons (b). The bar over a quark symbol identifies it as its antiparticle.

plausible constructions shown in Figure 7.6. An examination of these patterns shows that the symmetry observed above for the mesons and mentioned for the baryons is observed relative to charges and strangeness.

Nucleons

Matter is made of neutrons and protons. Let us look at the resulting arrangement. In the quark model the proton is made of two up quarks and one down quark. The neutron is made of two down quarks and one up quark. This combination works fine for the electrical charge. It is readily verified with the help of Table 7.3 that the resulting neutron has no electric charges and the proton has a charge of +1 e.

Each of the quarks has spin 1/2 and they are maintained within the nucleons by a strong force. Quantum mechanics applies to those new particles and the new particles are fermions obeying Pauli's exclusion principle. Within the baryons and the mesons, they cannot occupy the same energy state nor have the same quantum number. Although that condition does not appear to cause problems with the proton and neutron in the assumed configuration, the arrangement for some other particles, such as the Δ and the Ω formed of three quarks of identical flavour, would violate the exclusion principle. How can that problem be solved? Just invent a new quantum property: the strong force acting between the quarks is assumed to have a property called 'colour charge' (Greenberg, 1964) for the lack of a better word. Like the name flavour, it is imported from our macroscopic world without any connection to its normal meaning. Three colour charges are generally identified as red, blue and green. The colour charge is associated to the strong force holding those particles together within the given hadron. Like colours repel. Hadrons are neutral relative to colour. Consequently, neutrons and protons are made of quarks with three colour charges as shown in Figure 7.7 and an ensemble containing the three colours is said to be colourless.

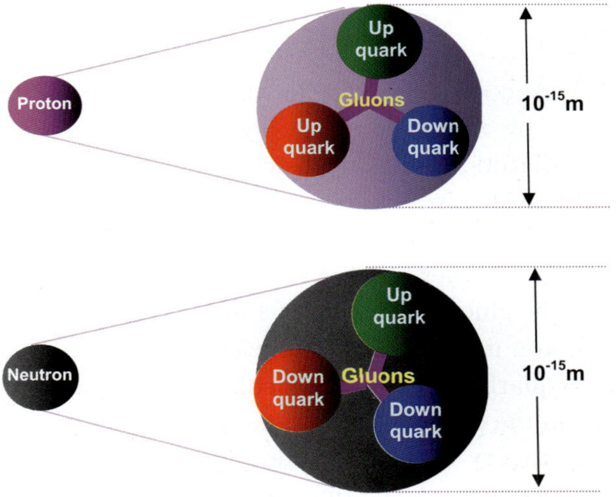

Figure 7.7. The proton and the neutron are made of three quarks, up and down flavours, characterised by three different colour forces. Gluons are the mediators of the strong colour force.

We note also that two of the quarks must have opposite spins, summing to zero, with the other quark providing the necessary spin 1/2, since protons and neutrons have a spin 1/2. They are fermions.

Mesons

They are simpler in a sense. The configuration found to satisfy the conditions of symmetry made explicit in the hexagonal figure of the eightfold way, including the quantum conditions such as charge, strangeness and spin, is a simple assembly of two quarks. This makes sure that the spin may be 0 or 1. The mesons are bosons and can occupy the same energy state. It is concluded that they are made of a quark and of an antiquark as shown in Figure 7.6b. At this point, it is worth saying something about the quantum property of strangeness introduced previously to divide particles in certain groups with somewhat similar properties. In the construction of the symmetrical hexagon, it is readily observed that the strangeness is simply measured by the number of strange quarks contributing to the mesons or baryons. It makes the whole picture symmetrical and coherent and it appears that the properties of all known particles can be visualised with the model.

The strong force

Following the success obtained in the modelling of the electromagnetic force by means of gauge symmetry leading to massless photons as mediators, it was suggested that the strong force holding those quarks together was also mediated by a particle. As already mentioned, this particle is called a *gluon*, a rather inspiring word for sticking objects together. It is the excitation of a gauge invariant gluonic field. The gluon has a spin of 1 and is thus a boson. As in the case of the photon it is massless. However it carries the colour charge from quark to quark acting as the mediator of the colour force. This is different from the photon, which does not carry any electric charge.

A whole theory has been developed around that model based on the concept that quarks, attached to each other by gluons, form hadrons and mesons. The theory is called 'Quantum Chromodynamics' (QCD). It deals with the behaviour of the strong force, the colour charge, hence the name 'chromo-', and the properties of the gluons mediating that force. The formalism uses the concepts of quantum field theory in a similar way as in the case of quantum electrodynamics, but is much more complex using advanced mathematical tools borrowed from group theory. However, it is possible to have a qualitative idea of the theory by direct comparison with quantum electrodynamics. In that last case, the electromagnetic interaction, originating from the electron field, is mediated by the photon and is represented by a diagram of the electron emitting a virtual photon that interacts with nearby electrically charged particles such as protons or other electrons. The strength of the coupling is given by the electric charge. In quantum chromodynamics the interaction is much more complex because the quark fields are three component vectors originating from the three quark colours and the force is mediated by eight types of gluons. This number of elements complicates the analysis, which is handled then through gauge symmetry and leading among other subtleties to the handling of 3×3 matrices. Similar Feynman diagrams representing the interaction can also be drawn. The strength of the coupling is given by the colour charge. At this point it is logical to question the range of the strong force as propagated by the gluon field. A most amazing result is waiting for us. No free quarks or gluons have ever been observed. Even with present accelerator technology, it is not possible to detach quarks from hadrons. The interaction is so strong that the quarks transform into other composite particles before they can be observed as free particles. It is a really shy behaviour, worse than that of the electron and photons that at least can be observed individually. That behaviour appears to be another layer of complexity of our world. The existence of quarks and gluons is limited to situations where they are in interaction with each other.

That property, very unusual in our world, is accepted as a basic characteristic of quarks and gluons. They can exist only within hadrons, as baryons or mesons. A model was developed that assimilates the quarks to entities held by loose elastic bands that get very stiff when the distance between those quarks reaches a dimension of the order of 10^{-15} m. The quarks are essentially free within that limiting border. It may be said that the protons and neutrons are perfect jails with walls so high that no internal components can escape. The model is called 'asymptotic freedom' (Wilczek, Politzer, Gross, 1973). Within the baryons the quarks are free to move and they do move violently, space available being very limited and the force involved being extremely large at the border of a volume whose size is of the order of that of hadrons. The quark motion is quantized and as in the case of atoms that have energy levels, baryons exist in specific energy states.

How about Beta Decay?

Is it possible to visualise in the quark model the radioactive decay of some nuclei by the emission of an electron, that is to say beta decay, in some isotope of radium for example? We have already looked at the process. In the decay, a nucleus emits an electron and a neutron is transformed into a proton. In order to conserve energy and momentum it is assumed that a neutral, massless particle, called a neutrino, is emitted. The development of a theory explaining the phenomenon was a multi-step process and extended over more than thirty years. A first attempt at providing a theoretical basis for the phenomenon was made by Fermi in the early 1930s. He assumed, according to the knowledge of the time, that neutrons and protons had no structure and that beta decay was a four-fermion interaction. Later, based on the success obtained with quantum electrodynamics and Yukawa's model for the strong force a new approach based on concepts of particles mediating a new kind of interaction was developed. It introduced the concept of the electroweak interaction or force mediated by so-called vector bosons carrying electric charge and mass. The model was developed over many decades and provided a rather credible theoretical basis of the phenomenon (Klein, 1938; Schwinger, 1957; Glashow, Salam, and Weinberg in the 1960s).

In order to illustrate the process it is best to first assume the existence of the neutrino and imagine that it is in interaction with matter. The interaction is very weak, much weaker than the electromagnetic interaction. Neutrinos are known to traverse great thickness of matter without being absorbed. However, when in rare events an interaction takes place between a neutrino and a nucleus,

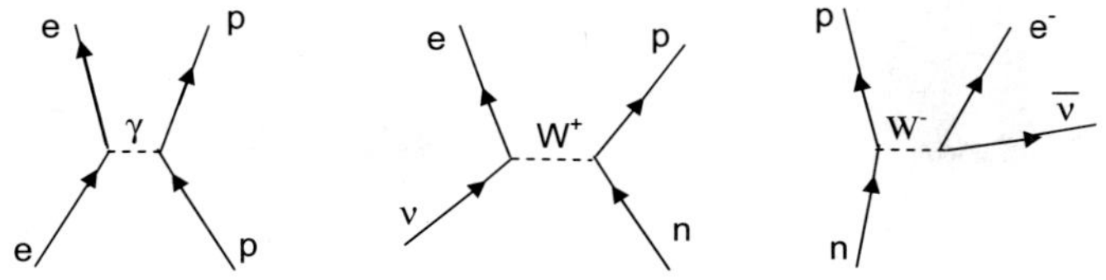

(a) electron-proton interaction (b) neutrino-neutron interaction (c) beta decay

Figure 7.8. Illustration of the similarities and differences between the electromagnetic (a) and weak (b) interactions by the respective exchange of a photon γ and a boson W^+. Figure 7.8(c) illustrates the reverse process of emission of an electron and of a neutrino (actually an antineutrino) by a neutron making explicit the weak interaction at work in beta decay.

the neutrino may transform a neutron into a proton. We can make a direct comparison between that type of interaction and the case of photon interaction as in Figures 7.8(a) and 7.8(b).

In the case of electrons and protons (Fig. 7.8(a)), electromagnetic interaction takes place by means of the exchange of a virtual photon and leads to Coulomb interaction. The photon does not carry charge, is massless and the charge of the electron and proton stays the same. In the case of the neutrino and the neutron it is assumed that the interaction takes place through the exchange of a particle called a W^+ boson as in Figure 7.8(b). That particle carries a charge $+1\,e$ and transforms the neutron into a proton. If we now go back to beta decay, which is the transformation we are interested in, we may think of it as a process in which the neutron is transformed into a proton by emitting a W^- boson as in Figure 7.8(c). We may wonder, however, about the process taking place inside the neutron itself. What is happening to the quarks? The conclusion is that the boson W^- is emitted by the down quark that has a charge $-1/3\,e$ and transforms itself into an up quark that has a charge $+2/3\,e$. Conservation of charge is maintained as well as baryon and lepton numbers. The W^- boson decomposes into an electron and an antineutrino in a time of $\sim 3 \times 10^{-25}$ s. To be more complete, it should be said that the model allows the existence of three types of weak force bosons: W^+, W^-, and $Z°$. The W^+ boson is then the mediator of the transformation of a u quark into a d quark causing positron emission as in isotopes ^{11}C, ^{13}N, ^{18}F and others.[6] The details of the processes are illustrated in

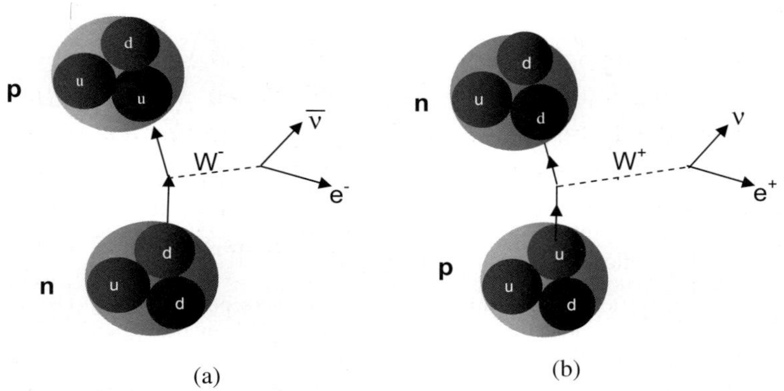

Figure 7.9. (a) Inside the nucleus, beta decay corresponds to the emission of a '$W^{\pm}$' boson by a d quark that transforms itself into a u quark. The W^- is short lived and decays into an electron and an antineutrino in about 3×10^{-25} s. (b) A similar process exists for the transformation of a u quark into a d quark with the emission of a W^+ boson that decays into a neutrino and a positron, the antiparticle of the electron.

Figure 7.9. The boson Z° is neutral and does not involve any change in the charge of the interacting particles and may be seen as the mediator of the force between a neutrino and a proton, for example, without affecting the charge on the proton.

The phenomenon is an extremely rich source of information relative to interactions taking place at the level of particle physics. Furthermore it appears to provide a confirmation of a phenomenon taking place at the cosmological level and the theory developed supporting it regarding the origin of mass. First it is obvious that somehow due to the transformation of the charge of the quark from a d to a u or vice-versa there is a coupling between the electromagnetic interaction and the weak interaction provided by vector bosons $W^{\pm}$. Electromagnetic and strong interactions do not allow such a transformation. The emission of an electron appears to provide a coupling between the charge on the electron and the charge on the quarks. The exact equality of the electron charge to the electric charge difference of the two quarks d and u appears to be more than coincidental. To be equal to that level, they must have something in common.

On the other hand, in classical physics, as discussed earlier, it appears that the laws apply equally well to the mirror images of objects in interaction. It looks normal that nature does not differentiate between right and left. This is called parity conservation. The transformation is represented by P, called parity operator. It was believed that the same rule should apply to particle physics. In fact, although analysis of data showed that parity was conserved in electromagnetism, strong forces and gravity, it had not been tested for the case of the weak interaction as pointed out in 1956 by Lee and Yang (Nobel Prize, 1957). They suggested experiments in which that symmetry could be verified and soon a group, under the leadership of C.-S. Wu, showed, using beta decay in ^{60}Co in a strong magnetic field, that parity symmetry was violated (1957).[7] There was thus a preference in nature between left and right in that particular process. On the other hand, it was believed that if charge conjugation was applied, that is to say, if a particle was transformed into its antiparticle, a process represented by the operator C, the symmetry combination CP would be conserved. This symmetry was shown, however, to be violated in Kaon decays (Cronin and Fitch, 1964; Nobel prize, 1980). Consequently, CP symmetry is violated in weak interaction. It was concluded, however, that in the Universe time reversal provided the added operation to preserve symmetry, a total operation called CPT. Thus, we may conclude that Beta and Kaon decays were instrumental in providing direct information on symmetry in the Universe.

In the weak interaction model developed by Glashow (1961), gauge symmetry used in analysing the weak force field leads to massless weak force mediator bosons. Salam and Weinberg[8] (1967) developed the model further using a hypothesis made by Higgs, Brout and Englert (1964) concerning the possibility of the presence of a vacuum field whose main property is to give mass to particles. As we will see later, the Universe, in the Big Bang scenario, developed from a primeval entity at very high temperature. At that energy or temperature it is believed that all forces were united under a single interaction with maximum symmetry. Upon expansion and cooling, a break in the symmetry took place and the forces of nature separated under different identities. Similar processes take place in our macroscopic visible world. For example, water in its liquid state has complete spherical symmetry in the sense that all directions in space are identical. Upon cooling, it freezes and form crystals with an evident loss of symmetry, since different directions do not have the same properties. Similarly, a piece of iron at high temperature has no specific magnetic properties in any particular direction. The material contains a large number of electrons with their spin magnetic moment oriented at random because of thermal agitation. It has full spherical symmetry. Below a certain temperature, called the Curie temperature, the agitation is reduced and the magnetic moments orient themselves in the same direction as their neighbours: they lock themselves in a state of lower symmetry as in Figure 5.5. It is thus concluded, as illustrated by means of those examples, that below a certain temperature, the original symmetry present at higher temperatures is broken. There are many other phenomena, in which physical states change abruptly under some specific conditions leading to a loss of symmetry. The whole Universe has been subject to those changes in its evolution and contains some hidden symmetries that simply wait to be discovered.

The vacuum of space is assumed to be a structured medium with the Higgs field present and the excitation of this field being identified as a boson in a similar fashion as the photon may be considered as a level of excitation of the electromagnetic field. It is assumed that a symmetry breaking took place in the Universe expansion and, upon cooling at a temperature of the order of 10^{17} K, patterns developed and some particles acquired mass. The $W^{\pm}$ and Z° bosons acquired mass while the photon gauge field remained massless. The two interactions, electromagnetic and weak, were then differentiated from each other. Beta decay is the evidence of this phenomenon.

The type of interaction just described is weak relative to electromagnetic interactions. It should be mentioned that the mass of the force mediating bosons, is very large, 80 GeV for the Ws and 90 GeV for the Z°, which lead to extremely short-range forces. A confirmation of the existence of such particles with the properties predicted was made by means of elaborate experiments organized and accomplished by Rubia, van der Meer and their team in 1982–83 (Nobel Prize, 1984).

It has also been found that gauge symmetry at the basis of the mathematical description of the three types of interaction introduced, electromagnetic, weak and strong interaction provides a unification of the representation of the forces. Each in turn sees its characteristics being described by symmetry groups of higher order than the previous one and the whole set of interactions appears to be unified and is given the name *Standard model of elementary particles.* This is now the accepted theory providing a description of the forces at all levels of interaction in the microscopic world. We recall that gravity, being a very weak force, does not play any role in that model.

As mentioned previously, the theory developed to provide an explanatory framework for Beta decay is far-reaching in the introduction of new concepts and addresses basic questions regarding the origin of electric charge, the origin of mass and symmetry breaking.

The Force Carriers

In the model just outlined, the world is made of elementary particles, quarks and leptons that interact by means of mediators or carriers of force. Three of those forces, electromagnetic, strong and weak and their mediators, the photon, the gluon, and the $W^{\pm}$ and Z° have been studied extensively by means of the powerful quantum field theory. An important conclusion of the model is that the W and Z particles have mass, which they acquire through a break in the symmetry of the physics involved and the introduction of a new field, called the Higgs field. We have raised the question of the origin of mass several times. The introduction of such a field, although its existence has not been confirmed experimentally, appears to address that question. In simple words, that field provides a state of vacuum, which acts as a viscous entity procuring mass to elementary particles as quarks and leptons. This effect is also responsible for the mass of the mediator of the weak force. The Higgs field is also represented by a particle called the *Higgs boson* that has mass. As of the year 2009 it has not yet been observed.

In the first chapters we have introduced the force of gravity. Einstein's concluded in his theory of relativity that gravity originates from a deformation of the geometry of space-time by the presence of mass. We first said that mass, at least a part of it, could be accounted for as energy through the relation $M = c^2/E$. Now, we just introduced the Higgs field that creates a state of vacuum through which elementary particles acquire mass. We can see immediately a difficulty in that. On the one hand, we treat mass in a classical context, with general relativity in the background, providing a mechanism for explaining gravity, a model accepted at large. On the other hand we introduce a quantum field to explain its existence. One approach is classical while the other is based on quantum field theory. Somehow, we should be able to make the two approaches coherent, that is to say, we should treat the gravitational field as a quantized entity. In such a case we would see appearing, as in all quantized fields, a particle, the *graviton*. As with the Higgs field and boson, the existence of the graviton is a contemporary problem. Theoreticians are still working actively on the development of a theory of quantum gravity and neither particles have been observed yet. The reason for the lack of success of such an enterprise is at opposite ends of the energy spectrum. In the case of the graviton it is its weakness that causes the problem; gravitational waves are expected to be extremely feeble and have not been detected yet. In the case of the Higgs boson, its mass is expected to be very high (many hundreds of GeV) and the possibility of seeing it as a result of particle collisions in accelerators has presumably been inhibited, among other things, by the limited energy of present instruments. One of the goals of the research presently pursued at the Tevatron and the Large Hadron Collider of CERN, in particular, is to observe that particle.

We have thus, in principle, the three families of interaction involving the carrier of force that we have encountered as required by quantum field theory to explain the behaviour of the microscopic world of particle physics. Underlying the whole landscape is the presence of that particular field, the Higgs field, giving mass to some of those particles. They are part of the Standard model of particle physics. Finally, we expect the gravitational field to be also represented by a carrier of force, the graviton.

All these particles have an integer spin and are bosons. We recall that such particles can exist in the same state, or in other words, can pile up on each other. Fermions, such as quarks and leptons have half integer spins and cannot exist in the same state. They cannot

Table 7.4. List of force mediating bosons assumed to exist in the standard model and gravity.

Force mediating bosons in the standard model and gravity						
Type	Photon	Gluon g	Weak force $W^{+/-}$	Weak force Z°	Higgs Boson	Graviton
Status	observed	Measured	Measured	Measured	Hypothetical	Hypothetical
Spin	1	1	1	1	0	2
Charge	0	0	(+/−) 1	0	0	0
Mass	0	0	80.5 GeV	90.2 GeV	>several hundred GeV	0

pile up at the same point. They give volume to matter. We can make a table of the force carrying bosons as shown in Table 7.4. We have added to the normal list the hypothetical Higgs boson and graviton.

As one readily sees, the situation has become rather complex since Newton's introduction of the laws of mechanics and gravity. The microscopic world is characterised by rules that are rather different from the rules of the macroscopic world in which we spend our life.

Are there more Fundamental Particles?

We have introduced the six leptons, given in Table 7.1 and then we have introduced three types of quarks: the up quark, the down quark, and the strange quark. They were the basis of the early model made of the constituents of nucleons and mesons. Are there other flavours of quarks? The answer is affirmative. In the development of the electroweak field theory, particles were grouped in doublets. Each charged lepton, electron, muon and tau, have their companion, an associated neutrino. On the other hand, we have introduced the up and down quarks as a doublet and we were left with the strange quark as a single entity. Furthermore, there are three classes or families of leptons. It appears natural, by symmetry, to expect also three families of quarks. It was a basis for predicting the possible existence of other flavours of quarks. When such a prediction is made an immediate search is started in institutes and laboratories at the forefront of knowledge in the field and large efforts are generally organized through specialized teams. Following these efforts

Table 7.5. Six flavours of quarks have been identified. Of those six only two, the up and down quarks, are required to construct all visible matter in our Universe.

Presently known families of quarks				
Flavour (mass)			Charge	Spin
up (1.4–4 MeV) u	charm (1.5 GeV) c	top (172.5 GeV) t	2/3 e	1/2
down (4–8 Mev) d	strange (80–120 MeV) s	bottom (4.2 GeV) b	−1/3 e	1/2

three new quarks were identified: they were called the charm (c), the bottom (b), and the top (t) quarks. The whole picture is summarised in Table 7.5. This forms a symmetrical arrangement of three generations and is our present state of knowledge on the number of flavours of leptons and quarks in existence and some of their properties. Could there be a fourth generation waiting to be observed? It appears that the present list forms a rather complete picture supported by a coherent theory, which leads to the conclusion that those are the basic particles of our Universe. However we don't know. Extrapolating from past lessons, Mother Nature may still be hiding some of her secrets.

How are Particle Properties Determined?

The reader may be intrigued as to how those particles are found, created experimentally, and studied? Or said simply: how is the theoretical quark model verified experimentally? In physics a theory is said to be valid if and only if it is in agreement with experimental observations. Symmetry, as identified earlier in the arrangement of the particles in pretty figures, is not sufficient for justifying totally the description made with quarks even if it is supported by an elegant theory. One needs experimental facts, otherwise it is abstract reasoning.

At one time the most energetic particles and their siblings that were available for study were those coming from space, cosmic rays. Unfortunately, we do not have much control on those. If one wants to study the characteristics of the sub-atomic particles in detail one should be capable of producing them at will in the laboratory. The only way to produce those particles is to make some of them interact strongly as happens in nature. This is what happens with cosmic rays. Particles come from great distances in space, travel at high speed with high energy, and hit the nucleus of an atom either in the intergalactic space or in the upper atmosphere. Individual events have a small probability but since the number of particles, atoms and nuclei entering in the picture is very large, a certain number of collisions take place and by chance the by-products can be observed on the ground or in high altitudes. However, this is a tedious task. The first idea that comes to mind then is why not accelerate some charged particles in the laboratory and send them on a target? Perhaps some new particles will come out providing information on the properties of those that created them. This is the idea behind the construction of particle accelerators: charged particles are accelerated by means of electric fields at speeds very close to the speed of light and directed by means of magnetic fields. They are then made to collide with a chosen target. Some components or by-products fly apart and analysis of those parts provides information on the original particles. This is really the hard way of finding the constituents of complex machinery. If you want to investigate the functioning of a watch, the last thing a jeweller would recommend is to knock two of them hard together and analyse the properties of the parts that fly in all directions. However, this is exactly what is done at present in this type of research. Why? There are really no other ways.

I remember the days of my undergraduate studies at University of Montreal when the physics department was very proud of its Cockcroft–Walton voltage multiplier used in an accelerator of deuterium nuclei. The energy involved was in the hundreds of keV. The director of the project was teaching us electromagnetism and most of the assignment and homework were concerned with calculating the field of the accelerating plates of the machine or of other components. At McGill University, where I made my graduate studies, the building next to our solid state physics laboratory was harbouring a Cyclotron with energies in the MeV range. Those machines were invented in the 1920s and provided the basis for the development of the more recent accelerators used nowadays. These can now accelerate particles nearly at the speed of light and provide energies million of time greater, in the range of TeV.

Progress in the field has been amazing, but at the expense of always larger instruments, involving contributions from several countries and employing a large staff with outstanding qualifications.

What is the basic idea used for accelerating electrically charged particles? We have seen that a negative charge is attracted by a positive charge: it is the Coulomb attraction. This is the principle used in accelerators: in a very simple arrangement, electrons emitted by a wire heated to a high temperature by an electric current are accelerated by a charged metal plate placed some distance from the hot wire. The plate potential is raised to a high positive value by an electrical source and becomes charged positively. It attracts the electrons. In scientific terms we say that the positively charged plate creates an electric field in the surrounding space, accelerating the electron according to Coulomb's law and Newton's second principle. A small hole in the middle of the plate allows the passage of a certain number of electrons that then form a beam. This idea is implemented in practice within a small structure with internal electrodes arranged to form a narrow beam of electrons. The structure is called an electron gun. As we mentioned earlier, this is the principle used in television picture tubes. A magnetic field created by a coil and an electric field created by metal plates can be used to alter the path of the electrons in the beam, which is deviated or scanned in desired directions. In a picture tube, the electron beam hits a fluorescent screen and produces light. This is illustrated in Figure 7.10.

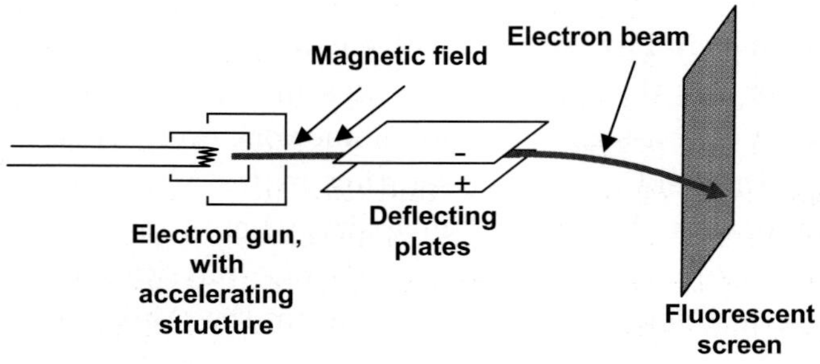

Figure 7.10. Basic idea used in the acceleration of electrons and their deflection by a magnetic field or electrically charged plates. The principle is used in television picture tubes. The image is formed by modulating the intensity of the beam and forcing it to scan the whole screen sideways and up and down several times per second.

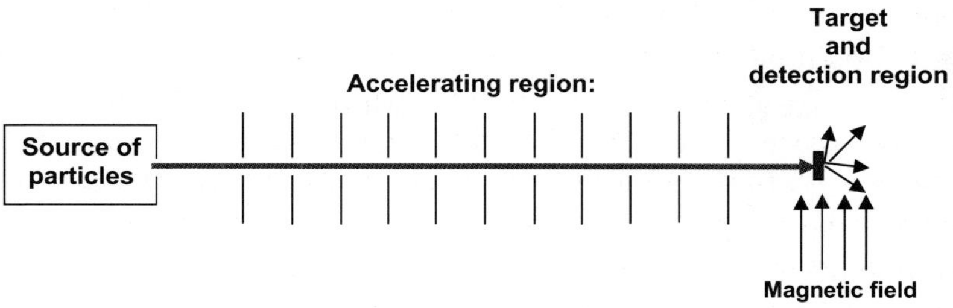

Figure 7.11. Acceleration of a beam of particles by a succession of accelerating plates. In modern versions the plates are submitted to alternative positive and negative potentials, providing increasing acceleration. In such a case the beam is not continuous and pellets of particles are accelerated. The particles may also be focused and deviated by strong magnetic fields in order to form a beam travelling on a circle and to reduce the space occupied by the device. The detector surrounding the collision region is a very important part of the ensemble. It uses various techniques, such as an applied magnetic field, location detectors and energy measuring devices, to identify the particles and measure their properties.

The same idea is used in accelerators. However, in order to attain high energies, the acceleration is done in several steps. As illustrated in Figure 7.11, the structure may consist of several stages of plates set at increasing voltages as in the Cockcroft–Walton design. A continuous beam is then accelerated towards a chosen target. In modern versions, pellets of particles cm in length and mm in diameter are made. They are accelerated by the successive plates whose potential is alternated. Setting the alternating frequency in phase with the passage of the pellet in the successive regions, it is possible to provide continuous acceleration of the charged particles forming the pellet. The process continues with the succeeding plates and in principle the acceleration depends only on the length of the path travelled and the number of accelerating plates. The reader will realise that as the speed of the particles increases, the time of passage between one plate and the next one will be shorter and synchronization will be lost. This can be compensated for by either altering the spacing between plates or increasing with time the alternating frequency of the potential applied to the plates. It is also obvious, however, that there are some practical limitations regarding size. This is the reason why some accelerators are circular. In those cases the pellet of particles is deviated by strong magnetic fields at given locations, travelling then in a great circle several kilometres in diameter. Furthermore, the frequency of the switching of the potential of the plates becomes very fast and reaches the microwave frequency region. In that case, cavities resonant in the microwave part of the spectrum are used. Fortunately, there is a limit in the speed of the particles. It is the speed of light. Consequently, when this limit is reached, the time of their passage from one region of acceleration to another is almost constant. The effect of the field consists essentially in increasing the energy of the particles by increasing their mass through relativistic effect, without altering their speed.

Recently, some large specialised machines were implemented using those basic principles. One of them accelerates electrons and positrons in a large circle in opposite directions and makes them collide head on. This is the idea behind the Large Electron Positron collider (LEP) that operated for many years at the *Centre Européen de Recherche Nucléaire* (CERN), near Geneva. It accelerated particles that resulted in head on collisions with energy of a few hundred GeV. It is being replaced by a similar type of accelerator called the Large Hadron Collider (LHC), that will accelerate protons to an energy of 7 TeV providing head on collisions with an energy of 14 TeV. The dimensions of such a machine are out of proportion with the size of the particles that are accelerated such as protons or other hadrons whose dimensions are of the order of 10^{-15} m. The LHC is situated underground in a tunnel and forms a ring having a circumference of 27 kilometres at a depth ranging from 50 to 175 metres. Other examples of such colliders are: the Fermilab Tevatron, with a circumference of 6.8 kilometres providing particles with energies up to 10^{12} eV (1 TeV, the origin of the name). This last accelerator has been the most energetic particle accelerator for a long period, but will be surpassed by the LHC by a factor of 7. The Stanford Linear Accelerator Center (SLAC) is an electron-positron linear accelerator, 3.2 kilometres long, buried about 10 m underground, which provides particles with energies up to 50 GeV. An accelerator called the relativistic heavy ion collider situated in Brookhaven (RHIC) has a circumference of 3.8 km and produces energies of 100 GeV per nucleon. It is readily observed that the size of these machines is dictated by the high energy that is desired.

The energy of the particles is so large that upon collisions, if not elementary, they are broken up into their constituents. Also, new particles are created out of the energy released. It is the siblings of those reactions that are studied in special dedicated detectors. Detectors in early experiments in particle physics could be cloud chambers and bubble chambers, which provided information on newly-found particles, their trajectories and their lifetimes. In cloud chambers, charged particles travelling through a cooled vapour would leave traces similar to jet airplanes flying at high altitudes. Bubble chambers detectors would use the property of a liquid to develop gas bubbles upon the passage of a particle. The modern detectors surrounding the collision region are large and complex, formed of several types of elements extending from solid state detectors including photodiodes, gaseous ionization detectors with multiwire chambers, as well as calorimeters. These detectors supply information to computers for analysis of the paths, lifetime, energy, momentum, charge, mass and other properties of the resulting particles. This is illustrated in the sketch of Figure 7.12. As an example, the solenoid in the Compact Muon Solenoid experiment

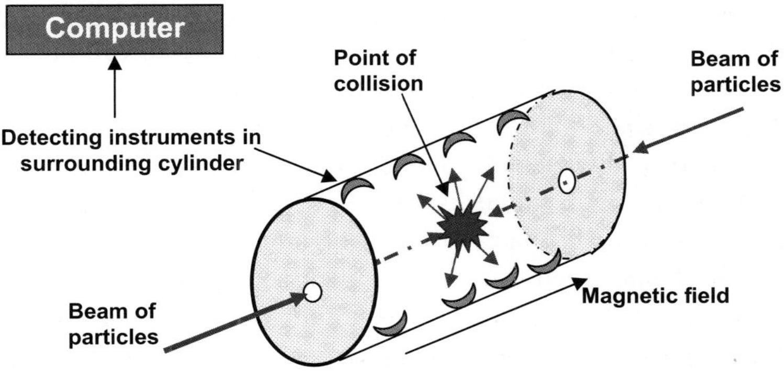

Figure 7.12. Sketch of the basic ideas behind the study of elementary particles by means of accelerated beams. In experiments such as those made at very high energies, as with the LHC and the Tevatron, the detector may be a rather large structure several metres in diameter. It may consists of regions filled with gas for recording the trace of particle paths, structures creating strong magnetic fields and forcing particles in definite paths, silicon detectors, calorimeters and other devices providing information on charge, mass, momentum, energy, lifetimes, trajectories of particles. Actually four large detectors are installed at the LHC for studying muons, quarks of the heavy type, antimatter, and the products of collisions involving ions that should result in quark-gluon plasmas.

(CMS) at LHC, CERN, produces a magnetic induction of 4 teslas, inside which are situated subsystems to measure properties of various particles such as those enumerated above, produced in the proton-proton collisions. The diameter of the instrument is 15 m and its length 21.5 m. Several other detectors have been constructed at the LHC such as ATLAS, a general purpose instrument with a primary goal of detecting massive particles such as the Higgs boson. Thousands of scientists from hundreds of institutions around the world are involved in the design, construction and operation of these detectors.

Obviously, the modern approach is very far from the technique used in early studies of cosmic rays with sensitive photographic plates sent on balloon trips at high altitudes. That technique relied on the random arrival of particles, and the results were obtained from analysis of the trajectories observed in the chemically developed photographic plates. Those were one-shot experiments that could not be repeated easily and the experimenters had to rely on a limited number of samples to draw their conclusions.

But one of the first questions that comes to mind is why do we have to go to such high accelerations of the particles or to such high energy? The reason is that if we want to extract particles from a given structure, we have to apply an energy that is at least comparable to the energy involved in the interaction of the constituents created by the forces that hold them together. In the case of the electron attached to the nucleus the energy that is required is at the level of electron-volts. For example, in hydrogen an energy of 13.6 eV is required to detach the electron from the nucleus. The energy may be supplied by a photon in the vacuum ultraviolet, having a wavelength of about 100 nm. In the case of nuclei and quarks, the force involved leads to energies in the range of millions of electron-volts. Another reason lies in the equivalence between mass and energy given by the relation $E = mc^2$, with which the reader should be familiar by now. In the use of colliding particles for studying the subatomic world we rely on the transformation of part of the energy into mass. The mass of some of the resulting particles we wish to study, as predicted by theoretical analysis, is sometimes very high. For example for a W particle to appear as a result of the collision, with a mass corresponding to 90 GeV of energy, there must be at least that amount of energy involved to start with. The top quark requires at least 172.5 GeV. Furthermore, those particles are normally formed in pairs in the process, thus requiring twice the energy.

We mentioned above that quarks appear to be in the nucleus as if in a jail from which it is impossible to escape. If an attempt is made to pull one quark away from a nucleon, so much energy is involved that new quarks are created from vacuum; they attach to the original quarks forming new particles. It is like trying to remove one of the two poles from a bar magnet: a new magnetic pole is automatically created. It is from the properties of those new particles that information is extracted on the properties of the original quarks.

Is This Whole Quark Story Right?

The quark model and its supporting theory had to be verified experimentally. In the beginning, in the period following the introduction of the model, since no free quarks were observed, there was some belief that quarks were simply an abstraction. It was thought to be a fine intellectual exercise providing with the help of high-level mathematics and symmetry arguments, a framework to make an interesting model for describing the constituents of the observed particles known at the time, mesons and baryons.

It had been accepted for some time that the electron and its partner the muon were elementary particles and their properties were known. On the other hand, nucleons, protons and neutrons, had been considered elementary particles for many years. However, already in the early 1950s, in experiments using elastic collisions similar to those made by Rutherford, but with electrons having an energy of 400 MeV, protons were shown to have a size of the order of 10^{-15} m (Hofstadter, 1953). This was about three or four orders of magnitude larger than the evaluated electron radius. The proton was thus suspected of being a composite particle. If a particle has size, naively it should be possible to analyse the internal parts that contribute to its extension in space, the substance forming it, and may be it could be possible to get a sample of that substance. One early hypothesis was that the size of the proton originated from a cloud of Pions surrounding it and providing the nuclear force. This proved to be wrong because it did not give the right properties to the proton, in particular regarding its magnetic moment. Furthermore, it did not provide a model for the symmetry observed among the particles as shown above. This is the way science works: one makes a hypothesis and experiments either confirm it or deny it. It is not democracy. Experimental facts decide.

At the end of the 1960s and beginning of the 1970s the SLAC machine at Stanford University could accelerate electrons at energies of the order of 20 GeV. At that energy or speed, the wavelength of the electron, as given by the de Broglie's expression $\lambda = h/mv$, is sufficient to resolve distances of the order of 10^{-15} m. The electron being a lepton is not affected by strong nuclear forces and it is expected that with such a short wavelength the electron can effectively penetrate the nucleon and investigate the possibility that it has components. The electron being a negative electrical particle, its scattering should be affected by the internal charges within the nucleon as proposed in the quark model. The collisions that take place are said to be inelastic, the electron penetrating the nucleus, unlike the case of elastic collisions done with alpha particles or electrons of lower energy, whose paths are simply deflected by the nucleons acting as small centres of attraction or repulsion. In the first experiments in 1967, there was evidence that the proton, the hydrogen nucleus, was not an elementary particle but had internal parts. Too many electrons were scattered at large angles. The constituents were called 'partons' by Feynman. They were the particles that had been predicted by Gell-mann and Zweig and that had been called quarks. We now use this name. A series of experiments showed, amongst several things, that effectively the internal parts of the nucleons, the quarks, appeared to be point charges like electrons, but with fractional charges, and had a spin (Friedman, Kendall, Taylor, 1970). Experiments done at CERN added information that confirmed the SLAC findings. The presence of neutral particles called gluons was also confirmed. Furthermore, the experiments showed that quarks were essentially free inside the nucleon. Later, high energy experiments at Fermilab and SLAC showed, without any doubt, the existence of those other ephemeral quarks called charm, top and bottom. The standard model was verified experimentally and is now accepted as a description of the world of subatomic particles. We presume that we now have a complete picture regarding elementary particles. They are given in Table 7.1 and 7.5. The forces mediating bosons are listed in Table 7.4 along with some of their properties.

Our World

Our visible world is thus made of atoms containing electrons and a nucleus made of nucleons, that is to say protons and neutrons. According to the standard model, those nucleons in turn are made of quarks held together by the strong force mediated by the gluons. How are the neutron and proton held together, for example, in the deuterium nucleus? How are the nucleons held together in nuclei more complex than the first isotope of hydrogen? In the theory developed by Yukawa the attraction of the two particles is mediated by a meson, the Pion, and the existence of the Pion was demonstrated experimentally some time later, making the model legitimate. However, in the standard model, it appears that the strong nuclear force is concentrated inside the nucleons themselves. It is concluded that what holds the nucleons together is a force that leaks from their internal parts. It is a phenomenon that looks very much like the force that holds atoms together to form molecules. Although to start with, atoms are electrically neutral, when they get close to each other they polarize: their internal charges are displaced and form what is called an electric dipole. At short distances, each of the atoms feels the presence of the internal charges of their partner. The electric force leaks through and there is attraction between the atoms. These forces are called Van der Waals forces and chemistry relies on them. The same thing seems to happen in the case of the nucleus. The strong interaction mediated by the gluons leaks through and holds the baryons together. We may say that the Pions, being exchanged between the nucleons, then act as mediators of that interaction.

A most important point is the realisation that up quarks and down quarks are sufficient to provide the basic material required for the construction of nuclear matter in our visible Universe. With the addition of electrons we then have all what is needed to build atoms. Gluons and photons are then only required to hold the soup together. A legitimate question is what is the role of all those other particles that have been found to exist or that are created during strong collisions of particles? We will see that the neutrino appears to play a central role in the evolution of the elements in the core of stars for the making of heavy atoms. But what is added by the existence of all the others? Are they only a consequence of the fundamental physics required for the existence of the up quark, the down quark and the electron? They look very much like accidental by-products or the residues of a large manufacture of complex products existing in the present Universe. They would then be necessary evils that appear in our high energy instruments and that we have to live with.

In any case, it appears that with some imagination and high level mathematics we have arrived at some understanding of the physics behind the functioning of the Universe on the small scale. We don't know all the whys, but we certainly know most of the hows of its working. It seems that we have reached a level of understanding of the rules that is quite satisfactory and our tool, mathematics, enveloped in symmetry arguments, has provided us with sufficient confidence that we can very often predict the results to be obtained in new experiments and even use those results for our own well being through the implementation of useful devices. We have understood and used the fission phenomenon of the nucleus to provide immense quantities of energy that, if used properly such as for producing electricity, can improve our well being. Unfortunately, we have found that the phenomenon of fission can also be used as a powerful weapon, which, if acquired by evil minds, could cause catastrophe. Furthermore the use of fission for producing energy is not without consequences to the environment, like residues of dangerous radioactive materials. However, physicists should not be blamed for the turns and consequences that have resulted from their findings and inventions. In general, those have provided a great deal of improvement in our lives. Dynamite was invented by Alfred Nobel and has been a tool of choice in construction of all kinds. It can also be used for evil purposes. There is no doubt that such work has led to an improvement in our well being.

One of the main results of all that work, however, is a better understanding itself of the functioning of the Universe, a quest that has inhabited the human being from the moment he has been conscious of his environment, from the ground that supports him to the air he breathes and to the stars he admires. We can be proud of that achievement.

VIII

Reflections on Interactions, Forces, Acceleration, Mass

In previous chapters we have talked about laws that appear to govern the behaviour of our Universe, from the large scale, the visible world with its objects with large masses, to the small scale, the atomic world. To make a proper description and provide some understanding of the dynamics of those two worlds, we have introduced basic models and theories such as Newton's principles, gravity, relativity and quantum mechanics. Overlapping the landscape, we introduced the theory of electromagnetism, which provided a window that makes it possible to explore the whole scenery. Finally, we found out that the window was itself part of that landscape. We have attempted to find connections between laws and have found that in some cases we could use a common language to describe various phenomena, a language that seems to provide the link to unify our comprehension of the whole. We have introduced concepts such as interactions leading to forces, fields, mass, energy, electric charges and several others, and have found that we could use them to develop those theories. It was done through the use of experimental observations and their interpretation in terms of mathematical expressions that then led through abstract reasoning to magnificent structures that we proudly call theories. However, do we really know and understand what force, energy, mass and all those concepts are?

Interactions and Forces

We find that objects somehow interact with each other, electrically, or through some other means such as gravity. We say that forces are entering into play. But what is the meaning of the concept 'force'? For example, in day-to-day activities we usually talk about it as a quality of a person that has strong muscles. That person can deploy great forces and handle large weights. In that context we feel that we know what the concept force means. That approach with such an example is surely a fine way to visualise a concept.

In the present text our first encounter with the concept force was with Newton's principles. We introduced it through the second principle: a force when applied to an object is said to accelerate that object and the resulting acceleration 'a' is said to be inversely proportional to the mass 'm' of the object. The more massive an object is, the more difficult it is to accelerate it and the greater the applied force must be. We think we know what acceleration is: it is a change of speed with time and we also think we know what speed is. It is a change of position in a given lapse of time. We wrote $F = ma$, a simple relation that in fact introduced also the concept of mass. Force is also involved in Newton's third principle that says that if a force is applied on an object, that object reacts by applying an exact opposite force to the first one. If you push with your hand on the wall, the wall reacts by pushing back on your hand, otherwise it would fall down. This is usually written as: action = reaction. This applies also to the rocket flying into space. The gas ejected backward is counter-balanced by the rocket accelerating forward. Recoil of a rifle on the shoulder is the reaction to the bullet sent forward. We note also that if an object has no mass, the second principle has no sense, because the force would be divided by zero leading to an infinite acceleration. But we conclude also that a particle with zero mass is moving in space at the speed of light. This is the speed of the photon and is the highest speed that can be reached. No object can exceed that speed. Consequently a photon cannot be accelerated by means of a force. There are no handles with which we can grab the photon and it always escapes the experimenter. We could say that force is defined by the second principle. But this approach is more or less turning in a circle since we also use the relation to introduce mass as the entity that opposes to acceleration when a force is applied.

As in the case of energy, when we introduce a concept such as force we have a difficulty in defining it exactly. Energy does not appear to exist by itself unless viewed as associated with another entity. We often read of 'pure energy' in science fiction novels. But what is pure energy? Does the question make sense? Energy can materialize as mass through the relation that we now know very well: $m = E/c^2$. Energy can take the form of motion, we call it kinetic. It can be gravitational, as latent energy waiting to be used. It can be interpreted as electromagnetic radiation, contained in the electric and magnetic fields forming a wave. It can be interpreted as the energy of the photons representing the excitation level of that wave. It can be chemical binding, essentially electromagnetic, holding atoms together to form a compound. Energy takes many other forms that appear as soon as we have interactions and forces entering into play. This is because when we have a force acting on an object work is done and alters the properties of that object. In some way energy is spent in moving the object. Interestingly enough, in the case of gravity the interaction energy is negative. What does that mean? It is a convention, but it is required to be coherent with other forms of energy. If we take an object from the floor, and place it on a shelf higher and more distant from the surface of the Earth, we do work: we have applied a force and displaced the object up in the Earth's field. We have given energy to the object: we call it potential energy. It is a convention to say that if the object is taken very far, say to a distance such that the interaction with the Earth is negligible; its potential energy is then zero. This makes a lot of sense since in that case the object being far does not know about the existence of the Earth and it would not be appropriate to give it a characteristic relative to an object, the Earth, with which it does not interact. If we bring the object back to its original position on the surface of the Earth we recuperate the energy we spent when we brought it very far away. That energy comes from the interaction of the object with the gravitational field of the Earth. If we call positive the energy that we get and can use, that energy must come from somewhere and then conservation of energy says that the gravitational interaction energy introduced by bringing the object closer to the Earth must be negative for the sum to be zero. There is thus a tremendous amount of negative gravitational energy in the Universe, originating from all the interactions of the celestial bodies, from planets to stars and galaxies and masses within those objects. Energy can thus take many forms and it is through such a description by means of examples that we finally get a feeling for the meaning of the concept.

It is similar with force. The second principle is too general to make us appreciate the full meaning of the concept and Newton gave a first example of force, that of gravity giving rise to the gravitational interaction that we just discussed. The attraction towards each other of two objects with mass is an example of force. The force is equal to the product of the two masses and inversely proportional to the square of the distance between them. The constant of proportionality is G, the gravitational constant. It is one of the identified forces known as *fundamental*. It is outstandingly small. One might say that this statement is exaggerated since we are essentially glued by that force to the Earth's surface: the size of the rockets that we have to construct to escape it can be used as evidence that it is large. But the force is large only because the mass of the Earth is large. As we mentioned, it is most fascinating to observe that a small magnet can beat the attraction of the whole Earth by lifting a paper clip from its surface. We can conclude readily that the attraction of the whole Earth is much weaker than the magnetic force of the tiny magnet that we have in our hand and that lifts the paper clip. Thus electromagnetic forces as we mentioned earlier are much stronger than gravitational force.

That reflection about the strength of magnetic forces brings us naturally to a second example of fundamental forces, that introduced by electrically charged and magnetic objects. The attraction or repulsion of electrically charged particles has been known since antiquity, but it is Coulomb who first really described the phenomenon on scientific grounds. It is very much like gravity. The main difference is that with electricity we may now have attraction and repulsion. We have thus assigned arbitrarily symbols to the charges in order to reflect that property. We chose the symbol (+) and (−). Charges with like symbols repel each other while charges with opposite symbols attract each other. The force is proportional to the product of the size of the two electric charges and inversely proportional to the square of the distance of their separation: it is called electrostatic force. In the international system of units, the constant of proportionality is $1/4\pi\varepsilon_o$, as we have seen previously. The constant ε_o like G acts without support, in vacuum. They are called vacuum constants. It appears that there are no other types of charge such as, for example, a $\times$ charge. As we mentioned, we could imagine that such a type of charge could accelerate other charges sideways relative to each other rather than towards each other. In nature there is no charge with such a property. When two electric charges are at rest relative to each other the only important force present between them is repulsion or attraction. However, if they move relative to each other, another force is developed that induces them to do exactly that type of motion, that is to say, to move somewhat sideways. It is called a magnetic force. It is most interesting to realize that this force is developed only when the charges are in *relative* motion at speed v. If they move together in a parallel motion at the same speed the only important force present between them is electrostatic, although they can exert a sideways force on a third particle moving in another direction. A most beautiful mathematical theory, based on these experimental laws, describing the dynamics of electric and magnetic phenomena was developed by Maxwell. It has succeeded in describing all macroscopically observed phenomena and has not been confounded yet. It is the classical theory of electromagnetism and all classical situations involving magnetic and electric phenomena can be handled by the theory.

Other forces? Yes, of course! As we said in the previous chapters we have identified four of them. We have seen that nuclear forces are extremely strong, in a world of their own. To explain the behaviour of the nucleons, the neutrons and the protons, we had to assume that they were made of internal components, internal gears, and we introduced a new type of particle, the quark. Quarks are attracting each other by a new type of force, represented by the colour charge. The interaction of quarks on each other takes place at very short range, within the size of a proton. The quarks cannot escape the nucleus, but their presence may be felt from a nucleon to another, causing the binding of those nucleons together, thus forming the large nuclei of heavy atoms.

There is finally the weak force that enters into play in the transformation of a neutron into a proton by the emission of an electron, and similarly in the emission of a positron by a proton transforming itself into a neutron within the nucleus. The emission of the electron or of the positron is accompanied by the emission of a neutrino. The interaction of the neutrino with nuclear matter is very weak and it is that interaction that is called 'weak'. Neutrinos can traverse thicknesses of kilometres of matter without being absorbed. We are traversed continuously by a large flow of neutrinos originating from the sun without, we believe, any consequence. It is indeed a weak interaction.

A basic question, however, is how are the forces transmitted, how does one particle know the presence of the other? In one approach, we write equations based on the presence of a field, assumed in a sense to emanate from the object, and we calculate a force. Naively, we may assume that the particles do the same calculation and react according to our equations. In the case of magnetic forces, the concept of field is almost evident. We just have to look at Figure 3.17 to conclude the presence of a field that forces the granules of iron to take a direction and make a pattern. And we say that the field follows those lines and creates the force. We even call them the lines of force. From empirical laws based on experiments we can construct a whole theory based on those fields and explain the behaviour of electric and magnetic charges in all circumstances. Are these fields real? A theory, called quantum electrodynamics or QED, the development of which started in the 1930s and continues up to this day, addresses that question in a rather intriguing way: an electron continuously sends out a so-called virtual photon, showing its presence and in a sense inspecting its surroundings. If nothing is present, the photon is re-absorbed by the same electron in a very short time. The process at first appears to violate the principle of conservation of energy since the energy of the electron is not altered by the emission of the photon. In quantum physics, however, uncertainties in energy and time allows energy to be borrowed from vacuum, for a very short time, as long as the relation $\Delta E \times \Delta t = \hbar$ is not violated. That photon, if it is highly energetic, may exist for only a very short time. It will not be able to travel very far in the time allocated for its existence by the uncertainty condition: the range of action is short. If it is a very low energy photon, however, it may exist for a long time and have effects at very large distances.

This is another way of looking at the interaction of electrically charged particles. QED introduces the photon as an instrument of interaction. The photon carries along electric and magnetic fields that interact with charged particles and is said to be the *mediator* of the electromagnetic force. In the atom, an electron is continuously exchanging photons with the nucleus. They are energetic photons, the distance is minimal and the time of travel at the speed of light is short. Our world is made of atoms interacting with each other, exchanging photons and forming chemical compounds. The whole of space is thus filled with photons travelling at the speed of light and mediating the force between all the electrically charged particles in the Universe. What a world! Vacuum is no vacuum at all! It is filled with energy. A similar type of process takes place in the case of the strong and weak force. Gluons (strong forces) and bosons $W^{\pm}$ and Z_o (weak forces) are the particles that are exchanged and act as mediators of those forces. That approach leads to some kind of unification making the theories behind electromagnetic, weak and strong forces somewhat symmetrical and represented by mathematical concepts that draw from concepts borrowed from common mathematical tools. It leads to the Standard model of elementary particles.

What of the transmission of force in the case of gravity? How does it work, what is exchanged in the process of attraction? In the classical context it is action at distance and in Newton's theory the carrier of the force is not specified. Newton did not elaborate on the nature of the force acting at distance. To him the transmission of a force without agent did not make much sense and he left it to the reader to interpret it. To provide a step in the description of such an agent, Einstein introduced the concept of the curvature of space-time. Space-time may be understood as a four dimensional canvas, a grid, that is distorted by existing objects with mass. An object, in constant motion in that canvas, when close to a massive object, appears to deviate from a straight line in our three-dimensional space because of the curvature of the canvas caused by the presence of the massive object: we conclude that it is attracted by the object. In that context the force of attraction is reduced to a geometrical property of space-time. As we have shown, it is like two persons moving north on a large sphere following two separate, apparently straight, meridian lines: they approach each other as they go north until they meet at the pole. If the persons do not know that the surface they are travelling on is curved it will look to them very much like mutual attraction.

However, in practice, attraction may also be interpreted as the effect of a field created throughout space by an object with mass, a so-called gravitational field that is sensed by all other objects having mass. As in the case of electrical charges, the gravitational field is one of the most useful concepts with which trajectories and orbits of cosmic objects can be calculated. Force is interpreted simply as the product of the object's mass by the gravitational field existing at the site of the object. On the surface of the Earth we interpret that force as weight. An important question is immediately raised in connection to that field: what is its speed of propagation through space? In Newton's theory, the effect of gravity is instantaneous. However, this is not so. Drawing from a similitude to the concept of electric field and the deduction we made of its property, could we not think of gravitational waves in a similar way as we do with electromagnetic waves? General relativity provides the framework for the existence of such waves. What would be their nature, what would they look like? What would be their effect on our environment? Our understanding of gravity through general relativity, as we just said, is that it is a deformation of the space-time grid. Such a deformation propagates on the grid at the speed of light. Such a wave, when passing by, would deform the space-time grid and an object on its path would also be deformed, stretched or compacted. However, owing to the weakness of the interaction the effect on length would be very small, corresponding to a fractional change of the order of the 10^{-21}, very small indeed. Could we measure that? As we have seen, the concepts of length and time measurements are very strongly coupled and we could rely on the precision of the technique developed in the field of atomic clocks to realize an instrument that would measure changes of length of that order of magnitude. LIGO, (for Laser Interferometer Gravitational wave Observatory), under development in USA by groups at MIT and Caltech has such a goal. Similar instruments such as VIRGO (Italy) and GEO (Germany) have also been implemented. The systems are designed such as to be sensitive to waves in the range of tens of Hz to several kHz. LIGO is an interferometer with two branches orthogonal to each other and several kilometres long. The structure is very stable and light waves propagating within and reflected by end mirrors can be used to measure the length of the arms which, in principle, would be altered by the passage of a gravitational wave. Changes in the pattern of the interference between the recombined waves propagating in the two arms would be an indication of the relative change in length of the two arms. The arms are made long in order to amplify the effect. It may also be mentioned that scientists in Europe are developing a space system similar to LIGO, called LISA (for Laser Interferometer Space Antenna), but of astronomical size (millions of km). It would thus be much more sensitive than LIGO and at the same time would avoid all the interferences produced on Earth by seismic activities to which LIGO is sensitive and could be interpreted as gravitational waves. That work follows the pioneer work done by J. Weber in the 1960s, who used a large suspended resonant metallic cylinder that would have responded to such waves but with much less sensitivity. No such wave has yet been detected directly by any of those instruments. However, their existence has been inferred from the study of a neutron star. Those stars are residues of star explosions (supernovae) and are essentially compact spheres of neutrons 10 km in diameter. They will be described in the next chapter. Those stars emit radiation in a preferred direction. That radiation appears as a beacon in the sky upon rotation of the star on itself and this radiation is detected on Earth as a pulse with a period of extremely high stability. In a particular case (PSR B 1913) the period is 59 milliseconds and was found to fluctuate slightly with time. Those fluctuations were explained by the presence of another close-by neutron star, the two of them orbiting each other with a period of several hours. However, that period was found to decrease by 76.5 microseconds per year. The cause of that decrease in period meant a loss of energy that could only be explained by the emission of gravitational waves (Hulse and Taylor, 1976; Nobel Prize, 1993).

Could we not interpret also the gravitational force as an exchange of particles as in the case of electric charges? In the limit of extremely high energy, within Planck's scale, in principle the field would be quantized, in which case we could say that a new hypothetical particle would mediate the force. Such an approach is considered and attempts are made to develop a theory called quantum gravity that would include that concept. The mediator of the force would be the *graviton*, an obvious choice of name. The graviton of course would have no electric charge. It would have an integral spin of 2 and thus be a boson as is the photon for electromagnetic interactions. Unfortunately, attempts at developing such a theory have so far been unsuccessful. It appears that the smoothness of the field in general relativity is not compatible with the granularity encountered in quantum mechanics. However, progress is made in the direction of a quantum gravitation theory by means of the so-called string theory based on the hypothesis that the world is made of basic elements called strings that can be either open or closed on themselves and whose modes of vibration would give rise the properties of objects surrounding us.

Acceleration and a Mysterious Force

Have you ever thought why, when travelling in a car taking a sharp curve, you feel thrown on the door opposite to the direction of the curve you are taking? Probably, but most of us, being asked about the nature of that force, since there must be a force that wants to throw us out of the car, would answer that it is obviously the centrifugal force that appears in any rotational motion. But what is a centrifugal force? We have found that there are only four types of force in nature. Did we forget one? Or is one of them acting without our knowledge? The way it is introduced in elementary texts on physics is through the third principle. It is a reaction. To see how it is introduced we will simplify things. Instead of using the complicated situation of a car with its passengers travelling on a curved road let us just think of the case of a sling, a small stone attached to a thread that is

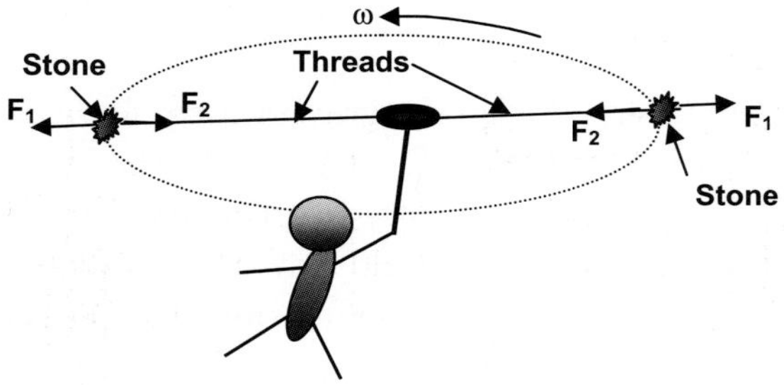

Figure 8.1. The rocks at the end of a sling appear to be in equilibrium due to opposing forces: the centripetal and the centrifugal force. F_2 is the centripetal force towards the centre provided by the threads, while F_1 is the centrifugal force.

swung rapidly at the end of your hand above your head. You can even have two of those attached to each other with the same length of string and stones with the same mass. With some agility you can get them to rotate above your head one rock on each side with the two threads tight in a horizontal plane as in Figure 8.1. Don't let go, we have just started. The stones are held by the threads. Since their motion is circular, they must be accelerating towards the central point, otherwise they would travel on a straight line. There is thus a force that acts on them and keeps them on a circle above your head; otherwise you know very well by experience that they would fly away. It is called the centripetal force. It is provided by the threads acting on them, holding them in place: it is an action on the part of the threads. On the other hand, since they are at the end of the threads and those threads are tight, there must also be a force pulling on them somehow, otherwise the threads would hang loose with the stones at the end falling on your head. The third principle tells us that this force must be present and it is the reaction. What is that force pointing away from the centre and keeping the threads tight? This is what is called centrifugal force. We observe that force in objects in rotation. For example, you may imagine yourself inside a large cylinder turning rapidly on its axis like a carousel, as in Figure 8.2(a). If the rotation is fast, aside from probably feeling sick, you will be glued to the outside wall by the same type of force. Sometimes, producers of science fiction films design their space ship as large wheels rotating freely in space

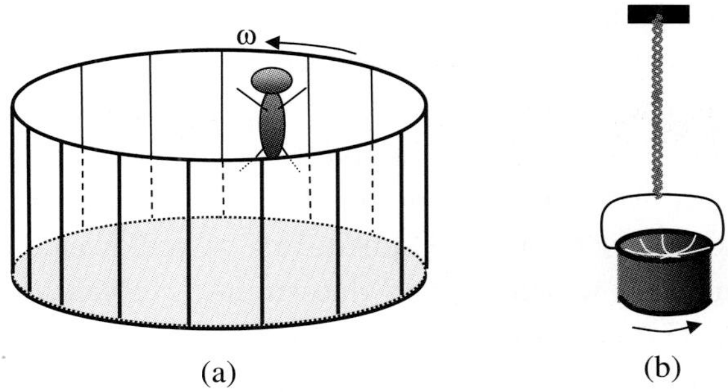

Figure 8.2. (a) In a carousel turning at large speed a passenger is thrown onto the outside wall. (b) In a rotating bucket, the water surface is seen to take a concave form as if it wanted to escape through the sides.

with the astronauts walking on the outside rim of the wheel. With a proper rotation rate, the centrifugal force in such a case would simulate the force of gravity and provide a comfortable environment. Newton raised the problem with a similar kind of experiment, a pail filled with water at the end of a string as illustrated in Figure 8.2(b). If the string is turned on itself until it is tight and then let free, the pail starts rotating. First the water is flat, not rotating, but friction in the water forces the water to follow the pail rotation and its surface becomes concave, as if water wanted to escape through the sides of the pail. This is something that does not surprise us: we are used to the phenomenon. We do not think much about it. We simply say again that it is centrifugal force. But where is that force coming from? When sitting in an airplane gaining speed on the runway to take off we also have a sensation similar to the one in the carousel: we are thrown back on our seat. It appears that all those phenomena have something in common, they are caused by acceleration.

Acceleration is either a change of speed or a change in direction. In both cases a force must be present. The water in the pail goes in a circle, changes direction, thus is accelerated. The rocks at the end of the threads are going in a circle, they change direction and are accelerated.

We have discussed at length motion of objects at constant speed. When two objects move relative to each other at constant speed, it is not possible to decide which of the two objects is at rest or in motion. We said that uniform motion is relative and that there is no way we can determine the actual state of movement of a closed system, like a closed vessel travelling freely in space, by doing experiments within the system only. We are not sensitive to constant motion or constant speed. We do not feel constant speed, but we feel acceleration. What is the reference by which we can determine acceleration and makes us feel it? In the case of rotating objects who or what is turning? When the skater turns on the tip of her skates in a stadium, who or what is turning? She or the crowd in the stadium looking at her? In the case of the bucket of water, is the bucket rotating or the whole world rotating around it? However, it seems that there is an asymmetry. For example if the skater by some means increases her rotation rate, her arms would tend to be stretched, spread apart from her body, pulled by an invisible force. There is no such effect on the crowd watching her. If turning at constant rate with her arm extended she decides to bring her arm close to her body, she immediately starts to turn faster, without any involvement from the crowd looking at her. The question of the actual reference in the whole affair, pail of water, carousel and skater with arms stretched — acceleration in general — has haunted the physics community for centuries. We may have an answer but arguments are still taking place.

In Newton's model of the Universe, space and time were independent and absolute. Consequently, in that model rotation is taking place relative to that absolute space. Space exists by itself without the objects in it. It is interesting to note that in that absolute space one can determine acceleration in an absolute way but speed cannot be determined otherwise than relative to another object. This is very different from the space envisaged by Leibniz that we mentioned in the introduction, which is simply a relation between objects: no objects, no space. It looks as if in Newton's world the objects were placed in a pre-existing space or vacuum. In Leibniz' world the vacuum seems to be an abstraction without real existence.

In the 19th century, a scientist-philosopher, by the name of Ernst Mach, gave serious thought to the question and proposed essentially, although it was not stated clearly, that the rest of the objects in the world was the actual reference for an object in rotation. When the pail rotates on itself the first reference is of course the walls of the room in which the experiment is done, the surface of the earth underneath. If one looks above in the sky the whole world looks rather stable and seems to be a good reference also for the rotation. Consequently, in that model if those objects are removed from space, there is no longer any reference and it is impossible to say if an object rotates or not. All the effects observed in rotation, for example the concavity of the surface of water in the rotating pail, disappear. The pulling of the rocks on the threads of the rotating sling, as well as the sticking to the walls of the carousel in rotation, disappear. In that scenario, without reference the threads of the sling become loose and the man in the carousel floats around.

One question is raised immediately. What is then the force that causes the water to rise on the walls of the pail in our world? What is pulling on the rocks of the rotating sling? A possible explanation would be that the gravitational attraction of all the other objects in the Universe is responsible for that phenomenon. Similarly, it would appear that acceleration is felt because of the presence of all the masses in equilibrium in the Universe, causing inertia. 'Mass there, is inertia here'. Consequently, if all the objects with mass were removed from the Universe and only space was left, the phenomenon would disappear. There is no longer an absolute space as in Newton's world. However, if it were possible to put one single star very far away and obtain a reference point for the rotation, would the effect reappear? It does not look right that a distant single object with minimal gravity would produce such an effect. Would it be a gradual increase of the effect with the addition of more objects in the sky? This kind of reasoning and explanation lead to a major problem as it does not lend itself to experimental verification. It is all done in the abstract.

In a previous chapter we examined the relation between gravitation and acceleration. Einstein[1] had an inspiration that he believed was the happiest thought of his life: gravity and acceleration were similar. In space, far from any gravitational attraction, a passenger in a room in a rocket floats around freely. If the rocket is accelerated at 9.8 metres per second, the passenger can stand on the back opposite to the direction of acceleration as if he were on the surface of the Earth. If he has an apple in his hand and leaves it free, it will look as if it were falling to land near his feet with the same acceleration as on the surface of the Earth. Everything is the same. So gravity is similar to acceleration. Now, suppose that you are in an elevator that falls freely in the Earth's field. What do you feel in that elevator? You are in free fall and objects around you are also in free fall: they float around you. It feels like being in free space and all the laws of physics apply as if you were not accelerating. More than that, you may consider your fall in the Earth's field as a free fall in the field of all the stars, galaxies, nebulae and other objects in the Universe since their gravitational field, although weak, is felt also at your location. Actually, anywhere in space, if no force is applied, you are in free fall and all the laws of physics that we know should apply to objects around you.

According to special relativity the proper frame of reference is the space-time frame. With the inclusion of gravity the space canvas is curved, distorted by the presence of all the massive objects in the Universe. In your free fall you follow the curvature of that frame. It looks like the ultimate reference frame and in that frame, although you do not perceive motion, you can perceive acceleration relative to that grid. It looks like the absolute reference we were looking for. However, it is dynamic: it changes all the time since objects in space are moving and their mass affects the distortion of the grid. This looks also very much like an idea associated with Mach's primary idea, stated in another way, that all the objects in the Universe are the agents relative to which we can refer in order to determine the presence of acceleration. As was done above, let us assume that all masses are removed from the Universe; can we say that the grid remains and that acceleration could be referred to it? Would it be flat space-time? We have no way of knowing. The stars are there for a long time. We can say that in all rotating examples that we have introduced, the rotation is relative to space-time in a Universe filled with mass providing the absolute reference. The skater is the one rotating and not the crowd. She is the one who feels the acceleration in her rotating exploit and her reference is the curved space-time canvas.

Mass in that Context

The second principle and the law of gravity both use mass as a basic quantity. The mass in the second principle is called inertial mass while in the case of gravity it is called gravitational mass. Is it the same in both cases? We have looked at that and we examine it also in Annex A. Let us look at it again. Does the force of gravity act on the same entity called mass in the second principle? There is no a priori reason why it should be the same entity. Einstein concluded that gravity on the surface of the Earth is a force that is equivalent to the one that would be felt by an observer being accelerated in free space at the same rate as that produced by gravity, $9.8\,\mathrm{m/s^2}$. This obviously leads to the conclusion that the mass in acceleration under gravity is the same as the mass under the same acceleration in free space. This is what Galileo is said to have tested at the Leaning Tower in Pisa. This is one representation of the equivalence principle, basic to the theory of relativity and it has not been falsified yet. Consequently, it is assumed that inertial mass is the same as gravitational mass.[2]

But again, what is mass? Although in Einstein's special relativity mass is a materialization of energy, it is not evident how that concept helps in providing inertial mass even in Mach's hypothesis. Inside the neutron and the proton, the three quarks interact strongly with each other through the strong force; they may exist in states with large energy, like electrons in atoms that are excited to states with high energy well above their ground state. The quarks move around within a very small volume restricted by the strong force. They are quantum particles and they obey Heisenberg uncertainty principle. The cosiness of their habitat forces them to have a large momentum. That means that they have a lot of energy and that corresponds to mass. In fact, it is shown that the mass calculated from that interaction can represent 90% of the mass of those particles. And since the mass of protons and neutrons accounts for most of the mass of visible matter, it would appear that the concept, mass = energy/c^2, is sufficient to explain most of the mass of our visible Universe.[3] You may say that there is a lot more interaction taking place in the Universe, for example in chemical compounds. However, this is very small, in the range of electron-volts, compared to the millions of electron-volts in the case of the interaction of sub-nuclear particles.

There is nevertheless that 10% left in mass that is not accounted for in those calculations and as mentioned by Wylszek, mass, as obtained in that calculation, appears to come suspiciously out of nothing. It is like magic. Furthermore, the electron, without internal parts that can move, has mass!

In the previous chapter we have introduced a new field that appears to address that question. Let us look at it again. The idea consists in the possibility of existence of a field throughout space, treated in the context of quantum field theory, the so-called Higgs field, which could act as the 'generator' of mass. This is, however, a rather strange entity. Normally one would expect that the lowest state of energy, which is the state desired by nature, is the absence of anything, total vacuum. The presence of any entity such as a field, whatever its kind, should raise the energy of vacuum. However, according to the model, in that case the lowest state is when the field is present. It is a rather odd entity in line with the strangeness that we see emerging in our progress towards finding the true original laws of nature. That field should act as a kind of molasses that embeds some particles such as leptons and quarks and interacts with them. It also interacts with the bosons W and Z that are carriers of the weak force, as we mentioned. It does not interact with gluons and photons. It opposes to changes in the motion of the interacting particles. Such a field is quantized as in the case of the electromagnetic field and it is thus expected to exist as a particle, the Higgs boson. In the quantum world, its interaction with the elementary particles would give them mass. As in a circular argument, the Higgs boson is expected to have mass itself, a rather large value of the order of a few hundred GeV. As of 2009, it has not been observed yet experimentally.

It thus appears that we can describe the functioning of all that we study and observe in the laboratory by means of four fundamental forces. And it is most satisfying to realize that we have succeeded in modelling our Universe with such concepts requiring, of course, a fair amount of imagination in view of the limited amount of facts that we have obtained from tedious experiments. And furthermore, we have succeeded in joining, by means of arguments on symmetry, the description of three of them by a single model called the Standard Model. The theory supporting that model has been developed over more than fifty years by leaps and jumps and has been very encouraging in our attempt to understand the basic principles that drive the mechanics of our Universe. The Standard model uses all basic theories developed in the first half of the 20th century: classical concepts, particularly the concept of energy and force, the concept of space-time, special relativity, electromagnetism, and quantum physics. It uses fairly advanced mathematics and in a sense shows that the functioning of the Universe obeys rules that are a reflection of the logic that we can uncover by means of that mathematics. This is quite appealing.

Is there any basic reason why gravity could not exist by itself with general relativity providing the framework in which objects evolve in space-time, their motion looking like attraction? It is most interesting that our quest for understanding has always led to the discovery of new models for the description of the underlying mechanism of an observed phenomenon. Gravitation does not escape in that quest. The three forces, electromagnetic, strong, and weak, appear to be unified, although this unification may be qualified as a mathematical unification by means of symmetry groups. It is nevertheless a model the mathematical description of which provides a very symmetrical description. The present consensus is that all forces at very high energies could collapse into a single entity. Consequently, the present state of theoretical affairs with general relativity standing by itself and all three other forces being unified under quantum field theory, does not appear to be totally satisfying.

An ideal situation would be to have a single integrated theory, using some basic postulates or assumptions that look evident to our logical mind and from which all the laws that we are now using are derived. This would be the theory of everything (TOE). Would we be better off with a single unifying theory? We can certainly continue our life and progress without such a theory. We are exploiting rather well the Universe resources and we are certainly successful in improving our way of life, although we are now questioning our influence on our environment. But the human being is curious by nature and we have learned that the more we know, the better our life becomes. For these reasons we will always question our knowledge and understanding of the functioning of the Universe and we will attempt to understand more the laws of its functioning in an attempt to obtain a satisfying answer to the existentialist question of our origin.

IX

Where Does It All Come From? Where Is It Going?

There we are, with a modelling of the constituents of the Universe. We have indeed a model for the small scale. It is based on a rather exotic theory, quantum mechanics looking more like science fiction than reality. However, that theory, extended into a quantum field theory that relies on symmetry and mathematical concepts not directly connected to the visible world we are familiar with, describes very well the behaviour of all particles at the atomic and subatomic levels, from photons to electrons, to the nucleus and its constituents, quarks and gluons. On the other hand, on the large scale, we have Newton's principles describing the behaviour of objects with mass and introducing the concept of gravitational attraction between those same objects. Finally, we have Einstein's theory of relativity, a new approach to the description of the macroscopic world, a theory that models the behaviour of objects as a geometrical property of space, based on the constancy of the speed of light appearing as a natural limit, and on the equivalence of gravity and acceleration. Within proper approximations, the conclusions of general relativity reduce to Newton's model based on gravitational attraction. Interlaced in all that is the presence of electromagnetic radiation, which, when properly handled with quantum mechanics, becomes itself a model for the interaction of matter with light or radiation in general. It couples the small and the large and provides a means of observing the whole Universe. It is so well integrated to the ensemble that it infiltrates the whole.

We have atoms, grouping to form molecules through electromagnetic and quantum interactions. Those molecules group into larger ensembles forming liquids and solids under the form of crystals and amorphous objects. Upon increasing their size by sticking together, objects start to feel gravitation more strongly and form larger objects to finally create celestial bodies. They form stars, which in turn form galaxies. Around the stars evolve smaller objects, the planets that follow paths that we can model and predict rather accurately. We have catalogues where many of those objects are compiled and, with powerful telescopes either in space or on the ground, we accumulate more data on them every day. We identify more and more far away stars and galaxies. We find in some regions of space agglomerations of atoms, of molecules, and clouds of dust particles called nebulae, where stars are still in formation. We see stars in our own galaxy, the Milky Way, thousands of light years away and we see galaxies millions of light years away. At first sight, it seems without end, infinite and not made to our scale.

We have thus a fairly good description of the constituents of our Universe and we have uncovered the basic laws that appear to govern the interaction between those constituents. But how was that whole Universe formed? Why are there individual stars? Why are they not simply grouped together, under gravity, to form a very large body of matter, dark and cold? This is a question that was raised by Newton in correspondence with Richard Bentley, a 17th century scholar who discussed Newton's physics in the framework of theology. Gravity draws matter together. In a nebula, for example, particles in random encounters tend to stick to each other, forming larger particles. Accumulating mass, those agglomerations are attracted to each other by gravity. The attraction, as we said, is very weak and the process of concentration is slow. However, nature is patient and there is plenty of time. We deal with millions and billions of years. The particles finally form large dense structures that become compressed under gravity, developing high internal pressures. Gravitational energy is transformed into heat, raising the temperature of the ball being formed. The core gets so hot that nuclear particles travelling at high speed collide with each other and nuclear reactions start taking place releasing enormous amounts of energy partly as electromagnetic radiation. 'A star is born!' Fusion of particles and emission of light takes place when the size of the accumulation of matter is sufficient. But we may wonder why the agglomeration does not take place at the level of the Universe as a whole? If the Universe were finite with a hypothetical 'boundary', all masses on the edge should be attracted towards those objects closer to the centre and the whole structure should collapse after some time into a single large entity. Why does it not happen and why do we see stars that appear to be rather stable in the sky without motion to the naked eye? Do they not attract each other? If we wait long enough should they not collapse into a single immense blob? In that scenario, however, it should also be realised that a Universe with a boundary raises another serious embarrassing question: what would be outside the boundary?

On the other hand, if the Universe is infinite, there is no 'centre'. Everywhere is the centre and there is no preferred direction. Except for local inhomogeneities that can result in clusters as large as galaxies, stars on the whole are attracted equally in all directions. In such a case we could say that, except for local fluctuations, the gravitational potential is constant everywhere or the gravitational field, originating from a variation of the gravitational potential with space, is zero on the average. This condition would require a perfectly isotropic Universe. On that basis and from the observation that individual stars exist, we could deduce that the Universe is infinite, a conclusion that sends us back to where we started. However, that does not answer completely the question. It is only a hypothesis, a speculation. A conclusion of such a nature, originating from the single abstract premise that the Universe appears static and infinite without further experimental proof, is far from being scientific. We need some more evidence from experimental facts. Does nature tell us something more on that matter? Does nature provide facts that can elucidate that question?

Another problem is raised with the hypothesis of an infinite static Universe that would have existed for all time. In such a situation there would be stars out to infinity all emitting light. What would the night sky look like? Very bright, because there would be, everywhere along our line of sight, a star in the way, which although very far from us would nevertheless emit light that reaches us. Since there would be an infinite number of stars, there would be a light ray that would appear to originate from every individual point in the sky and at night the firmament would shine from all directions as brightly as the Sun. This troubling effect is called Olber's paradox. Since the night sky is dark and since we see individual stars, there must be another interpretation regarding the size and nature of the Universe.[1]

When Newton and later Einstein were attempting to form a credible model of the Universe, knowledge of the properties of the world was rather limited: the Universe essentially consisted of the visible stars in the Milky Way and some nebulae within. Furthermore, that Universe seemed rather stable. Nothing appeared to move much in it. Einstein was fully aware of the problem of the stability of the Universe realizing that gravitational attraction should pull things together. As we have mentioned, he introduced arbitrarily in his theory a constant, called the *cosmological constant*, which exactly counterbalances the effect of gravity and at first sight makes the whole thing stable. That solution, although rather satisfying mathematically, looks like assuming off-hand that only local motion exists. The introduction of that constant seemed to be the only way out.

This is the way scientific progress is made. Scenarios, models, and theories are based on knowledge of the day. And when we examine the evolution of ideas attempting to describe the Universe and its evolution, we have to place ourselves in the context of the knowledge at the time those ideas were formulated. In 1915–17, at the time of publication of Einstein papers on general relativity and applications, which were the results of a decade of reflection and hard work, the capability of instruments available to study the sky was rather limited. The solar system was well known and stars appeared to exist in a very large number. But those fuzzy objects, nebulae, some of them having the shape of a disc, caused a problem. Their exact nature was questioned and in particular their distance from us was unknown. Were they within our own group of stars, our galaxy, the Milky Way? It is only in the 1920s that serious advances were made in addressing that question.

The progress made in that decade both theoretically and experimentally with the available tools of the time is astonishing. It was a glorious time and is a testimony to the genius of those scientists who participated in it. The following description will be concentrated on some key findings and key aspects of the ideas that were developed. We will limit ourselves to essential concepts that have resulted in the present model of the evolution of the Universe.

Distance Measurements

As just mentioned relative to nebulae, one of the main problems in astronomy was and still is the exact determination of distances of objects in the sky, particularly far ones. If all stars were identical in size and luminosity, our Sun included, it would be simple to determine their distances. We would need only to measure the amount of light emitted by a star and reaching us. It would give us what we call its apparent luminosity. With that number, knowing that the energy radiated by the star is spread equally in all directions and knowing that its intensity drops as the square of its distance R, since it is spread over the surface of a sphere, which is equal to $4\pi R^2$, we would determine easily its distance from us. We would need only to determine the distance of one of them and the distance of all others would be determined through luminosity comparison to that one. However, nature has not made things that simple. Stars come in many sizes, they do not emit the same amount of radiation power, and their colour varies showing that the wavelength at their maximum of emissivity varies since they are not at the same temperature. Nevertheless, we can make some intelligent guesses. Celestial objects need to be of a certain size, a certain mass, to become a star. Our Sun is an average star not too large, not too brilliant, and not too small. Many copies should exist in the sky. We could use it as a standard. We can make averages and attempt to evaluate, at least roughly, distances of other objects on that basis. This is what was done in the early years of the last century. For example, the distance of the Andromeda disc nebula was evaluated by means of an averaging technique of that sort by assuming it was actually an object made of stars similar to those in our own

Milky Way. Reasoning on stars count, luminosity, evaluating total mass, and assuming that stars like our Sun are probably spread throughout our galaxy, it is possible to obtain a rough number for the ratio between mass and luminosity of the Milky Way. Making the same assumption for the Andromeda disk nebula and evaluating mass from a measurement of the local velocity of the disk rotating about the massive centre as measured from Doppler shift of light emission, it was possible to compare the results to those obtained for our own galaxy and deduce its distance from us. That distance was evaluated to be 1.5 million light years, which means that it is so far that its light takes 1.5 million years to reach us, a respectable distance (Opik, 1922). The presently accepted value, obtained through better measurements, is ∼2.5 million light years. The number obtained in those days, however, was a triumph: it situated clearly the nebula outside our galaxy, which is of the order of 100,000 light years in diameter. The door was opened. Our galaxy was not unique and the disk of the Andromeda nebula appeared to be made of stars very much like our own Milky Way. It was a galaxy on its own right. That changed totally our perception of the dimension of the Universe.

Distance measurement of objects in the sky is indeed a major problem and considerable effort has been given to it over the years. As just mentioned above, evaluation based on average luminosity and mass alone may lead to large errors. Better methods are needed. Methods based on different physical phenomena applied simultaneously to the same object would be best, since they would provide a correlation of the results obtained in one technique with results obtained in a totally independent one. This is what has been done over the years, providing a method for cross checking one result against those obtained by other independent means. The whole process is called the 'cosmic distance ladder'. First, a geometrical technique is used to evaluate not too distant objects, for example the closest stars in our Milky Way, directly in terms of measurable geometrical parameters on Earth. Then, another technique in which measurements is made, based for example on the luminosity of stars of the same class in close and distant galaxies, can be used. Overlapping of the two techniques for the closer objects allows comparison to the measurements obtained by the direct technique and provides a basic standard for calibration. Other methods, capable of making measurements of specific properties over larger ranges make possible distance measurements over larger and larger distances. It is similar to a ladder where techniques having various ranges overlap and use some of the same rungs. It is obvious that errors made in the first rungs, however, propagates up the ladder and accuracy at larger distances is lost in the process. This is the reason why so much care is taken in correlating results obtained by various independent techniques.

The most direct technique that applies only to relatively close objects is the use of parallax, a visual effect caused by geometrical properties of space in which the apparent position of an object relative to objects situated far in the background is altered by the motion of the observer. To explain the technique let us use it on the surface of the Earth as in the diagram shown in Figure 9.1. In that technique, the observer situated at point A observes the position of the object relative to an object situated behind it very far in the background. He then moves to point B along line AB, counting carefully his steps. He measures the angle α between the new line BC pointing to the object and the line to the distant object in the background, which if far enough would not appear to move. A straightforward calculation gives the distance d as a function of the angle 'α' and the number of steps he made from walking from point A to point B. Knowing the length of his steps or measuring accurately the length 'a' with a calibrated tape he can evaluate distance 'd' in the units he desires. The distance to the object is simply d $=$ a/α

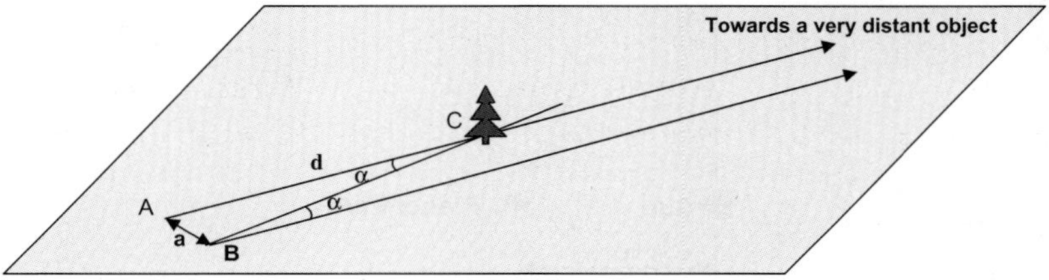

Figure 9.1. Illustration of the parallax technique in distance measurement. The distance 'd' of the tree at point C from line AB having length 'a' is measured by means of geometrical properties, using an object that is very far in the background and that does not appear to move when the observer moves from A to B.

where α is expressed in radians. The smaller the angle α is, the farther the object is. It is to be noted that the result may also be obtained from a simple reasoning on the properties of the sides and internal angles of triangle ABC.

It would appear that this technique when applied to stars from the surface of the Earth would be inaccurate because of the disproportion between the distance 'd' to a star, measured in light years, and the distance 'a' on the surface of the Earth, measured in kilometres, although possibly in the thousands. The angle α would be much too small to allow an accurate measurement. However, the Earth in its trip around the Sun situates itself every six months at opposite points of its orbit. The distance 'a' becomes very large. This is illustrated in Figure 9.2.

In that case, a very distant star is chosen as reference and the angle α resulting from the motion of the Earth is measured. We note that the distance d is much larger than a, even for the

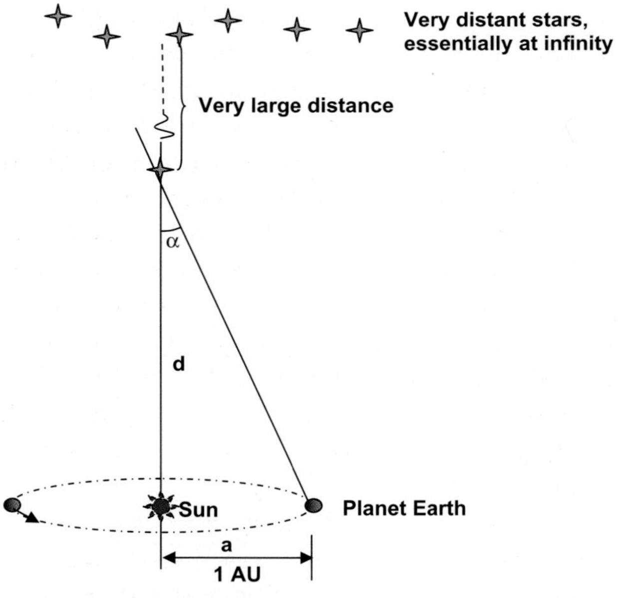

Figure 9.2. Illustration of the technique of parallax used to measure the distances of close-by stars. A measurement of only two parameters is needed, the radius 'a' of the orbit of the Earth around the Sun and the parallax angle 'α'.

closest stars. The star distance is then as above simply d = a/α. Fortunately 'a' is rather well known, its value being obtained by means of measurements of the motion of the Earth relative to other celestial objects including planets such as Venus. It is 149,598,000 km. Being the first rung of the ladder it is very important and has been given the name Astronomical Unit, AU. From the figure, if α is an angle of 1 second of arc, then the distance d is called a *parsec*, written symbolically as 'pc' (parallax of one arcsecond). The distance, 1 pc, is equivalent to 3.0857×10^{13} km, 206,265 AU or 3.26 light years. Most dimensions in relation to astronomy are expressed in terms of parsec. The closest star in our galaxy, Proxima Centauri, leads to an angle α equal to 0.77 arcsecond and its distance from us is thus 1.29 parsec or 4.2 light years.

Measurements of that nature have been expanded over the years to further and further distances through improvement of the measurement technique. The relatively recent use of a dedicated satellite by the European Space Agency, Hipparcos, has extended the measurements to stars at a distance of the order of 1,600 ly. The resolution of the method allows a determination of α in the millisecond of arc or 1 ly at the largest distances. A catalogue of over 100,000 stars has been assembled. Those measurements thus provide a very large number of first rungs in the cosmic ladder.

Other phenomena, such as proper motion of distant objects relative to the background, makes also possible the use of the parallax technique in the determination of the distance of farther celestial objects. For example, the expansion of the residual gas cloud of a star that has exploded (supernova) provides information that can be used to determine distance through the parallax phenomenon. In that case the expansion velocity is determined by means of the Doppler shift of the radiation emitted by the cloud. This last technique can be applied to very far objects. Unfortunately, star explosions are not frequent events.

Most of the other methods use the brightness of stars to determine their distance, as was mentioned above in the case of early distance determination of nebulae. The use of that technique relies on a model of celestial objects, or classes of objects, that have similar properties. It is then based on the definition of a standard luminosity

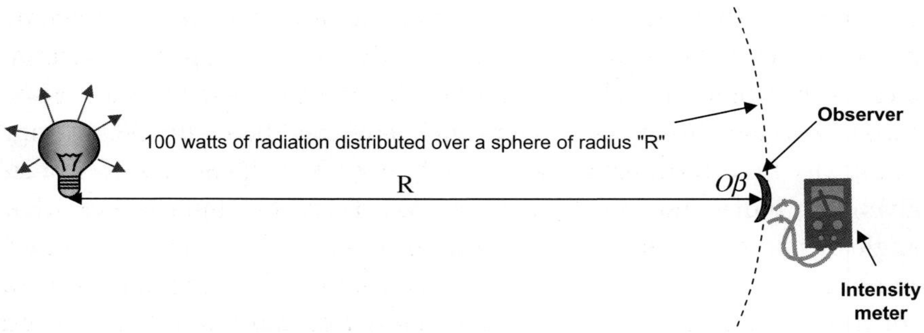

Figure 9.3. Light emitted by a 100 watts light bulb is spread over a sphere, which has a radius R at the site $O\beta$ of the observer. That surface is $4\pi R^2$. The luminous intensity, as measured by the detector at $O\beta$, can thus be used to obtain the distance R by means of simple algebra.

for the class in question. The technique is illustrated in Figure 9.3 be means of a common object.

For example the distance of the observer $O\beta$ from a light bulb B of known intensity can be evaluated from the measured light intensity at the observer's site. In such a case the light bulb acts as a standard and may be called a *standard candle*, a reminder of the use in the past of candles made of specific material as a standard of luminous intensity. A difficulty arises due to the fact that the detector may have an efficiency that varies with the wavelength of the radiation. However, it can be calibrated with known sources. Consequently, 100 watts light bulbs fabricated the same way form a class and could be used as standard candles for determining their distances. Their energy spreads in space on the surface of a sphere. The relative intensity of two of those light bulbs situated at two different distances is then simply inversely proportional to the square of their distances. Using the relation in reverse, distance is determined. We often use that idea in determining mentally, but grossly, the distance of various illuminated objects at night, like cars headlights, extrapolating in our mind to the original intensity of the light source, which is approximately the same for most cars. The astronomical problem is thus to find classes of objects in the sky that, in principle, have the same luminous intensity. Fortunately, there are many possibilities such as red giant stars, Cepheids, supernovae and many others for which a physical model can be made and a standard candle can be defined. For example a Supernova 1a occurs when a particular star, a so-called white dwarf, accretes material from a companion. When it reaches a critical mass of ~1.4 Solar mass, internal radiation pressure and so-called electron degeneracy can no longer counterbalance gravity and it collapses: it becomes unstable and explodes. We have a credible model for the evolution of such an event and all Supernovae 1a explode when they reach the same mass. Their luminosity is the same. Thus, Supernovae 1a form a rather homogeneous class, are great standards in the sky and provide standard candles for determining distances of galaxies when such supernovae are found in them. Another interesting group is a class of stars called Cepheids. These are stars whose intensity varies periodically with time from days to months. The pulsation is believed to originate from a cyclic phenomenon that reflects a dynamic equilibrium between the star outer layers helium gas gravitational contraction-ionization and expansion-recombination. The two states of the gas have different absorption properties. In 1912, Henrietta Leavitt observed that there was a direct relation between the period of fluctuations of those stars and their brightness. It is relatively easy to measure periods and, consequently, those stars provided a new class of standard candles that could be used in the cosmic distance ladder. The space telescope Hubble makes possible the use of that technique to over 50 million light years distances. Numerous other techniques relying on such physical phenomena have been developed over the years.

As mentioned above, the accuracy of distance determination by means of an individual technique is often questionable but comparison between techniques provides a comfortable level of confidence in the results. Thus, nowadays, there are many independent techniques to measure distances in the sky and astronomers can evaluate rather accurately the distance of most objects, stars, nebulae and galaxies, except the very far ones.

The Universe Canvas is Expanding

We can thus measure the distance of objects in the sky at least up to a certain limit. And the results obtained show us that some are extremely distant. Furthermore, since the speed of light is limited, we know that what we see is not the present state of the observed object. An object at a distance of 50 million light years appears to us as it was 50 million years ago since light has taken that time to reach us. We thus travel in time, backward of course. Looking through telescopes we obtain what is called a world picture. It is a picture that we take of the world by registering the light that reaches us, but started to travel thousands and millions or years ago depending on the stars and galaxies we record on film. We are part of the world and embedded in it. Our instruments give us that kind of information.

On the other hand, we may question the stability of that Universe and the actual position of a given object. Is it stable as Einstein seemed to assume in the solution of his equations? Where is that object now? If we were able to look 50 million years in the future, would it be at the same place? This is the question that was answered by Hubble in the late 1920s by means of tedious and patient studies with the Mount Wilson 100 inch telescope (2.5 m), assisted by his collaborator Humason. In those days, techniques of distance measurements as described above were not developed to the present level. However, some were known to Hubble. He could evaluate sufficiently well distances of objects in the sky from luminosity measurements and Cepheids periods to draw credible conclusions on their actual position relative to us. His observations led to the confirmation of the conclusion drawn by Opik that many of those observed nebulae were in fact galaxies of their own, formed of billions of stars and existing outside our own Milky Way galaxy. However, Hubble made a most important observation: he found that those galaxies, aside from a few exceptions, are all drifting away, recessing from us, at tremendous speeds whatever their observed position in the sky. And, most importantly, the further they were, the faster they were drifting away. Could we be privileged and be at the centre of that agglomeration of galaxies? It did not make sense. Only one conclusion could be drawn: all galaxies of the Universe are recessing from each other.

How did Hubble determine that? By examining carefully the wavelength of the radiation emitted by the elements forming the body of those distant objects and comparing those wavelengths to standard wavelengths of the same elements that we can generate here on Earth in the laboratory. Thus, we assume that the same elements exist in those galaxies as in our own, the Milky Way, and in our laboratory. The fingerprint of those elements is the wavelength of their emitted radiation, looking like a bar code. We have seen that hydrogen emits radiation with specific wavelengths. For example, a most characteristic radiation is the ultraviolet radiation Lyman α at a wavelength of 121.6 nm and many others at shorter wavelengths, all in the ultraviolet. They form a so-called series of spectral lines with very well determined wavelengths. It is called the Lyman series. Those wavelengths unfortunately are so short that radiation of that nature does not travel in air; the radiation is absorbed in the atmosphere. That range of wavelength is called vacuum ultraviolet since it propagates only in vacuum or in a very limited class of material. However, hydrogen emits also visible radiation consisting of spectral lines forming the Balmer series, of which the best known line is radiation H_α at 656.3 nm, a red line. Several other lines of that series fall also in the visible range, such as H_β (blue) at 486.1 nm, H_γ (green) at 434.1 nm. Hydrogen is very abundant in the Universe. The red Balmer line is the radiation that gives a beautiful reddish colour to some of the nebulae, showing us that their main constituent is hydrogen.

The element at the origin of the observed radiation can thus be identified by means of spectral lines that the element emits forming well known series. The observation made by Hubble was that the wavelength of radiation received from distant galaxies was longer than the wavelength of radiation generated in the laboratory from the same element, confirming earlier measurements made by Slipher. The conclusion was that the difference in wavelength originates from a phenomenon similar to Doppler effect, which we have described earlier in some detail for the case of sound waves and have applied to light waves. However, there is a subtlety in the present process. Who is moving? Is it the far away galaxy or us with our laboratory? The effect is the same in both cases. However, if all the galaxies are static, do not move relative to each other, and we are moving, we should see that the shift is different depending on the direction of our observation relative to the direction we are moving. This is not what is observed: it appears that for all galaxies, aside from some close ones that have a peculiar velocity, the shift is towards longer wavelengths, which mean that they are all moving away from us. Are we at the centre of the Universe? No. If we set ourselves on a long trip as the one made in the first chapter, we could reach another galaxy. We would find the same phenomenon. All galaxies would appear to move away from our new home. The Universe is isotropic; it looks the same in all directions and in all locations. It looks very much as if *space between galaxies is expanding.* This is a subtlety with important consequences. It appears to give space a personality of its own, an existence as something real, not simply vacuum or, to use a common word, not simply 'nothing'. Space is a reality on its own. This is illustrated in Figure 9.4 for two galaxies.[2]

Consequently, radiation emitted by a galaxy travels through an expanding medium, space. The wave travels locally at the speed of light as dictated by special relativity. When it reaches the observer in the other galaxy the wavelength has stretched in proportion to the space expansion in between during the travel time. Consequently, radiation received from a distant galaxy, which appears to recess from us, has a wavelength longer than that emitted by a source consisting of the same element at rest in our laboratory. As mentioned above there is no doubt on the actual elements in play as they are well identified by their characteristic series of spectral lines. Furthermore, for close by galaxies, the wavelength change is small and there is no ambiguity in the identification of the emitting element in question. We then define a factor z, called *cosmological redshift*, as the difference between these two wavelengths divided by the wavelength in question.[3] It has no units. It is called redshift since it is a measure

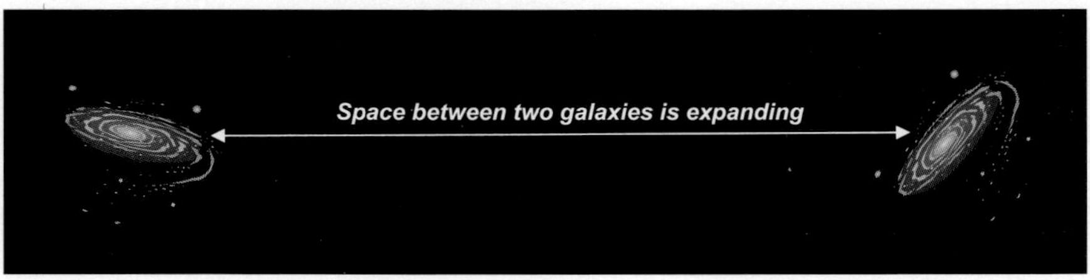

Figure 9.4. Galaxies appear to be moving away from each other. The interpretation is that space between them is expanding or space is created between them.

of the displacement of the spectral lines towards longer wavelengths, that is to say, towards the red when the spectral lines happen to be in the visible part of the spectrum. When the source does not appear to move, then z is zero, since the two wavelengths are equal. If the source appears to move very fast, say close to the speed of light, the detected wavelength becomes very large and z becomes large also.

The most important discovery made by Hubble is that the measured redshift for all sufficiently distant galaxies increases with their distance. The further is the galaxy, the larger is the redshift and the faster the galaxy appears to recess from us. It is very much like a raisin cake baking and expanding in the oven. All the raisins, like galaxies, are moving away from each other and no one is privileged. We can make a better picture as follows. Picture the Universe as if the galaxies, represented by small pieces of cardboard, were distributed on the surface of a balloon that is being blown up. This is illustrated in Figure 9.5(a).

The distance between galaxies increases as the balloon is blown up and increases in size. It is also easily observed that galaxies that are further apart are recessing from each other faster than those in close proximity. As another way of imagining the phenomenon, take an elastic band with spots printed on it as in Figure 9.5(b). Upon stretching the band uniformly, the space between the spots increases and since the space between spot C and spot A is twice as large as that between spots B and A, spot C recedes from A at twice the speed spot B recedes from A. This is the same thing with galaxies at various distances from each other. The conclusion is thus that *the whole of space is expanding*: space appears truly to have a personality of its own.

The world picture concept we introduced previously takes now all its meaning. When we look in our telescope we do not have a picture of the world as it is now since galaxies are recessing and have been recessing in the past. The present world should be very different. It is then more appropriate to represent the present state by means of a so called world map like a city street map. It is a picture of the world that would be obtained from outside, looking at the whole ensemble in its evolution. This conclusion has effects on the determination of

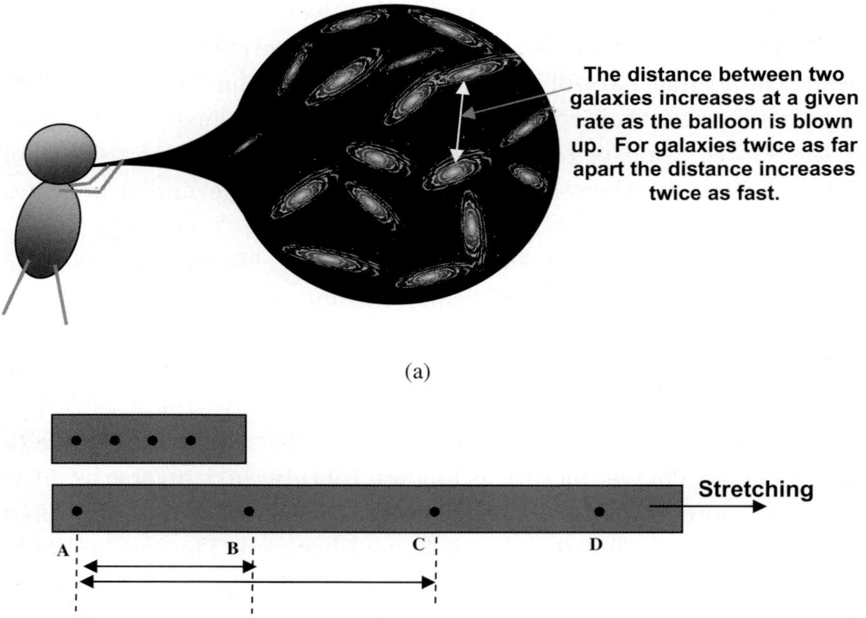

(b)

Figure 9.5. (a) Illustration of the Universe expansion by means of a two dimensional model in which the galaxies are distributed on the surface of a balloon that is blown up. (b) Illustration by means of a stretched elastic band of the increase in recession speed for points further apart.

parameters such as expansion rate, for example, since it could have varied with time. This is an observation that has to be taken into account in the analysis of the effects just mentioned and we need a model of the Universe evolution.

We may wonder, however, what happens to stars inside a galaxy or to planets within a planetary system like our own? Does the solar system expand? No. The planets around the Sun are held by gravity. The size of their orbits is determined by gravitational force and their own particular speed holds them at a well known distance from the Sun. The distance between Earth and Sun is not affected by the expansion of space. Gravity wins. The same thing happens within galaxies. The components, the stars, are held together by gravity, revolving around the central bulge of the galaxy, its core. They act as a whole, unaffected by the expansion of space. It is also the same on a larger scale where galaxies form groups. We are part of a so-called Local Group formed of over 35 galaxies that attract each other. Local gravitational attraction in fact gives peculiar velocities to close by galaxies such as Andromeda that is approaching the Milky Way at a speed of about 300 kilometres per second. We must realize that this is a very large speed when compared to familiar speeds reached on the surface of the Earth. However, it should be mentioned that even at that speed Andromeda will not collide with our Milky Way soon; it will take a few billion years.

Another important aspect of our illustration by means of a balloon, is that there is no edge to the two dimensional Universe formed by the external surface of the balloon. A two-dimensional being, travelling from galaxies to galaxies on that surface, would never reach an end, but would come back on its path. We note, however, that a surface that is curved on the large scale appears to be almost flat on the small scale. Can we say something like that about the three-dimensional space we are living in? As we discussed earlier this is more difficult to visualise. We have an advantage with the sphere: we can see it from outside and see the surface being curved as observed from our three-dimensional world. We obviously do not have this advantage for an object in our three-dimensional world. We are part of that world, living inside it similar to Flatant living in her two-dimensional world on the sphere. The ant is not aware of that third dimension unless she makes measurements on the properties of two-dimensional geometrical objects. Similarly, our three-dimensional space could be curved. We would have to make measurements on three-dimensional geometric objects to verify that curvature. In a curved three-dimensional space,[4] as in the case of the two-dimensional space of the blown up balloon, there are no edges: a continuous path would bring us back to our starting point. Consequently there is no centre to such a space and there are no edges in sight. We will come back to this point later.

Hubble was the first to quantify experimentally that important observation and to obtain a number for the rate of expansion of the Universe by evaluating both redshift and distance for many distant galaxies. In early work, his results on 46 galaxies were characterised by great uncertainties. As mentioned by Weinberg,[5] it is difficult to understand how Hubble could have arrived at the conclusion he reached about expansion when we examine the data he collected in those early measurements. However, later measurements made also in several other observatories provided greater accuracy in the measurement of distances by means of the cosmic ladder we mentioned earlier. We now have a number characterising the expansion of space, H_o, rightly called the Hubble constant, in which we are quite confident. At present, it appears that the Universe is expanding, or the galaxies are recessing from each other, at a rate of

$$H_o = 72 \text{ km/second per megaparsec of distance between the galaxies.}$$

This means that two galaxies that are 1 megaparsec distant (3.26 million light years) from each other are moving apart at a speed of 72 km per second, or rather space between them is increasing at that rate. The number is known with an uncertainty of a few per cent. Such an uncertainty reflects the difficulty in evaluating distances of objects in the sky. We take note that a galaxy distant from us by 326 million light years would appear to move away from us at the rate of 7,200 km per second, a rather large velocity that is not familiar to us on the surface of the Earth. Rocket speeds are expressed only in thousands of km per hour, not per second.

It is important to insist that galaxies are not seen as moving through space at that speed although locally they may move with a smaller peculiar speed due for example to local gravitational attraction. We can imagine space as a super grid with a fixed coordinates system. Galaxies are attached and fixed to that grid. The grid expands and space between the coordinates increases everywhere while the galaxies attached to it do not change coordinates. It is somewhat like longitudes and latitudes on a sphere. When the sphere radius is increased like a balloon that is blown up, coordinates, that is latitude and longitude, do not change but distances on the sphere increase. Similarly, since space expands everywhere, the further galaxies are from an observer, the faster they look like flying away although they do not move relative to the calibrated grid. Some galaxies very far from the observer could even be looking like recessing faster than the speed of light. Locally, in their own reference frame, they do not move and consequently they do not violate the basic principle of special relativity. This brings us to the heart of the space expansion concept. We can imagine a large sphere with a radius such that two specific regions of space are delimited. Inside the sphere galaxies are recessing at a subliminal speeds, that is to say speeds less than that of light, while outside galaxies are recessing at superluminal speeds, that is to say faster than the speed of light. That sphere is called the Hubble sphere. Each observer in its own galaxy has his own Hubble sphere and is at its centre. Light rays outside the sphere appear to recede from the observer and we could conclude that those galaxies would never be visible. However, galaxies may be decelerated for example by gravity and the Hubble sphere may overtake the receding galaxies, which then become visible. This phenomenon has obvious effects on the size of the visible Universe.

The conclusion relative to expansion was not without turmoil in the scientific community and the philosophical and religious world at large. It transformed radically our theoretical approach to understanding the functioning of the Universe and our existential position in that world. It provided a basis for the development of theoretical models leading to the identification of the origin of our Universe. One of the first obvious conclusions is that if two objects appear to be flying away from each other, they must have been closer to each other sometime in the past. If two galaxies one megaparsec apart are recessing from each other at the rate of 72 km per second, it is easy to evaluate the time when they were overlapping or on top of each other. It is simply the inverse of that rate of expansion expressed in the proper units by transforming km into light years or megaparsec. This exercise gives a number of about 14 billion years. Since this applies to all components of the Universe, we can conclude in principle that all galaxies were on top of each other at that time. That would be a rather crowded world and, as we will see below, such a world would contain so much energy and would be so hot that it could not support the presence of such objects as galaxies. Lemaître, a catholic Belgian priest, studying the matter in the light of evidence of the expansion demonstrated by Hubble, proposed that the Universe expanded from an infinitesimally small object called the 'primeval atom' containing an extremely large amount of energy. That energy would be the source of all masses or other forms of energy present now in the Universe. That idea was not accepted by all. For example, a number of physicists under the leadership of the astronomer F. Hoyle would not accept such an extrapolation and proposed a model of continuous creation of matter between galaxies providing, in a sense, the engine for expansion. That Universe would be eternal. They caricatured the approach proposed by Lemaître by giving it the name 'Big Bang', as if a big explosion had taken place some time in the past. Interestingly enough, as will be seen below, Lemaître's model won the case, and the name Big Bang invented by adversaries survived. It is somewhat misleading because it seems to say that somewhere in space and sometime in the past a large explosion took place. This is not so: in the most plausible scenario, space did not exist at that moment. Time did not exist either. The Universe does not originate from an object that has exploded at a certain time in an already existing medium that we call vacuum. That vacuum did not exist before the Universe was born. We cannot even say that there was 'nothing', because that word makes reference to the existence of something. Vacuum or nothing is relative to a state where something exists. Space or vacuum appears continuously at the same time as the evolution of the Universe takes place. In this interpretation, vacuum is as real as objects floating in it.

The Big Bang model that we will examine below was developed further over the years from ideas initiated by Gamow and his collaborators. The model could describe the actual steps of the evolution and could even make predictions on the proportion of the elements in existence today and on the existence of a phantom background radiation that is now filling the whole Universe. Those predictions have been found to be real and the model has now received general acceptance in the scientific community.

Gravity in Einstein's Universe

Several years ago, I was in Torino, Italy, working at the *Istituto Elettrotecnico Nazionale* (IEN). It was one of those privileges associated with the function of University professor to have the possibility of going away for a year, taking a so-called sabbatical year, with the goal of creating collaboration or reactivating old ones, as well as, presumably, broadening one's spectrum of knowledge. I had decided to go to the IEN in Torino to help one-time colleagues in the development of atomic clocks using caesium. I was invited at the same time to give lectures in that specialized field at the University of Pisa. What a delightful trip it was to go from that northern industrial city of Torino to the calm peaceful city of Pisa, realising that I was going into a very special place on the planet. Every time I travelled on that train with an imaginative colleague of mine, who was courteous enough to come with me and provide feedback on my lectures, I felt privileged to be in that country, which had been the cradle of the Renaissance and particularly to go to Pisa, the centre of a revolution in physics.[6] It was intimidating to go to that University to deliver lectures on the physics of atomic clocks, not only because of the notoriety of those attending the lectures but because of the proximity of the site to where Galileo had done his own experiments using a suspended lantern in the Duomo as a pendulum, essentially his first clock. The Duomo is also just besides the Leaning Tower where, as mentioned earlier, he is said to have proved experimentally a most basic principle of physics by dropping two objects of different weights from one of the highest level of the tower: the idea was to prove that they would arrive on the ground at the same time. Once, I went up the tower, (yes, it is really leaning) and, although rather busy watching my young son's wanderings to avoid a catastrophe, I could not resist thinking of the big show Galileo must have made from that height in front of the group that accompanied him. It must have been impressive. Galileo was a most brilliant scientist and when you think about it he must have been well prepared, knowing the result beforehand from previous experiments. It is said by some that the event is a tale and never happened. I know a lot of my contemporaries, including myself, that would not have hesitated to do it and I also tend to believe Galileo did it. In any case, he must have done something similar since we are still talking about it: a revolution in science was born.

The equivalence of inertial and gravitational mass that we discussed previously several times is probably one of the most basic principles of physics. This is what Galileo had proved. The gravitational attraction of the Earth exerts a force on objects proportional to their mass. A larger mass experiences more force, thus, should accelerate more than a smaller one. This mass is called the gravitational mass. However, according to the principle of inertia, when a force is applied to a given object the resulting acceleration is inversely proportional to one of its property that is also called mass. This mass is called the inertial mass opposing to changes in movement or acceleration. Nothing says that these two masses are the same. Galileo proved it experimentally throwing overboard at the leaning tower concepts that had been encrusted in the way of thinking from Aristotelian times. Einstein made a principle of it and it is the basis of his general theory of relativity, which is a geometrical theory of gravitation. It is most surprising to realise that although gravitation is the most common phenomenon around us and that we feel its presence all the time through the force that it exerts on us and on all objects around us, its origin remained a mystery for centuries. Even though Newton could put it in mathematical form, derive its property, and calculate the orbit of planets, he confessed that he could not understand that such a force could take place between objects without anything being exchanged between them. He could not address the question on a scientific basis and left it unresolved. It was Einstein, several hundred years later, who provided a theoretical model and opened the door to an explanation of the origin of gravity as well as providing a new model for the Universe as a whole, a model that has survived as the best one to the present times.

Following his work on special relativity, which was concerned with the motion of objects at constant velocities, Einstein addressed the question of objects under acceleration, in particular acceleration by gravity. We have seen that special relativity led to the concept of space-time, a world where time and space form a single entangled entity of which the velocity of light is the only absolute property. In the context of general relativity, Einstein developed a new geometrical approach for the description of the Universe in which the presence of mass produced a curvature of space-time. We have examined the geometrical property of curvature in a previous chapter. The space geometry introduced in special relativity led to the concept of the so-called 'space-time metric', a frightening concept in appearance, that for the purpose of explanation may be said to mean simply a measure of a distance or interval in the space-time grid. In our three-dimensional space represented by three coordinates, the metric is essentially the distance between two points. In special relativity, we use the metric introduced by Minkowski, a quantity formed of four components, three space coordinates and one time coordinate forming a 4-dimensional grid, space-time. Neither space nor time is absolute: they are irremediably entangled, interlaced.[7] We live in that world travelling both in time and in space with its three axes, continuously exchanging space for time and time for space depending on our velocity. If we travel fast, at a speed that is a good fraction of the speed of light, we have seen that space is heavily traded for time, which gets dilated: seconds get longer, a phenomenon verified experimentally. Without matter around, that space is flat. This means simply that the supporting grid is flat and regular, like the floor made of square tiles in a bathroom. What do we mean by that in geometrical terms? It means that the sum of the internal angles of a triangle on that surface is 180 degrees. It also means that if we lived in a two-dimensional world, and made a circle in that flat space the circumference of that circle would be the product of its diameter by π. As an example, we saw a very nice representation of a curved surface in the case of a two-dimensional canvas that was the surface of a sphere. The sum of the angles of a triangle was greater than 180 degrees. The circumference of the circle on that curved surface was not π times the diameter (Figure 3.3). Similarly, in a flat three-dimensional space, the surface of a sphere is 4π times its radius squared. This would not be so in a curved three-dimensional space, but it is more difficult to visualise.

Einstein developed a new geometry of space-time containing mass, a rather complex geometry based on curved four-dimensional space-time utilising advanced concepts developed previously by the mathematician G.B. Riemann. The metric in that geometry could no longer be expressed as in the flat Minkowski metric: mass caused a fundamental change to the geometry, introducing a curvature of the reference grid. In that model, objects near massive objects like stars and planets travel on a curved canvas, which seen from our three dimensional space look like movement on curved paths. Mass is said to affect the geometry of space-time and the geometry of space-time affects the path of moving objects having mass: in our space, gravity appears as acceleration because objects are deflected from a straight path and appear to be attracted by mass and thus accelerated. Again it looks very much that space is not simply a vacuum and space-time possesses a reality of its own: its properties are affected by the presence of mass.

That theory, based on intrinsic geometrical properties of space, represented by a so-called metric that includes the effect of mass and energy on its properties, leads to a set of general equations, called Einstein's field equations. At first sight those equations appear to be somewhat simple when written in proper notation. The application of these equations to practical situations and their solution, however, is rather involved because of their internal complexity. Such a statement may surprise the reader. I am sometimes told that if we have an equation representing a physical phenomenon, which may then be a physical law that says 'this is equal to that', then it should be obvious that one side is equal to the other. We should then have only to put numbers in it and the answer should come out. It is not as simple as that and the Annex showing some of the mathematics used in the description of the Universe addresses comments of that nature. In that Annex it is made obvious that applying basic equations to practical problems and solving them to obtain a description of the behaviour of an object in a particular situation is not always a simple thing. Just think of the law of gravity. It says that two masses attract each other with a force inversely proportional to the square of their distance. If we know the value of those masses and the distance between them, we can calculate the force and then their acceleration towards each other by means of Newton's second principle. However, this is very limited information. If you wish to know the trajectory of a mass thrown at an angle above the surface of the earth, the problem is much more complex, and, in applying the equations of attraction and inertia, conditions have to be set for original speed and direction of motion for example. It should be mentioned, along those lines of thought, that even with the simple equations involved in Newton's model, a solution can be obtained exactly only in the case of two objects. When another massive object is involved the exercise becomes a three body problem and cannot be solved under the form of a closed form mathematical expression. In any case, Einstein field equations cannot be solved in the general case to provide a closed form expression for the trajectories of objects submitted to gravity, but can be solved in particular simple situations that provide information on the dynamic behaviour of the Universe.

We have seen earlier how gravity, being assimilated to acceleration, could affect the rates of clocks. This could not be verified at the time Einstein developed his theory. However, the advent of the atomic clock in the 1950s and 1960s, with its unsurpassed stability made such a verification possible, particularly with the H maser as was done by Robert Vessot[8] in the 1970s. However, at the time the theory was completed, Einstein applied it to specific cases and could explain a particularity of the orbit of Mercury and could make a prediction relative to the effect of gravity on the trajectory of a light ray. Mercury makes an elliptical orbit around the Sun. It had been known for some time that the orientation of the axis of the resulting ellipse was changing with time relative to the axis of the orbit of other planets. In other words the axis was precessing with time by a small amount, actually once around the full circle in 3 million years. The planet is close to the Sun, feeling an intense gravitational field, and travels at great speeds. The effect could not be entirely explained with Newton's gravitational theory. General relativity provided the full explanation. The theory predicted also that if a light ray coming from a distant star passes close to a massive object, such as the Sun, it would be deflected by a small amount, but measurable. However, how can you see a star in full sunlight? The suggestion was to wait for an eclipse of the Sun by the Moon. A succession of photographs during the time that the eclipse takes place while the Sun moves across the path of the light ray should provide the data for a verification of the prediction. It required perfect synchronization since an eclipse lasts only for a few minutes. An expedition was organized by Sir Eddington to an equatorial region where an eclipse was to take place. The effect was measured and the result agreed within experimental error to the value predicted. Such a measurement was difficult, the deflection being of the order of a second of arc. It was a triumph for Einstein. In physics, there is no better proof of the validity of a theory than the verification of a prediction.

Einstein's Static World and Friedmann's Expanding Universe Models

The field equations developed by Einstein may be applied to the totality of space containing matter and energy under whatever form it could take. But a quick look at the sky tells us readily that this is a very complex structure that contains planets orbiting a star, our Sun, a structure that contains other stars that, as we now know from recent observations, have their own planetary system, stars that are themselves regrouped in galaxies, which in turn collect themselves in larger clusters and so on. It is so vast that it overwhelms our imagination. One thing transpires from this whole structure. Seen on a very large scale it appears to be very isotropic and homogeneous. When galaxies are represented as points on a hemisphere, as observed from Earth, they appear on a large scale, say a cube with sides of a few hundred million light years, as distributed homogeneously with an average density of matter varying by less than 0.1% from cubes to cubes. The properties of all such cubes, mass content, density, radiation content, are the same on average. Furthermore, if we travel in any direction, space looks the same. The recession of galaxies looks the same in whatever direction we go. The whole thing is isotropic. We are not at the centre of that Universe and no other place can claim that role. This is known as the *cosmological principle.* It may be seen as an extension of the principle introduced by Copernicus. Before his input on the question, the Earth was assumed, essentially under religious credo, to be the centre of the world. Following his findings and reasoning, the centre moved to the Sun and then, soon after, the centre disappeared completely in the vastness of space as a non physical entity, every location having the same properties and looking the same. In other words, the Universe looks like a uniform distribution of objects, somewhat like atoms or molecules in a gas contained in a very large container.

This is the basic idea that is used in attempting to obtain solutions to Einstein's equations applied to the Universe as a whole: it is assumed to be isotropic and homogeneous. Matter and radiation, and energy in whatever form, are distributed uniformly in it. The whole Universe is represented by a fluid and two parameters characterize such an isotropic substance through a so-called equation of state: it is its density or mass per unit volume and its pressure. It may be surprising to consider pressure as a kind of energy that acts like mass regarding gravitational attraction. One way of looking at it is to consider a spring that you compress with your fingers. That spring exerts pressure on your fingers. However, in order to compress it you add to do work and the energy you spent is stored in it. Thus the energy of the spring is larger when compressed than when left free. Within the principles of relativity these two entities, mass and pressure are thus considered as energy and both cause gravitational attraction.

In the days Einstein developed his theory, the common belief was that the Universe was essentially limited to our galaxy and was static. Nothing seemed to move on the large scale. In solving his equations, Einstein arrived at a difficulty. The solution did not lead to such a static situation. In order to satisfy that condition, Einstein had to alter his field equations and he introduced arbitrarily a so-called *cosmological constant Lambda* (Λ) that would stabilise the whole system. That constant looks very much like a negative pressure causing repulsion rather than attraction. In that model, the whole Universe stability is relying on that single constant. It needed to be tuned exactly to make the whole thing static. However, that Universe was really at a point of unstable equilibrium. Since Lambda introduces a repulsive force, a small expansion would decrease the attraction force due the mass present, leading to an instability that grows with time. As pointed out by Weinberg it is hard to see how Einstein could have missed that elementary difficulty.[9] Nevertheless, the solution obtained by Einstein was the first global view of the Universe. Einstein's Universe was finite, but without border, closed on itself with a large radius of curvature. A tourist, travelling in an apparent straight line, would find himself back to his original point of departure, although certainly after a long time. We will come back to that concept below.

Soon after Einstein developed his fundamental relativity theory, other theoreticians started to apply the resulting equations to the Universe, making various assumptions. For example, de Sitter (1917) formulated a solution of the equations without matter in it. It showed that the Universe should expand. It was a direct consequence of the cosmological constant introduced by Einstein. Within the next decade, other solutions of Einstein field equations in the presence of matter were found for the same homogeneous isotropic Universe. They were not limited to a static state. In particular, Friedmann (1922) derived a solution that provided direct information on that expanding Universe.[10] Lemaître, later, obtained independently similar results and, as we mentioned, went much further by proposing that the Universe had a surprising origin, a state in which all matter was concentrated in some kind of primeval atom.

Parallel to that theoretical work, Hubble was accumulating data on the identification of galaxies outside our Milky Way and was tabulating their recession velocity by means of their redshift as a function of their distance from us. In previous work in the late 1910s, Slipher had also found the existence of a redshift in the emission spectrum of certain nebulae. However, no conclusions were drawn relative to the exact interpretation of that phenomenon. It is Hubble that interpreted the experimental data as evidence that the Universe is expanding. In 1930, Einstein visited Hubble at Mount Wilson and became also convinced of the reality of that expansion. He accepted it, rejected his own previous static hypothesis, and abandoned the concept of the cosmological constant calling it the 'biggest blunder of his life'.

The theoretical model we now have consists of a space-time metric with a time component the same as in the Minkowski flat space approach, but with a space component expanding with time. The metric is called the Friedmann-Lemaître-Robertson-Walker metric from the name of those who have made major contributions to its development.[11] Friedmann obtained an expression for the expansion rate from a solution of Einstein field equations for a homogeneous-isotropic medium representing the Universe. Contrary to Einstein's hypothesis, the medium is allowed to expand freely on itself and the solution of the equations says that such a Universe will do so.

It is certainly most interesting to observe that the expansion of the Universe was all along contained inside Einstein's field equations, and that it was just a belief of the time that incited Einstein to fix his solution in such a way that it would make the Universe static. This is understandable since at that time the Universe was believed to be essentially the Milky Way. It was thought that nebulae were within the Milky Way. It took the patience and keen sense of observation of Hubble and the theoretical studies of Friedman and Lemaître to establish the facts.

Evolving Ideas

In summary, we thus have experimental evidence that galaxies are recessing from each other at a speed that increases with distance between them. We also have a model that is based on the fundamental geometrical properties of space within the theory of relativity. The analysis incorporates relativistic principles based on the concept of Minkowski's space-time metric that has been successful in explaining all measurements made in the case of constant velocities.

When general relativity principles are applied to that metric it is found that the solution leads to the expansion concept of the Universe. There are also a few other predictions that we have mentioned previously and that have been verified experimentally. Is that enough to justify the acceptance of the general theory of relativity and to accept that the Universe is expanding now and has been expanding for a long time from a single primeval entity as suggested by Lemaître? As mentioned above, for many years after the observations made by Hubble, a competitive model in which matter is created continuously in the vacuum of space was supported by several scientists. The model provided the material for the expansion. There had been a continuous expansion of the Universe in the past and it was still in action. In such a context, one may thus assume that the Universe had always existed and is infinite. Since there was no direct proof of either scenario, the two camps were sticking to their positions. The conflict opposing the two scenarios lasted for decades. Other elements confirming one or the other were required and were soon to be found.

A model for the Universe evolution, from an original entity into its present state with its various chemical elements, from hydrogen to uranium, took shape over the last half century. In particular, a scenario was developed by Gamow and collaborators Alpher, and Herman in a series of articles around 1948–50, that followed the general lines of Lemaître's expansion proposal.[12] They assumed the existence of a primeval substance from which the present Universe evolved. The model was based on a very high original temperature — billions of degrees, with a high density of particles such as neutrons, bathing into a sea of blackbody radiation. Those neutrons either decayed into protons or combine with them by nuclear fusion to produce the elements that we now have in existence today. As the expansion took place, the temperature of the ensemble decreased at a rate that was a function of the 'size' of the resulting Universe. From the model, it was possible to calculate the temperature of the Universe expanding adiabatically, that is to say, without losing energy or gaining some, and to predict roughly the temperature of the residual radiation that should exist still today.[13] Although not exact in some aspects, the scenario was accepted, after some time, and refined by many others. In particular, it was realised later that in the first instants of the evolution, only certain low atomic weight elements could be produced due to the rapid expansion, reducing density and temperature to a level that was not favourable to nuclear fusion. However, the idea of the residual low temperature black body radiation remained, with an evaluation of a predicted present temperature ranging from 5 to 40 K depending on the assumptions made in the model used. First experimental observations were reported in 1957, showing some evidence of the existence of an isotropic background radiation in the sky at a temperature of about 4 K with an uncertainty of 75%. The radiation was then finally detected in 1965, although first as a nuisance, by two scientists at Bell Laboratories, Penzias and Wilson, working on a communication antenna.[14] They had found that their antenna had been plagued by background noise like static in a radio receiver when attempts are made to detect weak signals from a distant radio transmitter. All attempts to get rid of that noise, by optimizing all electronic detecting systems as well as cleaning the antenna structure, such as removing from the structure so-called lossy dielectric matter (droppings from pigeons) failed. The antenna was operating in the microwave range at 4,080 MHz and the radiation noise they were observing was evaluated to correspond to incoming photons with such energy as if they were emitted by a black body at a temperature of about 3 K. Its nature was readily identified by Robert Dicke and colleagues as the relic of the original radiation that existed at the first moment of existence of the Universe and is now known as the Cosmic Microwave Background (CMB). It is worth mentioning that Dicke and colleagues had been working for some time on instrumentation to detect precisely that CMB. When he was informed of the results obtained by Penzias and Wilson he knew immediately the significance of the observation and supported them through his celebrity by means of an independent article in the same journal enhancing the importance of the discovery.[15]

Our Present Large Scale Understanding

The existence of galaxies rather homogeneously distributed throughout the visible Universe, the redshift increasing with the distance of those galaxies, the existence and the properties of the cosmic microwave background, the observation of the proportion of the various chemical elements presently existing in the Universe at large, and the darkness of the night sky, are the main observations at our disposal for making an empirical model of the evolution of that Universe. During the last few decades, enormous work has been done on some of those items and fantastic progress has been accomplished in the understanding of the Universe evolution and the parameters regulating that evolution. In particular, the Cosmic Microwave Background, the earliest witness that can tell us about the state of the Universe some 380,000 years after the beginning of time, has been extensively studied. New technology now available, such as satellites, made that progress possible. It is worth mentioning the dedication of several scientists who did not hesitate to work for decades with remarkable patience in adverse situations such as facing large organization's bureaucracy accompanied by long delays in execution, and adapting to extreme meteorological conditions such as those encountered at high altitudes and at the South Pole, to obtain definitive data.[16] We have now a rather good understanding of *how* the Universe evolved to the present stage, although, as I said before, we don't know the basic *why* such a world ever came into existence. As we mentioned, we may never know, because that type of question is of a nature different from those answered by physics. The presently accepted so-called Standard Model, based on an expanding Universe that came to birth about 14 billion years ago appears to describe, up to a certain point, the reality of the Universe's evolution. Most of the elements of the model are well correlated to experimental observations, giving it great credibility. However, there are unexplained particularities in the proposed evolution model and it still requires fine tuning. A major problem in establishing a scenario of the evolution of the Universe resides at the start of that expansion. Extrapolation back in time, as we know it today, brings us to an entity that includes all galaxies, matter, radiation and energies under any form, and all of space we observe around us on Earth and in the sky below and above. It looks as if everything existing was compressed into a small volume, a kind of a ball that contained a large amount of energy and matter. As was originally imagined by Lemaître, it looks as if some catastrophe happened that forced the volume of the original entity to expand to what we have now around us. We will use freely, for the lack of a better vocabulary, the word 'entity' to identify that primeval object assumed to exist at the birth of the Universe. What it was is not known and we have to use what we now know to imagine an entity that makes plausible the existence of the Universe with all its properties in its present state. Such a situation without possible direct verification can obviously lead to speculations, but using our knowledge of the known laws of physics we can imagine a fairly credible scenario that became a model qualified as standard.

An important point to repeat and emphasise is that the primeval entity, whatever it is, did not appear some time, somewhere, in space.

There was no space around it and we might say that what we call time started when the entity started to expand. In other words, time, space, and energy, appear to have been *created* all together. We also use the word 'create' freely for establishing the presence of the primeval entity. In our time scale it is time zero. What existed before that moment does not even appear to be a proper question to ask since our laws of physics seem to stop at that point and time as we understand it did not even exist. We may quote Gamow who, being asked the question about what existed before the Universe started to expand, is said to have answered something like: 'Hell is reserved for people who ask such a question'. Nevertheless, there is some thinking done on the subject leading to speculations without possibilities of verification within our present knowledge. It often leads to popular articles and books that include a large amount of speculation.[17]

In the Big Bang scenario that is accepted at present as the Standard Model, some of the basic questions are: What did 'bang'? How did it happen? Did a ball explode? Was it an explosion like that produced by dynamite as we know it on Earth? What was the size of that ball? Why did it 'bang'? Those questions are not simple to answer. Since all matter in the Universe appears to have been concentrated in that ball, which in principle could be infinitesimal in our scale, the density of energy would have tended to infinity. Our mathematics is not made to handle infinities and the physical laws we have discovered do not apply in that case. On the other hand, we will see below that according to the results obtained with satellite experiments, such as COBE and WMAP, both studying the cosmic microwave background, the Universe, 380,000 years after the Big Bang, was homogeneous to something like parts in 100,000. The small inhomogeneities present at that time developed later, under gravitational attraction, in the kind of Universe we know today with galaxies and voids rather uniformly distributed on the large scale. We know how standard common explosions occur, throwing away chunks of material in all directions. The results are not very homogenous. It does not appear to be a very practical way of building a Universe as homogenous as the one observed with COBE and WMAP satellites. The homogeneity observed does not look like the results of a pseudo-dynamite explosion with heterogeneous parts distributed all over space or vacuum that then did not even exist.

A Solution to a Homogeneity Problem: The Inflation Scenario

As mentioned above the Universe appears to be rather homogeneous. In order to satisfy that requirement, parts had to be in some kind of contact in the past. Objects out of contact tend to develop independently and reach different states. Electromagnetic radiation is the only vehicle we know that can provide energy transfer by which the kind of homogeneity observed could be accomplished. However, the rate of transfer of electromagnetic radiation is limited to the speed of light. This condition introduces the question of horizons. What do we mean by that? When at sea, for example, due to the curvature of the earth, it is not possible to see objects at the surface of water further than a certain distance. The distance depends on local conditions of course such as height of objects observed. It is called the horizon. Something similar exists in space in the case of static as well as expanding Universe, although the concept is based on phenomena of a different nature such as the limited speed of electromagnetic radiation. We have seen earlier in Chapter III that due to that limit objects could not have a causal effect on our present state if they were in the past outside a region of space delimited by a so-called light cone (Figure 3.13) Objects have their own past or backward looking light cone and so-called world line. The world line is simply the vertical axis representing time for a given object in the diagram of Figure 3.13. If an object backward looking light cone does not intercept in the past the world line of another object, situated somewhere else, there could not have been communication between the objects in question. They were outside their horizons. It is shown that such a situation existed and that different parts of our expanding Universe could not have been in contact in the past.[18] It is thus difficult to see how they could be in equilibrium and have the same physical properties. This phenomenon is called the 'horizon problem'.

Furthermore, it is also found that the Universe we live in is flat or with little curvature. This means that if the Universe could be represented symbolically in a two-dimensional model by the surface of a sphere, like in the illustration with a balloon, the radius of that sphere would be enormous and the surface around a certain region would look extremely flat. Consequently the primeval entity and the nature of expansion must lead to a Universe whose properties are compatible with those observations. The standard model in which the Universe would start from a hot ball, with enormous curvature, expanding to the size it has today does not appear to address properly these two requirements although, as we will see, it explains well the Universe evolution at later stages.

A process that is assumed to have taken place at the very early instant of the expansion, called *inflation*,[19] offers a possible scenario providing an answer to the questions raised above. The scenario was developed by several researchers in the USA and the former Soviet Union.[20] The details of the scenario differ from authors to authors but the basic idea is essentially the same. The difference is in the details. We do not have access to the Universe early moments and we can only evaluate the quality of the details of the proposed scenarios on their ability to predict a world similar to the one we presently live in. However, in view of the ability of the inflation scenario to explain some general characteristics of the present state of the Universe (and possibly also from a lack of other sufficiently credible competing proposals),[21] it has been well received and is now essentially accepted as a component of the Standard Model.

We need a starting point. Earlier, we introduced Plank's natural units: they were obtained as an exercise in dimensional analysis using combination of fundamental vacuum constants. Plank's quantum constant 'h' was introduced in the picture and appeared to provide a kind of limit to our classical world. We obtained a Planck length, 'l_p', equal to 1.6×10^{-35} metre, a Planck time, 't_p', of 5.4×10^{-44} second, a Planck temperature, T_p, of 1.4×10^{32} K. We will round off those numbers to multiples of ten or orders of magnitude. These are most interesting as they appear to provide a limit under which we can expect rather extreme conditions. The world at that scale is quite exotic. We find immediately that the length l_p is the distance travelled by light in Planck's time t_P. That can be seen directly from the definitions themselves given in Annex B. Consequently, entities within the primeval entity distant by Planck's length would have time to communicate within Plank's time and we can conclude that the physical properties of those parts would be submitted to causality leading to an homogenous medium. On the other hand, at those dimensions, it is expected that the classical world we know loses its properties and that quantum effects are predominant. It should be a world of turbulence with particles and antiparticles appearing, disappearing in a very short period and colliding with each other at a frenetic rate and with enormous energy. It is also expected that at those limits, gravity passes from a classical Newtonian or relativistic description to a quantum description. However, as mentioned above, a theory of quantum gravity has not been developed yet and we can only suspect that at those dimensions laws very different from the classical ones did exist. Those numbers appear to be a sensible starting point in constructing our Universe, since we do not know anything beyond. At Planck's time due to the extremely high temperature (10^{32} K) and energy content of the entity assumed to exist, it is believed that weak, strong, electromagnetic and gravitational interactions were united into a single type of interaction. This should be a conclusion of a Grand Unified Theory (GUT). However, attempts at evaluating specific conditions at that time leads to speculations. For example, nuclear dimensions are of the order of 10^{-15} m. This is much larger than Planck's length (10^{-35} m) and consequently we need to think of a different state of matter existing at that time. This is where use is made of a concept introduced in a very successful theory that was developed in elementary particles physics within the context of Quantum field theory: this is the concept of the scalar field. Such a field is introduced, for example, in the unification of weak and electromagnetic interactions as giving mass to elementary particles and some bosons: the Higgs field.

A similar field may be assumed to have been present in the cosmos at the time where all interactions were united at or before Planck's time. In a scenario developed by Guth[22] that field could end up being trapped in a state of high energy. It is called the *inflaton* field. That field would be characterised by a negative pressure, a state called false vacuum. This is similar to the idea behind the cosmological constant introduced by Einstein to provide a mechanism for counterbalancing gravitational attraction in his hypothesis of a static Universe. Consequently, the energy in the large inflaton field would create a negative pressure and would cause expansion of the Universe. In the scenario, the process creates space that contains the field, which stays nearly constant maintaining a constant energy density that fills all space available. This is a self-sustained process, the inflaton creating expansion, the field filling available space and creating more energy. Such a process leads to inflation. It is an exponential growth. One may think of the process as one in which the volume of a given region of the cosmos is doubled at fixed time intervals.[23] The process would last until the field decays towards its minimum, either tunnelling from its meta-stable state or tumbling from its high value in the early condition. The energy in the scalar field reaching its minimum value transforms itself into other fields and elementary particles forming a brew of material particles and radiation. At that time, gravity becomes effective, inflation stops and the Universe starts expanding at a rate comparable to that predicted by general relativity: a recession rate presently estimated as 72 km per second per Mpc and slowed down by gravitational attraction between the parts. The high temperature at the end of inflation provides an environment favourable to the existence of essentially free quarks within the rule of asymptotic freedom, as in baryons. That temperature is too high for quarks to unite to form baryons and obviously for joining electrons to form atoms. Those would dissociate immediately and the existing temperature would only allow the presence of a plasma made of charged particles in interaction with the background of intense radiation created at the same time.

In describing the evolution scenario of the inflation, it is difficult to provide exact timing for particular events to occur. Periods when such events happen are thus given with the understanding that only general characteristics can be imagined based on our knowledge of the present laws of physics. The inflation is assumed to have taken place in a very short period extending from around 10^{-38} second to possibly 10^{-32} second and may have started when strong and electroweak interactions became dissociated. At the end of inflation the temperature dropped to a value of the order of 10^{27} K. During such a period, exponential growth would have increased the size of the original entity by an enormous factor, with speculations leading to an increase by hundreds of orders of magnitude. The process provided the bang of the Big Bang scenario. As is sometimes said, it appears that this process leads to a free lunch, energy appearing out of nothing, violating the principle of conservation of energy. However, as mentioned by Guth and some previous authors, Newtonian physics leads to the concept that gravitational potential energy is negative and this concept is still valid in general relativity. Consequently, during expansion, when space filled with positive energy was created, the increase in total energy was zero, since the positive energy created is compensated by the gravitational negative energy appearing when matter and energy were created. If the reader finds this difficult to accept, he may be comforted by learning that he is not the only one. Gamow reports in a book that he mentioned to Einstein one day that according to relativity a star could be created out of nothing, since its negative gravitational energy could cancel its positive mass energy ($E = Mc^2$). Gamow mentions that Einstein was rather surprised by such a statement.[24]

This inflaton field approach is very different from the explosion scenario in which all matter or energy present in our Universe needed to be concentrated in the initial infinitesimal primeval entity. One may say that we are back to the hypothesis of the expanding Universe in which matter is continuously created in the vacuum of space, an observation that did not escape Hoyle since it reached in its essence his own proposed continuous matter creation scenario. The scenario proposed in the inflation model, however, is displaced to an early period of the Universe evolution.

So, in summary, scientists have displaced the problem of the primeval entity containing all the galaxies and radiation to an entity that contains the seeds of the forces or interactions known to exist today, an inflaton field, and the creation of space expanding with a constant density of energy. In that scenario, inflation was the 'bang' of the Big Bang model. The expansion was dramatic, creating space, possibly doubling dimensions every 10^{-38} second for a period extending to something like 10^{-32} second. Inflation can be very efficient at creating space and an environment that can be the starting point of the expansion of the Big Bang model.[25] The expansion in the inflation period may also be thought of as an accelerating process in which objects were pushed apart, by the creation of space in between, eventually at an apparent speed greater than the speed of light. In that type of scenario, the objects are seen at rest in their own surrounding frame, and the principle of relativity, which prohibits objects from travelling at speeds faster than that of light, is not violated.

Sometimes, models are made to illustrate the overall expansion and a graphic that attempts to depict the history of the whole Universe evolution is imagined. However, such graphics must be used with caution and with reserve. The actual shape and size of the Universe is unknown. It could be infinite spreading in all directions. Or it could be finite. As mentioned above this last option brings in the concept of a border. However, there cannot be a border: as we said, what would there be on the other side of that border? We have to say that space was created at the same time as matter. But how can we have a finite space without borders? We have seen that this is possible for the two dimensional being on the surface of a sphere; Flatant cannot find a border on that surface. How can we transform that concept in our three-dimensional world? We can imagine in two dimensions a situation a little more complicated than that of a sphere.[26] Imagine for example a flat rectangular piece of rubber as in Figure 9.6(a). It can be rolled on itself as in Figure 9.6(b). We obtain a cylinder with two ends. Suppose we bend this cylinder and, stretching it slightly, we close it on itself. The rectangular plane closes itself along two directions and the two opposite ends become the same. We obtain a shape like a bicycle tubing. Our Flatant living on the surface of the tubing would not see any border in any direction, either around the short circuit of the cylindrical part or along the long circuit, encircling the central hole. The plane closes itself in two dimensions. This is a torus. Although more difficult to visualise, we can imagine such a situation in three dimensions. For example, we could have a cube whose opposite faces join themselves in an acrobatic way two by two. We end up with a 'hypertorus'. A

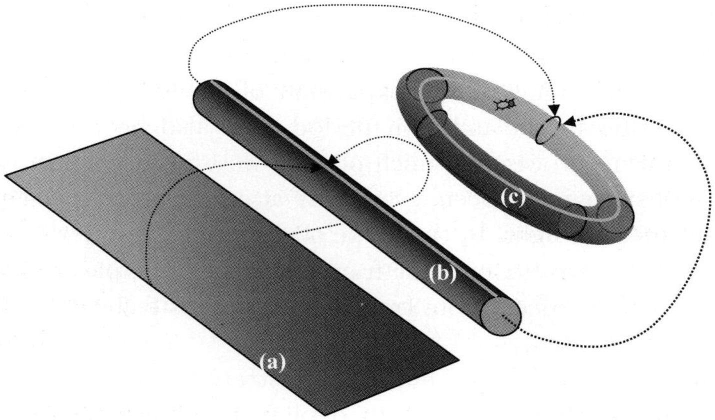

Figure 9.6. (a) A plane limited by borders in two dimensions. (b) A cylinder made of the plane in (a). The cylinder does not have a border on one of its dimension. (c) A torus made from the cylinder in (b), creating on its surface a two-dimensional space without borders.

hypothetical being living inside such an environment would traverse one face and enter the opposite face automatically. He would find no borders and space would appear infinite being closed on itself without borders. We can imagine a similar situation with any shape like a sphere, for example, associating diametrically opposite points two by two. We may wonder how to interpret the Big Bang in either case, that is to say in the case of an infinite space or a finite closed space without borders. If the Universe is infinite the Bang would have taken place everywhere, a situation difficult to imagine but not entirely ruled out. If the Universe is finite, the Bang gave rise to a closed Universe without borders. The radius of curvature of the resulting space may be very large, making the Universe look flat in our scale. In both cases our visible Universe, as observed by means of our instrument, would be a limited space like that enclosed in a sphere. Its size would be limited by its time of existence, the speed of light creating backward looking light cones and the expansion rate that would allow radiation to reach us. In such a space, light emitted by a galaxy could travel around coming back to its point of origin. Multiple images would be formed of the same object, depending on the number of times light has gone around. The times of origin of the radiation would be different depending on the length of the path travelled.

The geometry of such spaces is generally characterised by their curvature, which is easily visualised in a two dimensional space. The chosen example of the torus is similar to the case of the sphere: the geometry of the surface is said to have a positive curvature and the equivalent resulting volume in a three dimensional space is finite. A negative curvature may be represented in two dimensions by a saddle as in Figure 3.25; such a geometry is called *hyperbolic* and leads to an infinite space. A flat infinite space is said to have zero curvature.

Cosmologists have not reached a conclusion yet on the actual shape of the Universe. However, it looks flat, possibly having a very large radius of positive curvature. We may nevertheless make a graphic representation of the evolution of our Universe and its expansion with the understanding that distances represented on the graphic are connected to the expansion metric and not absolute size: they are distances between objects that appear to recess from each other due to the expansion of space between them. This is illustrated in Figure 9.7. Space in that figure expands from the original primeval entity.

In this context, inflation solves the main problems associated with the standard Big Bang scenario. The small homogeneous entity that may have existed at the beginning spreads extremely rapidly to form a homogeneous brew of particles and radiation, which acted as the starting point of the Big Bang expansion model creating a very large Universe with a very large radius of curvature and making space look flat.

For completeness it may be mentioned that the inflation scenario solves another problem relative to the absence of so-called magnetic monopoles in our Universe. As mentioned previously, in the early days of development of quantum theory, Dirac showed that the presence of magnetic monopoles provided a mechanism by which the electric charge would be quantized.[27] This phenomenon was introduced as a quantized singularity of the electromagnetic field. The presence of such entities is also suggested in such approaches as GUT and String theories. However, no monopoles have yet been

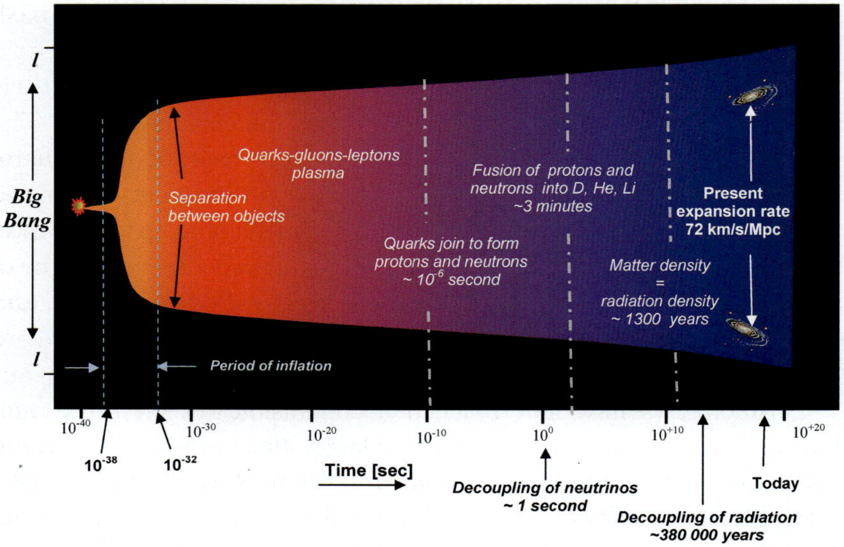

Figure 9.7. Imaginative illustration on a log-log scale of the expansion scenario of our Universe including inflation. The inflation period is very short but extremely efficient in increasing space exponentially. Events of special interest described in the text have been identified. Dimensions on the left-hand vertical axis must be understood as distances $l(t)$ between objects. The size of the Universe is not known and could be infinite.

found. It appears that the inflation scenario, through its large expansion, has caused so much dilution that the probability of detection of such a particle in our Universe is negligible.

On the other hand, according to our knowledge of quantum physics, quantum fluctuations should have appeared during inflation and should have created small spots of density higher or lower than the average in the original brew of particles and radiation present at that early stage. Those spots would not get homogenized during the expansion, it being too rapid to allow communication between the parts. In a time of 10^{-32} second, information, being transmitted at the speed of light, travels less that 10^{-23} metre, a distance much smaller than the expected distance reached between the parts, or in proper language, much smaller than the size that line elements in space-time had reached in that time through inflation. All those fluctuations affected the background radiation which interacted with the particles present in the brew that existed at that time. The effect is now visible on the microwave cosmic background. These small fluctuations in density have given rise to agglomerations of matter such as stars, galaxies and other cosmic objects that we observe today.

During the cooling process, scientists have identified regions of temperatures at which physical processes take place. An early process is, for example, nucleosynthesis, the fusion of protons and neutrons to form nuclei of certain atoms, such as deuterium, helium and traces of lithium. A later process is the recombination of electrons with nuclei, forming then an atomic gas of hydrogen and helium transparent to the radiation present. We can identify these temperatures by means of relations such as the equivalence of mass and energy ($E = mc^2$) and the equivalence of energy and temperature by means of Boltzmann constant. ($E = kT$). We can examine those various regions and correlate them to the evolution of the Universe. The whole expected evolution may then be illustrated grossly as in Figure 9.8.[28]

Various regions of the graphs are identified by a letter. The transitions are not sharp and there may be considerable overlap

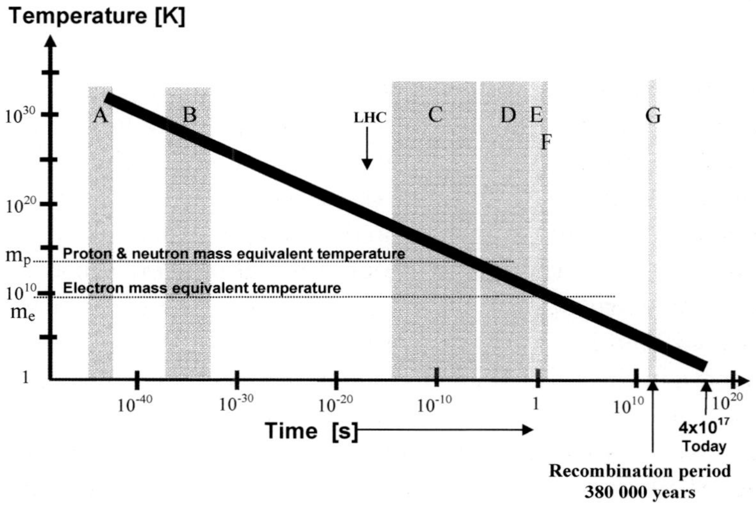

Figure 9.8. Identification on a graphic of events taking place during the expansion of the Universe. Periods during which events took place are identified by a letter and are explained in the main text. The graph shows on a log-log scale the bulk decrease of temperature with time believed to have taken place during expansion and does not differentiate between matter and radiation predominance.

between those regions. The following approximate steps, assembled from the various proposed scenarios could have taken place.[29]

A. This is the period before Planck's time, 10^{-43}. We do not know much about that region of the graph aside from the fact that the laws of physics as we know them today may not have been valid. But something must have existed: we may call it our primeval entity. Was it a quantum fluctuation? At that time, it is expected that gravity was quantized, and all the forces we know of were united under a single fundamental interaction. This is the region of temperature that is the dream of theoreticians who wish to proceed with a theory of everything. In the period 10^{-43} to 10^{-38} second, space starts to expand and it is believed that gravitational force, which according to relativity is a property of space, separated from the other forces, strong, weak, and electromagnetic.
B. It is believed that the period extending from 10^{-38} s to 10^{-32} s was a period of inflation. A scalar field, the *inflaton*, is assumed to be present. At a certain time, the temperature dropped to the range 10^{28}–10^{27} K where the strong force decoupled from the electroweak force. During that time, space expanded probably at an exponential rate e^{Ht} (early de Sitter model) where H is a very large Hubble constant leading to a large and flat homogeneous Universe. Energy is drained from vacuum and the inflaton field decays forming the basic matter of the Universe, quarks, electrons, neutrinos and of course, their antiparticles. There was also the possibility of creation of some other kind of substance, called dark matter, that we will describe below. The insert LHC shows the kind of energy that it will be possible to achieve in the Large Hadron Collider in CERN, in which collisions between protons will be created with energy of the order of 14 teraelectronvolts, corresponding to a temperature of 10^{16} K.
C. In the range of 10^{-15} to 10^{-6} second the temperature dropped to the range of 10^{15} to 10^{10} K. It is believed that all four interactions, strong, weak, electromagnetic, and gravitation are now independent. The fundamental particles, quarks and leptons formed a dense plasma. Temperature was still too high for quarks

to assemble and form stable nucleons except at the end of the period.

D. Between 10^{-6} second and 1 second the temperature was low enough for quarks and gluons to start to assemble and form baryons such as protons and neutrons. The mass of these particles is equivalent to an energy of 930 MeV, corresponding to an equivalent temperature of 10^{13} K. Below 10^{10} K the energy is low enough for them to be stable. Consequently, below that temperature, they can exist without being continuously dissociated by collisions. During that time the baryons recombined with antibaryons creating radiation. However, it is assumed that a break in symmetry caused an excess of matter over antimatter by a few parts in 10^9, leading to our world. This remains somewhat of a mysterious process. At the end of that period, the neutrinos decoupled from the plasma and became free. Their interaction with matter is very weak and they formed a very dilute neutrino background that exists up to the present moment. That background has not been yet detected.

E. Between 1 second and a few minutes, the temperature dropped to less than 10^9 K. The mass of electrons is about 0.5 Mev corresponding to a temperature of 6×10^9 K. Consequently, the temperature was too low for leptons to be formed and annihilation took place between leptons and antileptons as in the case of hadrons. A small asymmetry prevailed and there was an excess of leptons. In the Universe just created, radiation dominated over matter.

F. Between 3 minutes and approximately 20 minutes the temperature dropped to a value below 10^9 K low enough for fusion of protons and neutrons to take place without being continuously dissociated: deuterium and helium nuclei were formed with traces of lithium. That situation saved the neutron since it has a lifetime of about 15 minutes. When fused to protons it becomes stable. An analysis based on energy equilibrium considerations, shows that the process lasted for about 17 minutes after which the temperature was too low to accommodate fusion: this is the period of nucleosynthesis. Above the temperature mentioned, the process of fusion could take place but the resulting products were dissociated rapidly due to the high energy of interaction of the colliding particles. Upon cooling the products became stable, the temperature being below that required for dissociation. At that point the Universe was formed of a dense plasma of electrons and nuclei, essentially those of hydrogen (protons and deuterons, with one positive electric charge), and of helium (alpha particles with two positive electric charges) all bathing in a brew of photons. The plasma expanded for a long time without much happening. The photons were in continuous interaction with those electrically charged particles, their free time between collisions being extremely short because of the density of interacting particles. The photons were continuously scattered through a phenomenon called Thomson scattering. We have mentioned this phenomenon previously. Let us recall the process. Electromagnetic radiation consists of oscillating electric and magnetic fields. Those fields couple to the charged particles and set them in oscillation. The particles, being accelerated, in turn emit radiation. We say that the original photon, which caused the charged particle to accelerate, is scattered. This is the essence of Thomson scattering. The new photons are scattered again by other charged particles. The photons are continuously absorbed and new ones are re-emitted. It is an extremely foggy situation and the photons are essentially trapped in that dense plasma forming a very opaque environment. The situation lasted for several hundred thousand years.

Since the time of inflation, both matter and radiation were being diluted by expansion. The dilution, however, was different with time for each component because they obey different physical laws relative to temperature. This causes radiation density that was predominant in the early stage of the expansion to decrease faster than matter with time. At about 1,300 years after the Big Bang, radiation lost its predominance. At that time, when the temperature was of the order of 10,000 K, it is evaluated that both radiation density and matter density were equal. After that time matter became predominant.

G. After a time lapse of about 380,000 years of expansion, the temperature was about 3,000 K. This is a temperature low enough to allow electrons to attach themselves to nuclei and form hydrogen and helium atoms without being knocked out by collisions with other partners. This transition took place over a period of several thousand years. As we have seen in an earlier chapter, atoms are neutral and react with electromagnetic radiation only at very precise frequencies or wavelengths, corresponding to specific separations between the lower energy levels of the atoms. Consequently, the interaction of the background radiation takes place with hydrogen and helium over a very narrow range of very specific frequencies. Since the residual radiation background has a very broad spectrum, meaning that it extends over a very large band of frequencies, it follows essentially that most of the background radiation does not interact with hydrogen and helium atoms and the Universe became essentially transparent. The photons started to travel freely through the whole of space without being absorbed or scattered. At that point in time, radiation and matter separated completely. As we can readily see, the state of that radiation is a snapshot of the Universe at that particular instant. It contains information on the inhomogeneities that were implanted by the quantum fluctuations that took place in the early Universe. These inhomogeneities, consisting of fluctuations in density of particles materialized from the scalar field in the early Universe, gave structure to the background radiation by interacting with it. On the other hand, it is those same density fluctuations that under the effect of gravity attracted each other, grew up to become nebulae, factories of stars, and finally caused those stars to group together to form galaxies.

After the G period, the Universe continued to expand and cool to reach the present state. The background radiation stretched with expanding space, became redshifted through the expansion, and is now what we call the Cosmic Microwave Background that appears through the whole of space as blackbody radiation at about 3 K.

During that long period, between the time of recombination and the present, celestial objects as we know them were formed. We may describe grossly the formation of a star with its planetary system as follows. First, gas clouds or nebulae start to form from hydrogen and helium under local space curvature or in simple words under gravitational attraction. In such nebulae, further gravitational contraction of certain parts creates regions of high density. Gravitational contraction of such a region ends up in a central object of very high density and very high temperature. A central core is formed where nuclear fusion of hydrogen and helium nuclei takes place. Equilibrium develops in which the radiation emitted by the core causes internal pressure within the surrounding gas and prevents further collapse. Such an object becomes a star such as our Sun. During the whole contraction process, any original rotation accelerates, as in the case of the skater in Figure 4.3 since angular momentum is conserved. That rotation produces a flattening of the section of the nebula under gravitational contraction and a disk of material looking like a pancake, called a protoplanetary disk, is created. That disk becomes the seeds of planets that form through accretion from the surrounding material. In the solar system, formed some 4.6 billion years ago, we see the residue of that disk as planets orbiting the Sun all in nearly the same plane.[30]

Formation of stars is still taking place today in nebulae, as in the case shown in Figure 2.6. It is these stars, at various stages of their development, that group under gravity to form galaxies. These galaxies are found distributed in small groups forming long filaments and walls with immense voids between them, reflecting the small fluctuations present in the primary brew that expanded after the inflation period. That description is rather simplified. In fact, the process is complicated by the presence of so-called dark matter. As we will see below, the presence of dark matter is required to explain the dynamics of stars in galaxies. It turns out to be more important than ordinary matter (baryonic) and should play an important role in the formation of celestial bodies and their clustering. Since it does not interact with electromagnetic radiation, dark matter could have aggregated earlier than baryonic matter and could have played an important role in the formation of galaxies. However, we do not know yet exactly the nature of dark matter and it is difficult to evaluate the exact effects of its presence.

The above description is a concise summary of a possible scenario of the evolution of the Universe in the Big Bang model assembled from present published proposals. There is of course a certain amount of speculation in it, but most of it is supported by the presently observed state of the Universe and detailed calculations using the physical laws that appear to apply throughout the Universe. But most importantly, it is based on one of the most successful theories developed up to now: Einstein's General Relativity. In the beginning Einstein believed in a static Universe. Friedman and Lemaître showed that Einstein's relativity equations led naturally to expansion. Einstein, in front of the evidence provided by Hubble, had to admit the reality of expansion. Nowadays, the fact is accepted at large in the scientific community and the approximate scenario described above appears to be the most probable. It has become the so-called Standard Model of Cosmology.

In the Big Bang scenario just described, the laws of physics as we know them do not apply at time t $=$ 0: at that time one is faced with infinities and our mathematics and physics cannot handle such a situation. The scenario stops there and we do not know the cause of the presence of the primeval entity and of the inflaton field. Another scenario has been proposed and is based on the existence of a field called *dilaton* existing in the primordial epoch before time t $=$ 0 of the Standard Model. The properties of the dilaton are dictated by the evolution of the curvature of space leading to a kind of smooth transition at the origin of time leading to our Universe. It is somewhat like the cyclic scenario originally proposed by Lemaître. The new scenario is based on string theory and the relics of the pre-Big Bang dilaton field could possibly be detected nowadays as an accelerated expansion and the presence of gravitational waves created at that early moment.[31] String theory relies on the concept that elementary particles originates from one-dimensional strings or filaments of a very small size (10^{-32} m). The particles are then seen as materialization of those strings oscillating or vibrating in various modes. The size of the strings involved prevents of course their observation with presently available equipment limited to resolutions of the order of 10^{-15} m. String theory forms a field of its own and is studied actively by several groups offering a totally new approach to the description of the constituents of the Universe. Unfortunately, in spite of providing some promising theoretical results and sensible predictions, the new scenario proposed has not found yet experimental confirmation.

In that short summary we left many points outstanding, for example the origin of the elements, the actual information contained in the CMB, the constancy of the present expansion rate and several other phenomena. Let us look at some of the most important ones.

The Origin of the Elements on Earth and Elsewhere

On Earth we find natural elements extending from the atomic weight 1, hydrogen, to 238, uranium. We recall that the atomic weight is the total number of protons and neutrons in the nucleus of the atom in question. The charge on the nucleus is the number of protons, which is equal to the number of orbiting electrons, thus forming neutral atoms. It varies from hydrogen, with 1 electron to uranium with 92. Some of these elements exist in stars. We detect their presence through their fingerprints, the radiation they emit that provides information on their identity. Where are all those elements coming from? During the very early period of existence of the Universe, elementary particles, such as quarks, gluons and electrons appeared as materialization of radiation from pre-existing fields through quantum processes. When the temperature of radiation reduced substantially, that process stopped, and neutrons and protons were formed through synthesis of quarks and gluons. There was then left a mixture of neutrons, protons and electrons, those last two being charged particles and interacting with the residual radiation. At a stage of the expansion, a few minutes after the bang, temperature reduced to something like 10^9 K. At that temperature nucleosynthesis (joining of hadrons) could take place, the temperature being high enough to permit nuclear fusion. This means that, at that temperature or energy, deuterium, helium 3 and helium 4 nuclei were formed in various quantities from the fusion of protons, neutrons and deuterons. Deuterons do not have a large binding strength and are continuously dissociated through interaction with other particles at that temperature. On the other hand, free neutrons have a lifetime of the order of 15 minutes and transform into protons and electrons. Collisions between the just formed alpha particles and deuterons could take place producing small quantities of lithium nuclei with 3 neutrons and 3 protons. The same fusion could have taken place between helium nuclei, producing nuclei with 4 protons and 4 neutrons. Unfortunately, those nuclei are unstable, due to interaction between protons. Their lifetime is of the order of 1 microsecond. It takes the addition of another neutron to provide some stability to the produced nucleus. To be successful, the collision must take place during the lifetime of the newly formed nucleus. Such collisions are rare and it appears that a bottleneck was created. Furthermore the Universe expanded, cooled down, and the energy of the particles was no longer large enough to produce fusion of the existing particles. The whole nucleosynthesis stopped in about 17 minutes, which is 20 minutes after the Bang. This is an important moment. Equilibrium calculations show that the Universe consisted then of approximately 75% hydrogen nuclei, 25% helium nuclei with traces of deuterium (0.01%) and very small quantities of lithium and beryllium (10^{-10}).

In this model it thus appears that essentially none of the heavy elements were produced at the moment of the original nucleosynthesis minutes after the Bang, contrary to earlier original hypothesis.[32] Where were they produced? The accepted model is that they were produced in stars and still are today. The evolution of stars is rather complex. As we just described, they start from the compaction of material from interstellar gases that first form nebulae, containing mostly hydrogen, through the effect of gravitational attraction. Within a given nebula, some higher density region forms and compaction continues at those spots, always under gravitational attraction. Nebulae are manufactures of stars. Figure 2.6, the 'Pillars of creation', shows a famous example of such a mill. Upon sufficient compaction by agglomeration of material, the core gets extremely hot, billions of degrees, and nuclear fusion of light nuclei within it ignites. Low weight nuclei are transformed into heavier nuclei by fusion.

The density is high enough in such a compact space that very energetic and frequent collisions take place. The beryllium-carbon barrier is surmounted and heavier nuclei are produced. The complex process of nucleosynthesis develops to the iron nuclei which in turn produces a kind of bottleneck. As we have seen in Figure 7.3 this is the point where the assemblage of heavier nuclei requires energy rather than releasing some. However, the nuclear reaction taking place within the core produces a large number of free neutrons, which fuse to existing nuclei producing heavier elements. All these processes, although rather complex, have been studied in details and the whole subject is now well understood.

Matter produced finds its way in free space through a cataclysmic explosion of a star after it has depleted its fusion material. This can take place in several ways. For example, after depletion of its fusion material, the central core emits less radiation and the outward radiation pressure exerted on the outer layers of the star reduces considerably. The star collapses on itself, the outer layers being pulled in under gravitational attraction. The various intermediate layers forming the star rush to the core producing a shock wave that spread to the outer layers ejecting material. Calculation shows that if the mass of the core reaches a certain value called the Chandrasekhar limit ($\sim$1.38 times the mass of the Sun), the compaction is so large that such a star explodes in a violent cataclysm becoming what is called a supernova, spreading its internal fused material into space. This is the material that much later regroups again under gravitational attraction to form various objects, such as new stars and planets like in the case of our solar system, with all the elements and chemical compounds that we can dig up in the soil of our planet. We are the product of stars decays that took place a long time ago. We find our origin in star dust.

The implosion of stars as we have described is rather interesting and leads to other phenomena that challenge our mind and have only recently been explained. For example if a star is sufficiently large, say more than nine times the mass of the Sun, its collapse, after having been depleted of its hydrogen, may cause a sufficient rise in temperature to ignite again the core causing the fusion of helium nuclei into carbon nuclei. The fusion process may stop the collapsing process due to internal radiation pressure and the star may expand again. Other elements are formed. The process may repeat itself several times until fusion produces nickel and iron nuclei. Then, no more radiation is released and at that point the outward pressure disappears. For such a large star, the core mass exceeds the limiting Chandrasekhar mass and the core collapses on itself. The velocity of the falling matter may be of the order of 70,000 km per second or 23% the speed of light. The density and temperature rise cause photodisintegration of the elements present in the core. Released electrons and protons merge to produce neutrons. Residual electrons and neutrinos escape and the object that is left is a neutron core at a temperature of 100 billion K. Energy is shed away by the release of neutrinos. The neutron core is of the order of only 30 km in diameter. Consequently its density is extremely large: it is made of closely packed neutrons and looks like a very large nucleus. It cannot collapse further because neutrons are fermions and cannot occupy the same space. It is called a *neutron star*. Angular momentum is conserved in the formation of that object. Upon collapse the star starts rotating faster since any rotation velocity is amplified as its radius gets smaller, in a similar way as the rotating skater pulling her arms closer to her body as described earlier. The star may emit radiation from decay of the neutrons energy states or acceleration of particles near the magnetic poles. Since the magnetic poles and the rotation axis may not share the same orientation, the star may appear as a beacon in the sky and its radiation beam may be observed as a radiation pulse from any nearby planets, in particular the Earth. As mentioned previously, the object is called a *pulsar*.[33] Many such pulsars have been observed and the repetition rate of the beacon may be in the millisecond range. The stability of the measured pulse repetition rate reflects the importance of the rotating mass involved and is comparable for short periods of averaging times to that of atomic clocks described earlier.

What happens if the mass of the star that was at the origin of the collapse is still larger, say greater than 20 times the mass of the Sun? The core collapses into what is known as a *Black Hole*. This is one of the most mysterious objects that exist in the Universe. In the early days of the study of Einstein's theory of relativity by several theoreticians, a scientist named Schwarzschild found a particular solution of the equations for a non rotating spherical mass. He arrived at the conclusion that if a spherical object with a given mass has a diameter smaller than a certain value, the situation produces a singularity in the equation. The force of attraction within a certain radius is so strong that objects cannot escape the core attraction whatever their speed. In fact a horizon, at a certain distance from the centre of attraction, is calculated below which no objects whatever the velocity at which they may travel, can escape.[34] It is a point of no return called the event horizon. Even light cannot escape when crossing that horizon. The radius of the object calculated by Schwarzschild is small and the collapse has to be tremendous to reduce the core volume to a size that becomes a black hole. However, it is possible. It is believed that the central bulge of most galaxies is occupied at least by one such black hole, our Milky Way included. The mass of such black holes may be equivalent to that of millions to billions of solar masses acquired by means of gravitational capture of objects that have the misfortune of traversing their event horizon. Since they do not emit any radiation the presence of black holes is difficult to confirm. However, they can be observed through their effect on the surroundings and the emission of radiation by in-falling objects before those objects are swallowed by the black hole, particularly gamma rays. It is natural to think that the creation of black holes distributed over space could be the end of the Universe sometime in the distant future. As we will see, however, our Universe may have a different fate. It may also be mentioned that Black holes do not have an infinite life. Stephen Hawkins has shown that quantum mechanical effects taking place in the black hole environment could lead to the escape of some of the radiation at the event horizon and cause the black hole to evaporate, slowly of course. But we realise that there is plenty of time. The Universe does not evolve on a day-to-day scale but on a billion-year scale.

There were fears that the Large Hadron Collider (the LHC) in Geneva, in view of the large energies of the colliding particles, could initiate the creation of black holes that would swallow objects around themselves. Fortunately, on the basis of presently known well established physical principles this has been shown to be totally improbable.[35]

Another interesting property of an exploding star, such as a supernova, is that, as mentioned previously, it can be modelled rather exactly. Particularly, Type 1a supernovae have been found to behave rather reproducibly emitting a very well determined amount of radiation that can be used as a standard candle to evaluate distances in the sky. This is due to the fact that such a star, building mass through accretion, collapses at a very determinate size, the Chandrasekhar limit of 1.38 Sun masses. Unfortunately, events such as a supernova do not happen often, something like once every 50 years in a typical galaxy, and their use for distance determination, so much desired to evaluate the rate of the expansion, requires patience and diligence. It is not possible to predict such an event and when it happens and the event is detected, its study is limited by the scheduling of telescopes and their availability as well as the rapid evolution of the exploding star. Nevertheless, we will see below that with efficient planning of the available resources, those Supernovae have been studied and have recently revealed a new characteristic of our Universe: its expansion is accelerating.

What Does the Cosmic Microwave Background Tell Us?

The cosmic microwave background is extremely rich in information on the origin of the Universe. As we said, it is the residue of the radiation that was created in the early moments of the Universe. It is radiation that evolved during the early period when the Universe was essentially a plasma. The quantum fluctuations that developed in the early moments, during inflation, fluctuations that created small inhomogeneities in the constituents of the Universe at that time, became imprinted on the background radiation. They appeared as small inhomogeneities in that radiation. Consequently, the CMB is a picture of the state of the Universe when it became transparent to radiation. This occurred when electrons recombined with protons, deuterons, and alpha particles to become hydrogen, deuterium and helium atoms. Those atoms interact only at very specific wavelengths and only in a very narrow range of wavelengths of the existing radiation. Consequently, for all practical purposes, the Universe became totally transparent. The interesting questions are: what is the present wavelength of that radiation? How big are those imprinted fluctuations? What do they tell us about the Universe at that time? Is the Universe in its present state compatible with our understanding of the expansion that took place according to Hubble's law? After the first detection of that radiation by Penzias and Wilson, a detailed study was made of the characteristics of that radiation by several means. The main problem in the detection and study of that radiation is its nature, being at a wavelength that corresponds to radiation emitted by a blackbody at very low temperature ($\sim$3 K). It is like noise in receivers. Consequently, in order to detect that radiation without ambiguity, detectors have to be conceived in such a way that they would have as little internal noise as possible. One way is to cool such detectors to very low temperatures, which cause severe design constraints. Furthermore, humans produce noise on the surface of the Earth through their activities in communication and the use of various electronic systems; the atmosphere also produces noise, for example, through such phenomena as electric storms. The main problem in the detection of the CMB is thus to eliminate those sources of stray radiation. This was a major task for those involved and solutions were found in using balloons to raise detectors way up in the atmosphere far from human activities, in using high flying airplanes such as the U2 originally designed for spying at very high altitude, and finally, in using the ultimate, satellites in orbit around the Earth totally outside the atmosphere. The orientation of the detectors in those satellites could then be adjusted away from the Earth surface with considerable gain in differentiation of the origin of the detected radiation.

Two principal characteristics of the CMB needed to be measured. These were the actual spectrum of the radiation, or its intensity as a function of wavelength, and its homogeneity. The measurement of the spectral characteristics would determine if the radiation were blackbody radiation as was expected in the standard model and, if so, would determine the equivalent temperature. Black body radiation has very well known characteristics, regarding spread of energy over a spectrum of wavelengths and the wavelength at the maximum intensity. The spectrum of the emission was derived by

Planck who introduced the concept that light intensity increments are discontinuous. It is the radiation emitted by an object solely due to the fact that it exists at a given temperature. Its mathematical description is given in Annex B (section a). The hotter the object, the shorter is the wavelength of the radiation emitted at the maximum of emission. When we say that an object is 'red hot' we simply put in common language the fact that the maximum of emission of the radiation takes place at a short wavelength, in the visible. If the CMB were not of the nature of that emitted by a black body, the whole theory behind the Standard model would have to be revised. The other characteristic, the presence of small inhomogeneities, provides basic information on the size of fluctuations present at the time the Universe became transparent. They need to have a particular size to explain the size of the galaxies that we observe today and that originated from those fluctuations.

To measure those to a great precision, a typical detector was built in such a way as to measure differences between two given points of the sky rather than absolute values of intensities: it was thus constituted of two detectors oriented in two different directions (for example 60°) observing two different regions of the sky. This is called a differential microwave radiometer (DMR), invented by Robert Dicke in the 1940s. It allows the measurement of very small fluctuations of the background by means of a subtraction of the signals detected by the two detectors oriented in the two different directions. Furthermore, a rotation of the ensemble by 180° allowed also a correction for the difference that could possibly exist between the sensitivity of the two detectors. Nothing was left to chance. The experiment culminated from the U2 version to a satellite experiment called COBE for Cosmic Background Explorer. That experimental package was placed in a 900 km high orbit around the Earth. The satellite was also equipped with an absolute detector of the radiation called FIRAS for Far Infrared Absolute Spectrophotometer. The purpose of FIRAS was to verify the agreement of the microwave background spectrum with the predicted intensity of the radiation of a black body at the same temperature.

The results obtained over years of measurements by means of FIRAS were most successful and are a testimony of the talent, dedication, and perseverance of those involved in such delicate experiments. The results obtained were most rewarding. First, the nature of the CMB was determined to be black body radiation, the experimental points of intensity against wavelength fitting exactly a theoretical curve for a black body at a temperature of 2.725 K. Those results are shown in Figure 9.9. This is what is meant by the characterization of the temperature of space: it is the equivalent temperature of a black body that would emit an identical spectrum.

On the other hand, the results of the Differential Microwave Radiometer (DMR) caused a surprise. The microwave background showed the presence of a large so-called dipole in the overall picture of the radiation temperature. What does that mean? When the temperature of the background radiation was plotted for the whole surface of the sky from one horizon to the other, one end of the hemisphere thus plotted was slightly warmer than the other end. This is illustrated schematically in Figure 9.10. That observation meant that the wavelength of the detected radiation was longer when

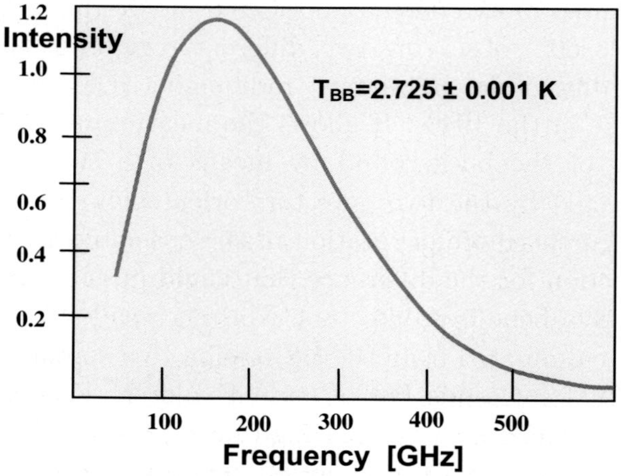

Figure 9.9. Results of the COBE-FIRAS instrument. Measurements were done at several frequencies. The theoretical black body emission spectrum at 2.725 K is drawn as a continuous line (Equation B.1 of Annex) over the hidden experimental points. Theory and experiment agree extremely well. The intensity is given in units of [10^{-4} ergs/cm^2 steradian sec.cm^{-1}]. (This graph is adapted from published data, by NASA).[36] We take note that a frequency of 300 GHz corresponds to radiation with a wavelength of $\sim$1 mm. Such a wavelength is often given in inverse wavelength, in cm^{-1}, through the relation $1\lambda = \nu/c$. 300 GHz corresponds to 0.1 cm^{-1}.

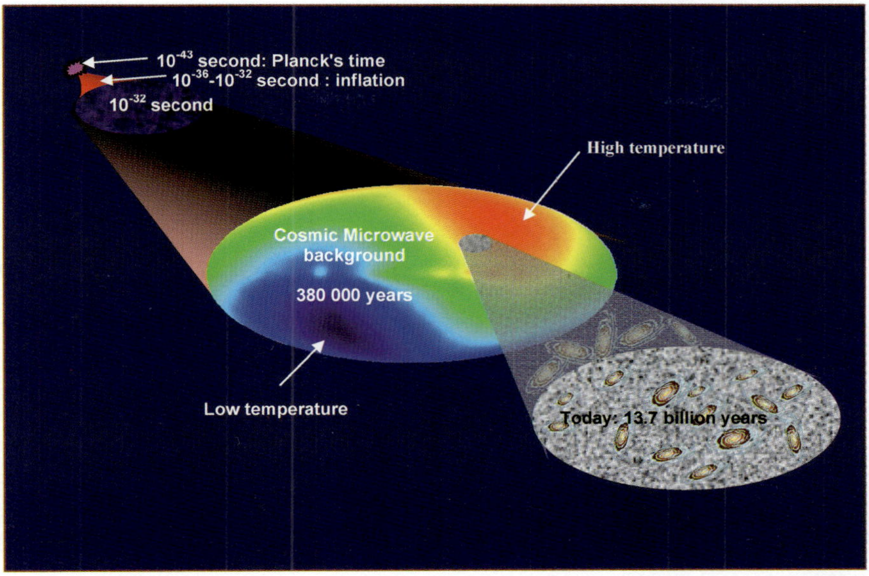

Figure 9.10. Integration of the map of the Cosmic Microwave Background (CMB) as observed by COBE into the Universe expansion. The temperature gradient across the sky, represented by various shades of the CMB, makes explicit the motion of Earth relative to the radiation background. Left hand bottom corner of the CMB is the lower temperature. The temperature differences between the regions are only a few parts per hundred thousands. The average temperature is 2.725 K (The microwave background cross section has been adapted from data published by NASA). The diagram is adapted from a similar one shown in reference 16.

observed in one direction than when it was observed in the other direction. The speed of light being the same in all directions it meant that the satellite was moving relative to the background radiation causing a Doppler effect. The colder region makes explicit longer wavelengths, which means that the satellite was moving away from those regions. You might say of course that this is not surprising. The satellite is moving, it is in orbit around the Earth. However, it is moving at a mere 8 km per second relative to the surface of the Earth, a speed that would not cause a visible effect on the background radiation. When calculation were made, it was found that the Earth trapped in the solar system, itself attached to the Milky way, was moving at a tremendous speed of about 600 km per second towards a point somewhere in the distant horizon. This speed is combined to the motion of the Earth around the Sun at 30 km per second and the

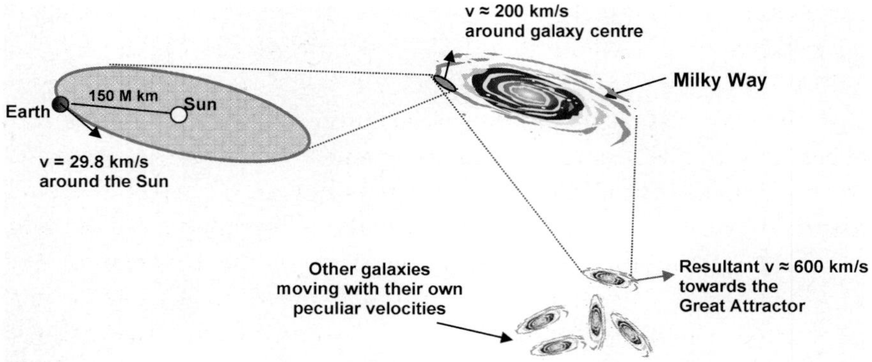

Figure 9.11. Illustration of the peculiar velocity of the Milky Way towards a massive object called the Great Attractor. That velocity is ~600 km/s as determined by satellite COBE.

Sun itself moving within the Milky Way at a speed of 200 km per second in its orbit around the central core, which last of the order of 250 million years. This is illustrated in Figure 9.11 making explicit those velocities.

What does that show about the CMB? You must remember the Michelson–Morley interference experiment where the velocity of the Earth could not be determined relative to the speed of light. The speed of the Earth did not add to or subtract from the speed of light. That finding confirmed the basic principle of special relativity, the speed of electromagnetic radiation or photons being independent of the speed of the source or of the observer or in other words of the frame of reference. But now we have the velocity of the Earth, Sun and galaxy relative to the microwave background. It looks very much like if we were given an absolute reference frame in the sky. The background radiation being isotropic acts somewhat as a reference grid in the Universe, an aether, but not quite. Radiation components at all wavelengths forming the background radiation travel at the speed of light in all directions through the Universe and the principle of relativity concerning the speed of light still applies. Electromagnetic radiation travels at the speed of light whatever the speed of the observer. The experiment does not provide a determination of the speed of the Milky Way relative to the speed of light. It does measure the Doppler shift of the radiation received by the observer on the moving Earth. This is very different: it is the method also used, as mentioned above, for determining the speed of recession of galaxies. The only difference is that the CMB radiation used for determining speed through Doppler effect does not originate from another galaxy but from the creation of the Universe itself. Consequently it is not an aether of the kind assumed to exist in first theories about the propagation of light waves. That radiation is propagating through the Universe at the speed of light. We could say that the Universe looks like a very large cavity holding radiation like a microwave cavity that we use in the laboratory. Such cavities can be excited in a multitude of modes like a violin string which can be excited at various harmonics of its natural frequency. In the case of the cavity, the vibrations are radiation modes having specific frequencies. The Universe, being very large contains an infinite number of modes, which then look essentially like a continuous spectrum. This is the CMB that fills the whole Universe that acts as a very large container at a temperature of 2.725 K. Why is the temperature of the equivalent black body radiation so low now? Are we not seeing the radiation when the Universe was only 380,000 years old? We said that, at that time, the temperature was about 3,000 K and that it was the moment when electrons recombine with nuclei to form hydrogen and helium atoms making the Universe transparent. The reason is that the Universe is now 13.7 billion years old, space has expanded 1,000 times since that time and temperature has lowered by the same factor. Through the expansion, the background radiation has suffered a stretching, which is seen as a redshift of the order of 1,000. This is the largest redshift directly observable. It looks very much as if the large cavity formed by the Universe is getting larger and that the frequency of the oscillation modes is decreasing with time. The background radiation has thus suffered a stretching in the same way as the radiation of the very far galaxies. From short wavelength radiation around the time the Universe became transparent, the background radiation has shifted to the microwave range.

The temperature of the CMB as mentioned above is now 2.725 K on the average. As we said it is not perfectly homogeneous. Small fluctuations expected to have resulted from the early quantum fluctuations are present. They have created our present somewhat inhomogeneous Universe with its galaxies. However, these fluctuations are so small that gross effects such as the dipole variation had to be removed by digital analysis from the COBE satellite data to see them. The fluctuations observed were of the order of parts in 100,000, very small indeed but large enough to be responsible for the forming of galaxies through the effect of gravity. The fluctuations were called 'wrinkles in time' by the principal investigator of the COBE mission.[37]

Several other systems were used to measure the CMB. The measurements were refined to a very high level by means of another satellite called the Wilkinson Microwave Anisotropy Probe[38] (WMAP) launched in 2001 and placed in a special point in space at a distance of the order of 1.5 million km from Earth and called the Lagrange point L2. Lagrange points are situated at privileged positions of the combined gravitational potential of the Earth-Sun system. A satellite situated at such a point rotates with the Earth-Sun system. The L2 position is advantageous in minimizing the amount of undesired radiation from the Earth, Sun, and Moon. The detectors in the WMAP satellite have a much higher resolution than in COBE and allowed a better determination of several properties of the Universe as we know it now. The result is shown in Figure 9.12. That CMB picture is really a rather high resolution snapshot of the

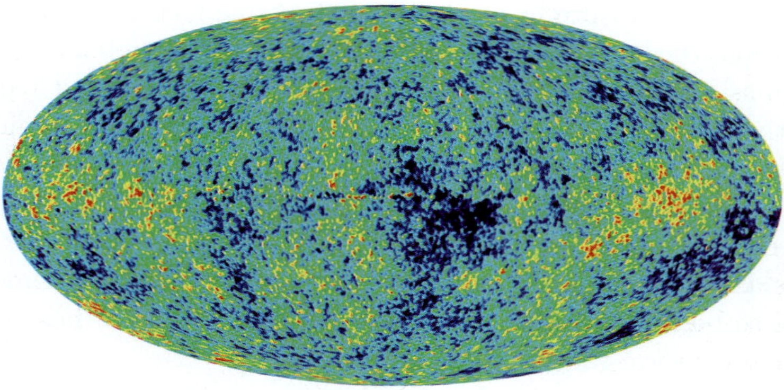

Figure 9.12. Snapshot of the Cosmic Background Radiation obtained with satellite WMAP, making explicit the minute fluctuations in the intensity of the radiation, which are of the order of a few parts in 100,000. (Credit: WMAP Science Team, NASA).[39]

Universe some 380,000 years after the Bang. It must be realised that this is a very short time span after the assumed 'bang'. The important information is contained in the spatial distribution of the fluctuations, their frequency, their intensity and their polarization. The model used assumes that sound waves were induced soon after the Big Bang in the electron-proton-photon plasma by density fluctuations. The fluid that existed at that time contained regions of high density of material. That material was made of baryonic matter that interacts with the electromagnetic radiation and so-called dark matter totally transparent to that radiation. As will be seen below, the presence of that dark matter is required to explain the motion of stars within galaxies. High density regions within that plasma tended to spread while the photons interacting with the baryonic matter cause internal pressure on the ensemble. Those phenomena created waves similar to sound waves in a gas and spread around at great velocities ($c/\sqrt{3}$). In a similar way as sound emitted by an object like a violin or another instrument contains information on the internal characteristics of that instrument, those sound waves present in the primeval plasma contained information on the properties of that photon–matter fluid. A theory was developed that provides a framework for the analysis of the properties of those sound waves that became imprinted in the distribution of photons at the time of decoupling. The spectrum of those fluctuations then contains information on the characteristics of the fluid before decoupling. An analysis of those fluctuations provides information on many parameters directly connected to the state of the Universe at that time, such as the size of the fluctuations that resulted in the present distribution of galaxies, the density ratio of baryonic matter to dark matter, the flatness of the Universe and several other parameters. Those results coupled with those obtained from other measurements such as redshifts, distance measurements, matter content, have provided a global picture of the evolution of the Universe in the Big Bang model. The Hubble constant is now known to be 72 km per second per Mpc with an uncertainty of a few percent. The age of the Universe is evaluated to be 13.7 billion years with an uncertainty of 1%. Some other parameters such as matter-dark matter content and curvature are also rather well determined from those results and provide some information on the future of our Universe. However, before discussing those results we need to introduce some new concepts resulting from recent observations that changed completely our understanding of the evolution of the Universe.

It may be said that the Cosmic Microwave Background is an element of the Standard Model that adds strong evidence of the existence of an event in the past such as what we call the Big Bang, in opposition to the continuous creation model mentioned earlier. However, observation of a background of gravitational waves could add an important element to the scenario, element that would provide information on the Big Bang itself.

Most of the Matter in the Universe is Not Visible

Spiral galaxies, like our Milky Way, are groupings of stars, nebulae, dust particles assembled under the influence of gravitational attraction. A standard approach to determine the mass of such galaxies or clusters of galaxies is to assume that mass is distributed according to the luminosity of its constituents. The total mass can then be inferred from the known ratio of mass to luminosity. Normally this is done in terms of the ratio for the Sun, M/L, where M is the mass of the Sun and L is its luminosity. For the Milky Way, for example, this ratio is about 3. In the 1930s, F. Zwicky[40] studied galaxies and clusters of galaxies as well as their motion within those clusters. He came to the conclusion that motion of those celestial bodies could not be explained from the mass deduced from the luminosity equivalence and that there was mass missing in the whole picture. Since that mass was not visible contrary to ordinary matter that would be seen through either emission or absorption of radiation, it came to be known as *dark matter*, a kind of new substance that was not interacting with electromagnetic radiation as protons and electrons do, making them visible or observable, their effect on the radiation taking place through either absorption, diffusion, or reflection.

Within galaxies, stars appear to orbit around the central core, the yellow nucleus clearly visible and formed of a large concentration of stars.[41] Somewhat recently, Vera Rubin measured the velocity

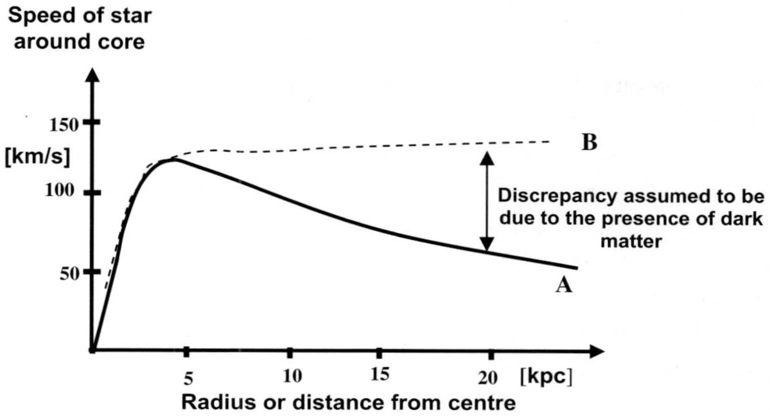

Figure 9.13. Illustration of the translation speed of stars within a typical spiral galaxy as a function of distance from centre. The continuous line (A) is obtained from a calculation assuming mass is distributed according to luminosity and Newton's gravitational law while the doted curve (B) represents experimental data. (adapted from one of the many graphs included in articles by Rubin and van Albada).

of stars as a function of their distance from the centre in several galaxies.[42] The velocity of the stars was determined by means of Doppler effect on the radiation emitted (H_α). The velocity measured that way was plotted as a function of distance from centre. When this is done a graph similar to that shown by the dotted line in Figure 9.13 is obtained. On the other hand, when the distribution of mass is evaluated as mentioned above from the luminosity of the contained stars, it is possible to calculate the star's velocity expected as a function of distance from centre using Newtonian mechanics. The result is shown as the continuous curve in Figure 9.13. The theoretical curve is made to fit the data at small radii. Obviously there is a discrepancy at large distances from the centre of the galaxy. Either there is something wrong with Newton's gravitation law or the distribution of mass is not what we think it is.

In view of the verification of the validity of Newton's law in all known circumstances the last conclusion is retained. To get agreement between data and calculation one has to assume the presence of added matter in the distribution, forming a halo around the galaxies. That matter is not visible; it does not emit or absorb electromagnetic radiation and does not appear to reflect any such radiation. It interacts with other visible matter through gravitational forces only. It is assumed to be the dark matter mentioned above. As mentioned several times in previous chapters, it is always important in science to confirm a finding or an interpretation of a result by means of independent types of measurements or observations. Gravitational lensing, predicted by General relativity provides such a confirmation regarding dark matter. When light is passing by a massive object it is deviated as was shown earlier. A massive object in front of a source of radiation, having gravitational properties, would thus alter the path of the light rays passing by and could act in a way similar to a glass lens. This lensing effect has been observed in the case of several galaxies and the same conclusion was drawn relative to missing mass to account for the strength of the lensing effect. In some cases, dealing with the collision of two galaxy clusters, it was possible to deduce the location of the dark matter relative to the normal matter, since both were separated physically in space after collision. In the collision process, the hot normal matter interacts electromagnetically and is slowed down while the dark matter, which does not respond to electromagnetic forces, but only to gravitation, keeps travelling faster than normal matter and finds itself isolated.[43]

The presence of dark matter in the Universe is a rather astonishing discovery. If this idea is accepted, we must conclude that the Universe we observe is not made only of protons, neutrons, and electrons, but of another type of matter that is not visible to us. What that matter is, we really don't know. It appears to have been created at the same time as the ordinary matter during the first moments of expansion and distributed throughout the Universe, playing a role in galaxy formation and surrounding them. At present we believe that dark matter is so important as to make 24% of all matter. There are proposals that neutrinos, if they have mass as it is now believed, although very small, not interacting with electromagnetic radiation, could be the source of that missing mass. However, since they travel essentially at the speed of light they do not appear to be the material required as halo to help in the formation of galaxies. Objects such as undetected black holes, neutron stars, small so-called brown dwarfs, objects too small to become stars on their own, or even large distant planets in the galaxies are also proposed as the source of the extra mass required for explaining the motion of stars within galaxies. They are identified as MACHOS for massive astrophysical compact halo objects. Some other candidates have been given a name such as WIMP for weakly interacting massive particles. The properties of WIMPs are similar to those of the neutrino; they would be interacting possibly through the weak force and gravitation but not through the strong force and the electromagnetic force. There are also proposals that for all fermions there could exist a corresponding boson and vice versa. This is the basis of the so-called Supersymmetry theory. It was suggested that those particles which are believed to be very massive could be responsible for the dark matter. None of those particles have yet been detected. Experiments performed with exotic detectors such as single crystal of germanium and silicon to detect the heat generated within by the interaction with a WIMP, have been unsuccessful. Similar experiments, using large xenon masses to detect a light flash when a WIMP interacts with a nucleus, have also been unsuccessful. Consequently, although dark matter is needed to explain phenomena that require additional gravitational masses other than those visible, its nature has not yet been determined and it appears that the question will need to be addressed with a lot of imagination and will require a major effort.

We are in Trouble: The Expansion is Accelerating

The rate of expansion of the Universe from its discovery to the present times has been a subject of great interest both in the public at large and in the scientific community. Its origin as the result of an explosion of a 'fireball', large or infinitesimal, has caused considerable controversy. Consequently, over the years it has always been submitted to scrutiny and to measurements of increased precision using more refined techniques. We have discussed above the question of identifying precise standard candles in order to evaluate as exactly as possible the distance of the galaxies, particularly those far away. All measurements have converged to the conclusion that the expansion assumed by Hubble to explain his results was valid and represented reality although the creation of space remains a phenomenon that we need to accept in order to make sense of the picture we have of that expansion. On the other hand, we have seen that in the very early stage of its existence the Universe may have expanded at an enormous rate called inflation. Since the expansion has not been constant over time, it is legitimate to investigate the constancy of that expansion over the more recent past. From our knowledge of the physics involved, it has been a general belief that gravitational attraction between the parts, stars and galaxies, should slow down the expansion. Why? When we throw an object up in the air, attraction from the Earth slows it down and, if we do not throw it with enough speed, it falls back. A rocket sent up in the sky is in the same situation. If it does not have sufficient speed it falls back on Earth. This is the case of so-called ballistic missiles. However, if its speed is large enough it could escape Earth's attraction and even the Sun's attraction and be lost in space within the galaxy. The same phenomenon should be present in the Big Bang scenario. All those galaxies recessing from each other, because of creation of space between them, still feel their mutual attraction. This gravitational attraction is the same agent grouping together particles to form nebulae, then stars, then galaxies. The attraction does not stop with the formation of those large objects. They still attract each other. Consequently, a fundamental question remains: at what level is the expansion affected by gravity? Can we measure a slowing down of the recession of the galaxies due to the attraction of all the others? From what we said about the difficulty of just measuring with precision the expansion rate, that is to say Hubble's constant, we can appreciate the difficulty of addressing that question. Actually, what we need is a mapping of the history of the recession rate to identify clearly the changes in recession speed, which is the same thing as evaluating Hubble's constant more precisely as a function of time. Fortunately, we can travel in time by looking back at galaxies away from us. The further the galaxies are, the earlier are our observations in the Universe history. We can thus, in principle, map the history of the evolution of expansion by measuring precisely the *speed of recession* with the *distance* of the far away galaxies. We know well how to evaluate the speed of recession. It is done through the redshift that we described earlier and the technique is quite accurate.[44] On the other hand, we must face the fact that accurate distance evaluation has always been a major problem. It seems, however, that Mother Nature is not always mean and sometimes provides natural instruments that can be used to address properly some questions of importance. We mentioned earlier that close to our planet we could use the parallax effect to measure such distances. Unfortunately, as we said, the technique cannot be used at large distances because the parallax angle becomes much too small. So we have to use another technique and the most appropriate one is using the luminosity of objects as we described it earlier. The problem is to find an appropriate standard candle that we can use with far away galaxies.

Mother Nature has provided us with such an object and we have mentioned it a few times before. It is a Supernova 1a, a very special type of exploding star among the larger class of supernovae. Let us recall some of the characteristics of such an object. The Supernova 1a has characteristics that make it ideal as a standard candle of very high reproducibility. It is recognized by the presence of specific spectral lines in its radiation spectrum, a characteristic that distinguishes it from other types of supernovae. It is also found that its evolution in going from accretion of material from a companion through the explosive shedding of its material through space can be modelled rather well. The main characteristic is that this type of star explodes when its core gets close to the Chandrasekhar limit of material accretion of 1.38 solar mass. The whole evolution of the star can then be modelled through its various stages and measurements on several of them found in various galaxies are found to confirm their reproducibility with rather high precision. When they are found in different galaxies, they can thus be used as standard candles by measuring their luminosity to determine the relative distance of those galaxies. The absolute distance can then be determined by means of the cosmic ladder described previously. They can be observed in very distant galaxies due to the strength of their explosions: their brightness may become equivalent to 5 billion times that of the Sun, thus almost equivalent to that of a whole galaxy. However, the problem is to find them. Explosions as we know them are rapid and the evolution of the star from the time of accretion to dispersion in space is a question of days. The whole process is over within about 40 days when the brightness has changed by 3 to 4 orders of magnitude. Furthermore, although a galaxy may contain billions of stars, it is not expected that such explosions take place more than once every 50 years in a given galaxy. When such a supernova is observed it may be too late to obtain the relevant information on its characteristics because of the long term planning required in the use of large telescopes. Nevertheless, astronomers have found appropriate scheduling at various international observatories, with observation of wide regions of the sky with special recording instruments. They have been able to observe a sufficient number of those supernovae in various galaxies with redshifts reaching unity, which means the time when the distance between cosmological objects were half their present distance.

What is found in that study is most surprising: the expansion of the Universe is accelerating.[45] Space is created continuously in the expansion process, but now space is created at an accelerated rate. This result is opposed to expectations. Without much thinking about it, we expect that the expansion should slow down with time due to what we mention before: gravitation should slow down the expansion and could even reverse it. What causes such an acceleration of expansion? It is not at all obvious. Cosmologists have introduced such an accelerated expansion phenomenon at the early evolution of the Universe. They called it inflation and invented a scalar field, the inflaton, whose internal properties acted in such a way as to cause expansion at an accelerated rate. Could we have a similar field that existed all along and that was hidden by the presence the gravitational field? Gravity would win in the early moments of expansion until distance between celestial objects has increased sufficiently to reduce it to a point where the expansion field would now win. It is possible also that the field varies with time thus changing the expansion rate. This scenario is different from the cosmological constant scenario, which is constant in time. The process has been called quintessence, like the fifth element in the antique way of describing the Universe, by means of fire, air, water, and earth. It looks that when we do not know something we go back to the basics.

On the other hand, as we said, the vacuum of space created by expansion may be imagined as a substance as real as matter 'floating' in it. In fact, as was inferred above, there is a certain cost in energy to create that space. We said it is borrowed from the gravitational potential. As we have done it several times before we may call upon the notion that energy is equivalent to mass, as per Einstein relation $E = mc^2$. Consequently, according to General relativity, the vacuum energy would act through gravitational effects. Since it is vacuum, not containing matter as such, it would be characterized by a negative pressure, which in General relativity is a cause of expansion acting in opposite direction to gravity. We can associate it to a cosmological constant, in a very similar way as Einstein introduced it to stabilize his model of the Universe and that he rejected when he was confronted with the evidence of expansion. We can simply re-introduce it on the other side of the equation to satisfy our requirement in a similar way as Einstein did and qualified as a blunder. What is really that vacuum energy? In quantum physics, zero-point fluctuations of fields are an important source of energy in the vacuum of space. Similarly, virtual particles created by quantum fluctuations could give energy to the vacuum. However, those phenomena lead to a very large number incompatible with the energy involved in the experimental observations of accelerated expansion. That quantum fluctuations energy is over 100 orders of magnitude too large and is almost an embarrassment to the physics community. The nature of dark energy thus remains a question mark and for the moment it is represented simply by a cosmological constant similar to that introduced by Einstein. As mentioned by Perlmutter,[45] 'Perhaps there was more insight in Einstein's blunder than in the best efforts of ordinary mortals'.

We thus have a world made of visible matter, objects we see in the sky, and of invisible matter required to justify motion of visible objects within the physical laws that we have concocted. Furthermore we have discovered that we need extra matter in the form of energy, which fills the so-called vacuum of space to explain an observed accelerated expansion. What is the proportion of those components when expressed in the same units of energy? It is frightening, because we have to come to the conclusion that visible matter constitutes only about 4.5% of the whole. Of that amount stars and planets constitute

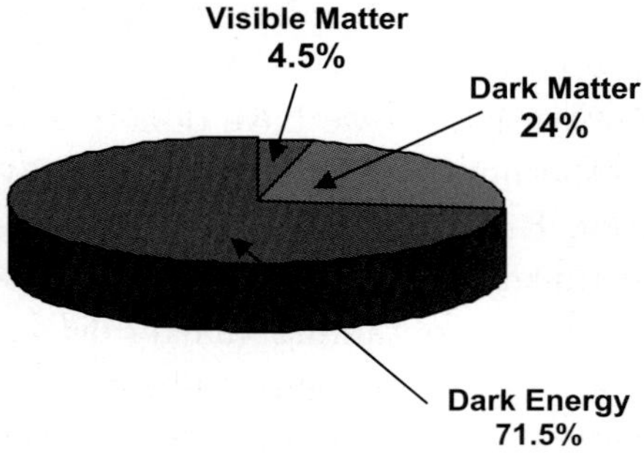

Figure 9.14. Our Universe is made mostly of energy and matter whose origin is unknown. Most of the visible matter is under the form of H and He. The rest is not visible. Dark Matter and Dark Energy is not accessible to us with present technical means.

only 0.5%, the rest being interstellar gas. This is best represented as a pie chart as shown in Figure 9.14.

This situation is rather humiliating. After having developed all those beautiful models and theories that appeared to explain rather well the behaviour of the Universe from the small to the large scale, we are obliged to admit at the end that the object of all that research constituted all along only 4% of the whole landscape. The Universe is again escaping our attempts to understand it. It looks almost as if it were plotting all along against our full understanding and inventing new scenarios to confuse us. However, this unexpected Machiavellian plot will not prevent physicists from looking deeper into the 'reality' and 'virtuality' of that world and attempt to understand more deeply its basic foundation.

In any case, this is the world we observe and live in: we need dark matter to explain the movement of stars within galaxies or other effects such as gravitational lensing and we do not know its nature. We need a special field or a cosmological constant to provide an explanation of the accelerated expansion of our Universe. We certainly are faced with two major questions for which we do not have an answer. It is a real challenge to the mind. However, this is not the end of the story. There are other questions raised by the Universe expansion to which we do not yet have an answer and they have a deep meaning regarding our own existence in that world. For example, where is the Universe heading? What is its fate? These are questions still under study, either theoretically or experimentally, and are most challenging.

The study of expansion based on Einstein general theory of relativity gives rise to the standard model called *Lambda-Cold Dark Matter Model* or, for short, the *Lambda-CDM Model.* The model is based on Friedmann–Lemaître's solution of Einstein's field equations and introduces the concept of expansion of the Universe. As its name implies, it involves also the last phenomena that we just introduced: Dark Matter and accelerated expansion through a cosmological constant, lambda (Λ). This model is used as an attempt to increase our understanding of the present state of the Universe and its future evolution. The solution involves concepts such as the critical mass density, the curvature of space, and the importance of the energy content associated with the various parameters such as cosmological constant, present mass density (baryon and dark matter), and radiation.

The results found in the study of the inhomogeneities of the cosmic microwave background radiation supports the inflation scenario at the creation of the Universe and leads to the conclusion that at the end of that episode the curvature of space was reduced to a negligible value. Our Universe in that scenario is Euclidean, or in common language, is flat. Inflation at the early stage of evolution has enlarged space so much that it has flattened it as, in a previous model, the surface of a balloon is flattened over a limited portion when it is blown to a large volume. Expansion of the Universe simply means that the balloon is continuously increasing in size.

In view of that expansion the most important question concerns the fate of the Universe. Several scenarios are possible. For the moment let us examine the question in the absence of accelerated expansion i.e. in the absence of Λ. It is convenient to introduce what is called the concept of 'critical density' and define a parameter omega (Ω) as the ratio of the present Universe density to that so-called critical density.[46] If the value of the present density is equal to that critical density, we have $\Omega = 1$ and the present Universe radius of curvature is so large as to make it look flat on our scale. In that case the Universe will expand forever. However, due to gravity the rate of expansion will continuously decrease with time. It is like the

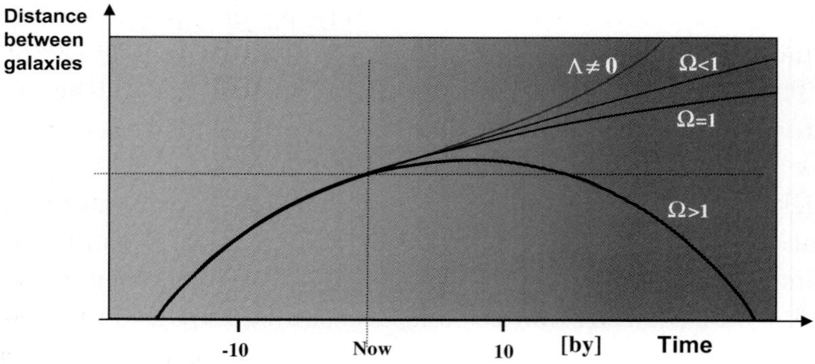

Figure 9.15. Possible scenarios for the expansion of the Universe. If the density is higher than the critical density, making Ω larger than 1, $\Omega > 1$ in the graph, the Universe could contract back on itself under the influence of gravity. If the density is assumed to be equal to the critical density, that is $\Omega = 1$, the Universe should expand to infinity. If Ω is smaller than one the Universe will similarly expand forever. If the expansion is such as to allow the presence of a cosmological constant, that is $\Lambda \neq 0$, the Universe expands at an increasing rate and will be ripped apart.

rocket that has been given just enough speed to escape the Earth and the Sun attraction: it drifts into space and never comes back. The scenario is illustrated in Figure 9.15 as a plot of the evolution of the Universe size as a function of time, and identified as $\Omega = 1$. There would be total dilution of matter after an infinite time leading to a very cold Universe.

Another scenario concerns the case when the mass is less than the critical mass. This is $\Omega < 1$. Again the Universe expands forever and is slowed down by the gravitational attraction but not quite as much as in the case of the critical density. If the matter density, however, is larger than the critical density, that is to say $\Omega > 1$, then at one point in time, gravitational attraction would win and the expansion will reverse to a contraction. It is like the rocket that was not given sufficient speed to escape gravitational attraction. It falls back on the surface of the Earth. In that case, the Universe is curved back on itself. A real disaster would follow, resulting in a so-called Big Crunch. That possibility has given rise to the Phoenix scenario,[47] imagined by Lemaître, in which the creation of the Universe is a stage in a possible cycle of repeated creation and destruction of Universes possibly very different from each other. These scenarios are also connected to the curvature of space examined in Chapter III.

In the case of critical density, it appears that the radius of curvature being very large corresponds closely to the case $K = 0$ for a flat Universe. In the case where the density is larger than the critical density it appears that the curvature is positive leading to a closed Universe.

The results found with the recent study of the expansion using Supernovae 1a as standard candles concludes that the Universe is expanding at an accelerated rate and the best way to take it into account is to include a cosmological constant in the model, in which case $\Lambda \neq 0$. This is the Lambda CDM model mentioned above. It also shows that the Universe is heading to a catastrophe, a so-called Big Rip, the expansion increasing continuously and heading to an infinite rate that would rip apart all the beautiful construction that we observe at night in the sky. What a disaster! This is essentially where we now stand. We know that the Universe is 13.7 billions years old. We do not know exactly the present density of matter in the Universe, but from the results of the CMB we can conclude that it is looking very flat. However, if it keeps expanding at an accelerated rate as we observe it now, the Universe will be ripped apart.

X

Some Last Reflections

Can we believe those models and theories that have been summarised in the previous chapters? They are what we think constitute at present the most credible description of the functioning of the Universe. The physical laws that we have enunciated, based on experimental observations, are used to make models and theories that are then used for a description of the evolution of the Universe we live in. To do that, we use logic and mathematics, rather efficient tools that have somehow been put at our disposal through our evolved brain. Some of the mathematics is described in the Annex as an illustration of the basis of the models that are assumed to represent as close as possible the reality of our world. The machinery that makes the laws of physics work is hidden to us. Then why does mathematics works? We don't know. Could there be some other tools that we don't know about? At present this is all we have. This mathematical basis is what makes the physics we know today so different from philosophy, which works logically but mainly in the abstract. The mathematics we use provides logic with numbers that are verifiable experimentally, directly, and quantitatively to several decimal places. I remember those years in college where we had to go through philosophy courses filled with assumptions, followed with logical reasoning, assembled in so-called thesis that appeared to satisfy the mind for the moment, but left us in a nebula afterwards, because we did not know what it really meant in practice and what to do with those theses. The conclusions could not be verified experimentally. They probably had some bearing on our way of thinking about life but they did not provide us with a real answer about how the world functions in practice. Religion is another approach to a description of the Universe from its creation to the arrival of man in its environment. The common belief is that what is written in the Bible was inspired from an 'exo-being' called God and is the absolute truth. This is one of the characteristics that make science different from religion. They are 'orthogonal'. Science is based on experimental observations and proofs, not beliefs. It is not absolute and is always ready to be revised relative to its fundamental conclusions when confronted with new evidence, while religion stands by dogma and very reluctantly adapts to new reality, as is shown clearly by history, a situation still prevalent today. On the other hand, philosophy works in a parallel path to science claiming that its level of analysis is at a higher rung on the ladder of understanding than anything else. This claim is questionable.

However, are we satisfied with all that knowledge of physics derived from observations and organised as logically as possible in models with a supporting theory? It is a great structure, very well organised, but obviously it does not answer totally to our desire of understanding our Universe and our origins. Tremendous progress has been made during the past century building and assembling our knowledge on the shoulders of giants of previous centuries, such as Galileo and Newton. However, physics is now facing a real challenge in its last steps towards understanding some of the basic behaviour of such a Universe that finally evolved to create beings like us, capable of asking existentialist questions. We have often attempted to make the leap between the *how* it works and the *why* it is the way it does. It seems that on occasion we have almost an answer to the why, but we find that it is only a question of a level of understanding that is more satisfying than the previous level, which looked like acceptance of facts rather than a logical deduction. For the first human beings on this planet capable of asking questions, the Sun rose every morning and it seemed a supernatural fact controlled by an external being. For many of those events that occurred naturally, gods were imagined and they ruled the functioning of the Universe and some other realities became calcified in the mind as 'natural' without explanations. Gravitational attraction was accepted as such a fact of life. Apples fall on the ground because of gravity and that was for a long period a satisfying answer. But a law was found for its motion and trajectory. That seemed to provide a 'why' the apple falls, but on second thoughts it was realised that gravity describes only 'how' it falls. Newton was fully aware of it because he raised the question of why there is such a gravitational attraction at a distance when objects are not in contact. It looked as if they were communicating and a so-called gravitational field could be imagined for doing that. Einstein came later and provided a model for that pseudo force of attraction. He invented the concept of space-time and introduced a description of gravity as a distortion of the space-time grid by the presence of mass: it seems like a final explanation and a true why. But what is mass? Although we know that energy can materialize on occasions under the form of mass and that mass may be transformed back into energy of another kind in special occasions, we are still looking for its origin. In other words, although we know rather well its property, we do not know exactly its nature. A theory exists that leads to the presence of a field filling all space, called the Higgs field, from the name of its proponent. The field is quantized and should be observed by means of its materialization as a particle called the Higgs boson. Major efforts are presently concentrated on the detection of such a particle. That field, through its interaction with elementary particles, would provide them with a property that would make them obey the physical laws that we know, from Newton's principles to Einstein's space-time grid distortion. That property is mass. As it is readily observed, this is another step in the ladder of understanding. We have now a field to which elementary particles are indebted for their mass, but that field is a property of space. And it goes on like that in all models we make. If we make a step forward we usually find another question that needs answering.

Let us look at some of the questions that do not appear to be solved in the present models, or at least, still require to be addressed at a higher level of understanding. First, we have to realise that all theories developed, from Newton's principles to the law of gravitation, even as expressed in Einstein's geometrical model, as well as the special theory of relativity, rely on the introduction of parameters that are arbitrary and called constants of nature. The same is true of other theories concerned with elementary particles and the proposed string theory, for example. Fundamental constants pop out like rabbits from a magician's hat. Let us think of the velocity of light. Long ago it was measured experimentally and found to be about 300,000 km per second. Maxwell made a beautiful theory unifying electricity and magnetism and found that it led to electromagnetic waves that travel at a speed of $(1/(\mu_o\varepsilon_o)^{1/2})$ in which ε_o and μ_o are two constants belonging to experimental laws found in static electromagnetism. They are electric and magnetic characteristics of what we call vacuum. And the number obtained from the exercise in calculating the speed of propagation of electromagnetic waves is c, the speed of light. That seems to explain why light travels at speed c. It is an electromagnetic wave. The constants involved are of static origin as if vacuum were a substance that was being stretched and compressed by the wave, with the two constants determining the importance of the resistance to those deformations. We did not think that vacuum, being 'nothing', could perform that kind of acrobatics. Physicists invented the aether and then found out that it did not quite make sense and abandoned it. Consequently, what is that vacuum and why does it react that way? It does not appear to be vacuum at all, at least not the vacuum that we interpret in our mind as the absence of anything. We really don't know the origin of those constants and the reason for their connection to give the speed of light. The relation comes out of the mathematics developed by Maxwell. Is it sufficient to the critical mind? The mathematics just follows logic but does not provide an explanation of why vacuum should be characterised by such constants and should provide a limit to the speed of light. Some theorists brush them under the carpet, make them equal to 'one', as they do with many other constants, but that does not answer the question. It displaces the problem. As well, to say that c is the inverse of the square root of the product μ_o by ε_o, a property of vacuum, is not a satisfying answer either. Could we formulate a theory that would say that vacuum is a state of the Universe, which, based on an evident hypothesis, would allow the calculation of those constants? We do not know. Maybe that theory would itself be based on an observation with its own constant of which we do not know the origin. However, in any case, all theories that provide a description of the functioning of the Universe are based on models requiring the introduction of some constants that we have to measure experimentally, such as Planck's constant, gravitational constant, and Boltzmann constant. Did all those vacuum constants materialize at the early moments of the Universe existence?[1] It is possible.

This leads us also to those other constants characterising elementary particles, in particular the electron with its electric charge, its mass, and its magnetic moment. Those parameters or constants are quantized. They are unique and cannot be split and appear to be constant with time. In the theory of chromodynamics the quark is introduced as an elementary particle with a charge being an exact fraction of the electronic charge, $-1/3\,\mathrm{e}$ for the down quark and $+2/3\,\mathrm{e}$ for the up quark. That hypothesis seems to be in exact agreement with experimental observations. Proper combinations of quarks make all atoms exactly neutral, the electric charge on the nucleus cancelling exactly the charge of the electrons enveloping it. Although the charge of the two particles, quark and electron, does not appear to originate from the same physical context, the balance is exact. If it were not so, our world would not exist the way it is. Another most interesting thing is the spin magnetic moment of the electron. Where does it come from? The electron is presumably a point in space without dimensions, but it has mass, charge and a spin. If it were spread somehow, as we said before, we could probably split it and analyse its parts. But no, it is quantized and we do not find half an electron in nature. It looks rather fundamental in the sense that it does not have internal parts. But it has a magnetic moment and, to our understanding from a classical point of view, the presence of a magnetic moment means that electric charges are moving around and rotating somehow in space. It is not possible for a point in space to do such acrobatics. We may reply that it is not classical, that it is a relativistic-quantum-mechanical object. The strangest thing of all is that if you couple special relativity to quantum mechanics, the spin magnetic moment pops out like magic. It is true that classically magnetic effects appear to originate from the relative motion of charged particles, but the conclusion that relativistic-quantum-mechanics gives a magnetic moment to a point charge particle is quite a difficult concept to visualise and understand.

This is Dirac's theory of the electron and is a most beautiful exercise in matrix mechanics. Does it really tell us where the spin comes from? Mathematically yes. Physically, I have not succeeded yet in seeing it. It appears, however, that Physics can live with that. It starts with a great concept that elementary objects are both wave and particle and that they are subject to the special rules of quantum mechanics and, at the same time, they follow the rules of relativity. But both are rather strange. As we saw, quantum mechanics is the strangest theory of all, with particles essentially obeying the rules of probability, with uncertainty being its basic rule of conduct. It also makes possible spooky action at distance, a most unreasonable property of entangled particles. It is a property that Einstein could not accept as real and attempted to explain as originating from our lack of understanding of the physics involved and as a result of an incomplete theory. Although we can handle the mathematics behind it, we have not really yet made sense of the physics involved. On the other hand, relativity is based on the concept that whatever you do, whatever the speed of the source or of the observer, the speed of light appears to be the same in all directions. These rules do not make sense to our simple mind. We simply have to accept them. It looks as if we are really limited and we have to accept that a conspiracy of nature hides things from our understanding.

Another peculiarity of the Universe is the supremacy of matter over antimatter. We can speculate that in the beginning of the expansion there was a break in the symmetry and this is what caused matter to win. But what caused that break is not clear. A possibility exists that parallel Universes developed, some entirely made of matter and others of antimatter. The whole thing becomes neutral regarding the matter-antimatter question. However, this looks more like speculation than a solid theory based on facts and observations. We have no way of verifying it. In the same line of thought we seem to have a perfect balance between positive and negative charges in the Universe. Quarks and antiquarks were created independently of electrons and antielectrons, but the Universe stayed neutral electrically in the process. During the catastrophe of matter annihilating antimatter, matter appeared to predominate by a small margin and won over antimatter. And presumably that process left exactly zero net charge, the number of quarks with $2/3\,\mathrm{e}$ and $-1/3\,\mathrm{e}$ left in the proper proportion to cancel the charge of residual electrons. This seems to be a rather improbable coincidence.

There is another question regarding the number of elementary particles that exist and all the possible combinations that can be realised with them. They do not seem necessary for our world and many of them have an ephemeral life, so short that sometimes their actual existence as particles is questioned. Up and down quarks, gluons, photons, electrons and neutrinos are entirely sufficient to construct the world we see and live in. When examined carefully, those particles seem to be dictated by some law of symmetry that we barely understand. Other kinds of elementary particles exist and many kinds of combinations are possible with those particles. Why it is so, we do not know, but the symmetry is there. Could the extra particles be required by-products of a planned larger construction that we do not understand well?

It also appears that the present state of the Universe is very special and that the fundamental constants play a very important role in its evolution to that state. The present Universe conditions are such as to allow our presence and it is claimed that this accomplishment is due to the exact value of those constants. If they had a slightly different value, the Universe would be very different and would most probably not lead to our existence. This brings us to the so-called anthropic principle. This principle is formulated often with different perspectives. One formulation is that 'we are here because the conditions in the Universe are such as to allow our presence', or 'the universe is the way it is because we exist'. On the other hand, some may argue that it is most improbable that a fine tuning of the precision required for the existence of life and our presence could exist by chance alone.

This kind of reasoning has lead to philosophical arguments leading to strong positions about our origins. In a theistic approach, an exo-being exists, creates the Universe with a very intricate space-time grid and with very special laws that we try to understand. In that argument we are desired as final product and the knobs on the console that control the fundamental constants were adjusted by that exo-being to such a value as to make sure that we appeared some time in the evolution of the Universe. We are really privileged. It almost says that we are unique and that most probably there are no others like us in that huge Universe. As it is sometimes said, it looks like an awful waste of space. We did not need that much.

It does not appear that we can address such existentialist questions with scientific arguments. We probably can only say that the fact that we are here is a proof that the fine tuning was possible and it happened. If it did not, we would not be here to talk about it.

That previous discussion is about fundamental constants and adjustable parameters of the model we have developed. How about the theories and models themselves? Are we satisfied with them? We are very proud of them and they certainly are a testimony to the talent of those who have developed them. However, those models are still rather compartmentalised in their own sphere of specialization. We have general relativity on one hand with a dynamic space-time grid as background that applies to macroscopic objects and quantum theory on the other hand that applies to the microscopic world. They both seem to work well in their own sphere. Even with all its absurdity we believe in quantum mechanics. It works so well. But we also believe in the general theory of relativity. It works very well also and we have given several examples above where it provides explanations of observations without ambiguities. It has also provided us with a fairly solid mathematical framework for the description of the evolution of the Universe allowing us to express Hubble's law in terms of physical parameters. Then, why can't we have a theory that describes at the same time the small and the large scale of that Universe? It is a real challenge and according to theoreticians the solution does not appear to be around the corner. It is a dream of those theoreticians to realize on paper a theory that provides a unification of all the forces of nature into a single representation.[2] We have presently a satisfying unification of electricity and magnetism as well as a unification of the weak force and electromagnetic force. The strong force is treated in a similar mathematical context as the weak force, but the gravitational force, however, remains well isolated.

There are other elements of our ignorance that are even more frustrating. From what we discussed earlier, it appears that to explain the behaviour of stars in galaxies we need to make the hypothesis that there is in our Universe a new type of matter that we cannot see. It is so important that it amounts to approximately 24% of the whole mass that is required to explain the observation of the behaviour of trajectories of stars in galaxies and the expansion of the Universe. And this is not a one-shot observation made on one case. It is verified by means of analysis made of a large numbers of galaxies. The dark matter that we infer from observations is necessary or we have to abandon Newton's theory of gravitation or Einstein's general theory of relativity. Nobody is ready to do that. The brave person that would attempt it would probably find himself quite isolated. We thus say that dark matter must exist.

The same thing happens with another observation. The Universe, the expansion of which should slow down under the influence of gravity, has been found to expand at a faster rate over the last few billion years. It is difficult to imagine a bigger blow to our ego. We make a whole theory that is based on gravitation, of which we think we know the effect, and we derive logically its main consequence — the expansion should slow down. We find that it does the opposite: the Universe expands faster with time and has done so for some time. Even worse than that, the energy involved in that expansion when converted to mass is 71.5% of the total mass forming our Universe. This is energy with a very special character that causes expansion rather than contraction. Where is that energy coming from? We don't know. Presumably it comes from that entity that we call vacuum. We are obviously forced to conclude that we really don't know what vacuum is. It is a seat of effervescence in which virtual particles are created for short periods of time obeying the uncertainty principle. Quantum electrodynamics leads to the presence of residual zero-field fluctuations and according to that theory, this effervescence is the cause of several effects that we measure with great precision. We were never able to disprove that theory. This is energy present in the totality of space. All those contributions make a lot of energy and actually are more than necessary. The energy involved is way too large to explain the accelerated expansion. We do not talk much about it. It is again a real blow to the ego. The concept of vacuum has to be revisited and revised. We raised the question in connection with the speed of light. When electromagnetic waves travel in materials, we find that they interact with the atoms of the material and their speed is altered. They travel more slowly. Thus, we know that something is happening to them: they interact with electrons in the material. What is then happening to the electromagnetic field in vacuum to set its speed equal to c?

In the concept of the expansion of the Universe we concluded that space expands, not the galaxies, but the space in between the galaxies. We concluded that light waves are so well connected to that space that their wavelength stretches, expands with it: we call it the redshift. Why is it not just the speed of the photons that is reduced? No, and it is coherent with Einstein's principle of relativity that the speed of light is the same for all observers whatever their speed. However, photons, interpreted as the level of excitation of the electromagnetic field, are altered somehow in that expansion. They lose energy and in doing so alter their frequency: the associated wave changes wavelength accordingly. It looks very much as if the photons interact with vacuum, which has a life of its own.

There is something about that Universe that is much deeper than our mind can dig. We appeared in it through evolution, although this is contested by some. But the evidence is there. We have evolved from beings of low intelligence that slowly acquired the cognitive faculty that we now possess. Are we the main actor in it? Are we the goal, as some theological arguments wish us to believe? Why would there not be in those billions of stars inside billions of galaxies another planet with a favourable environment that would end up with a being like us, or still more evolved? To answer 'no' does not seem to be rational. With Copernicus we accepted that the Earth was not the physical centre of the Universe. It would be strange to assume now that as thinking beings we would be at the centre of that new dimension: intelligence. At the present stage of our knowledge, with what nature has given us for developing that knowledge and partial understanding of our origin, we look much more like an accident in that whole landscape, a happening in a Universe that does not seem to be directed to anywhere precisely. In that context, that accident could very well have happened somewhere else. If it did not happen anywhere else, we are then very lucky to be here to ask ourselves those questions.

Einstein, one of the greatest physicists who ever lived, said at one point that he wanted to know if God had a choice in making the Universe. He also said, in discussing the fundamental uncertainty aspect of quantum mechanics that he did not believe that God played with dice. Analysts have attempted to deduce Einstein's personal relation to religion from those statements. However, Einstein's questioning may be simply related to desperation faced with observations that do not find logical explanations. What we find is that the construction of the Universe appears to follow in general some logic that we interpret with our mathematics. The dream of the physicist is to find that logic in its totality and the very first primeval rule that governs the whole. We keep adding to our knowledge and push back every day the frontier of that knowledge by means of new findings, experimental and theoretical. Will we find one day the ultimate rule in a theory of everything?

We have gone backwards 13.7 billion years to find out that our Universe was born essentially of nothing. Were the rules embedded in that moment or did they develop later? Were there any rules or 'obligation' for quarks and electrons to materialize as they did to form matter? Was it possible for other types of particles to appear? It seems that the possibility existed for antimatter but it soon lost the war to matter, which now dominates. There are also other quarks and leptons that do not seem necessary for the existence of our Universe. It appears that they are there because nature likes symmetry. There are many other particles than those that constitute our visible world, particles the existence of which is made possible from the proper combination of quarks. However, they have a limited lifetime. Consequently, they also lost the battle because time was also created with space and those particles were then ephemeral in that extra dimension. I.I. Rabi, when learning about the discovery of the mu lepton, commented: 'Who ordered that?' We could ask the same question about all those extra particles. They appear to be a by-product of a main construction plan, the residue of a manufacturing process. We also face a really humiliating situation that probably is the main challenge to our mind. We thought that we were getting to understand what the material that constitutes the world is: it was made of baryons and leptons, that is to say neutrons, protons, electrons, neutrinos. But we find that those constitute only 4% of the whole. We would certainly have been happier with 100% and obviously, like Rabi confronted with the muon, we may say that we did not need dark matter and dark energy. Things were complicated enough. As written by Weinberg, '... the more the universe seems comprehensible, the more it seems pointless'.[3] This may be an exasperated and exasperating statement that may irritate some. However, in the light of the comments just made, we can understand the frustration of the physicist before questions that cannot be answered. Is it a question of the capacity of our brain that leaves us unable to grasp the whole picture?

We might say that the Universe looks very much like an experiment that was done, that succeeded to a certain level and then turned sour, at least in part: then human beings appeared. Planet Earth was evolving with its minerals, plants, and animals giving rise to a great variety of species. Dinosaurs and close cousins appeared and existed in equilibrium with other living beings. Rules presumably written in the evolution process were followed. They existed for a hundred million years and seemed to live in some kind of harmony with their environment. A catastrophe happened and they all vanished. Sometimes, chaos seems to be the rule.[4] We appeared with a brain capable of reasoning, capable of using a minuscule part of the Universe to improve our well-being, and we started to transform it. We have been here with that capacity for a few thousand years. And we cannot live well with each other and with our environment. We certainly look more like an accident in the whole evolution rather than a goal.

This kind of reasoning leads us back to Einstein's question about God's possible choice in creating that Universe. We do not have an answer to that question. We do not have a theory with an equation that gives the recipe on how to construct a Universe like ours. We do not know if such a theory is possible. We do not have a logical reason why we should be part of the landscape. When we go back in time we find answers to some of our questions, one by one. It is a lot of work because our brain is limited. It is a similar process as getting rid of old gods that were invented in the past to explain natural phenomena. Will we find one day that final theory or that recipe? Physicists will keep hunting for it, that's for sure. What will we find at the beginning? It appears that it may always escape us. If the Universe emerged from a singularity, our physics and mathematics cannot handle the infinities that result in the primeval entity. Our mathematics fails and we cannot calculate anything when infinity arises in the equations. If the Universe is the result of a cycling process, a Phoenix scenario, information about the previous stage may be blocked by the passage through the narrow channel at the origin which may also lead to a singularity. At the risk of displeasing many, as it appears that we are a minor actor in the whole scenario, we may never know what that whole affair is about.

There are speculations that in the cycling process of the Phoenix scenario, Universes would appear in series in a Big Bang and disappear in a Big Crunch. They would be different from each other and could have different physical laws and fundamental constants. In that scenario we just happen to live in one cycle where the laws and fundamental constants are such as to be favourable to life and our presence. This would be compatible with the anthropic principle. A similar speculation proposes that there could be many Universes existing simultaneously in parallel and it just happens that we are in the right one. It could very well be that those physical constants and laws are totally interconnected and that it is not possible to change one without changing the others. This possibility would lead to a great diversity of those Universes, in series or in parallel and only particular ones would be apt to offer proper conditions for life.

When examined closely, life, looking like the consequence of complexity, looks inevitable in the complex Universe we inhabit. As is sometimes said, it would have taken divine intervention to *prevent* it from appearing. Can we say the same thing about intelligence? It appears so; it is a question of level or degree: evolution was a most efficient process to reach the stage we are in. It is very plausible that it could go further and evolve into something greater. If intelligence grew up in part with the size of the brain, it can certainly grow further, if not on this planet, perhaps somewhere else.

Where does that discussion lead us? At the present stage of our development, we do not seem to have the kind of tools needed to comprehend the whole. We will certainly find answers to many of the questions that prevent some of us from sleeping. We will find such things as what dark matter and dark energy are. We may find out why fundamental constants have those values, or at least we may be able to establish a higher level in our understanding. We may even find out what quantum mechanics really means. We may also find what the nature of the complexity that creates life is. But, being a part of that Universe, integrated in its space-time canvas, having a myopic view of the whole structure, it seems that we very likely may never find out *why* we are here, questioning ourselves on our origin and the origin of that world around us. There may not be an answer to it, inside the space-time grid: a real challenge to our limited mind.

Annex: Mathematical Support

In this annex, an elementary mathematical support to some of the concepts treated in the main text is presented. It is a very concise summary and no detailed demonstrations of formulas are made except in very simple cases. Its aim is to provide the mathematically oriented reader with a very concise mathematical overview of some selected physical theories that are referred to in the main text. It is not necessary for understanding the main text, which can be read without the details of this Annex.

A. Classical Physics

(a) *Newton's principles:*

- If no force is applied to an object its velocity is constant.
- A force F applied to an object of mass 'm' causes a change of its momentum, mv, given by:

$$\frac{\mathrm{d}}{\mathrm{dt}}\mathrm{m}\mathbf{v} = \mathbf{F}, \tag{A.1}$$

where $\mathbf{v}$ is the speed of the object and (d/dt) is the derivative operator. The parameters written in bold characters are vectors, entities that have a direction associated to them, like speed and force in the present case. The group m$\mathbf{v}$, called the momentum of the particle, plays an important role in the sector of physics that we call mechanics. If the speed is small, m may be assumed to be constant and the equation becomes:

$$\mathbf{F} = \mathrm{m}\mathbf{a}, \tag{A.2}$$

where a stands for d$\mathbf{v}$/dt and is the rate of change of the velocity or the acceleration.

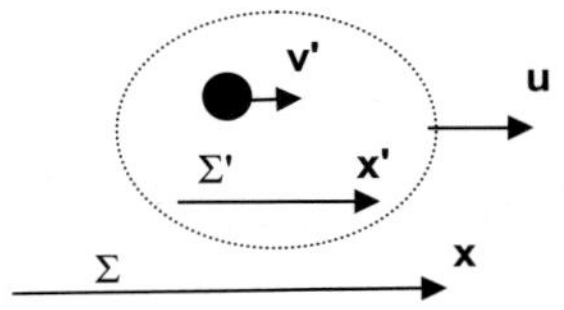

Figure A.1. Galileo's transformation for a reference system moving at speed u.

- Action is equal to reaction.

$$\mathbf{F}(\text{action}) = -\mathbf{F}(\text{reaction}). \tag{A.3}$$

An object A acting on an object B with a force $\mathbf{F}_{\rm AB}$ will be submitted to a force $\mathbf{F}_{\rm BA}$, equal an opposite to $\mathbf{F}_{\rm AB}$, object B acting on object A. In equilibrium, the net force, that is the sum of the two, is zero.

It is assumed that the laws of mechanics remain the same in reference frames in relative motion. For low velocities, the Galilean transformation leading to a direct addition of velocities in reference frames in relative motion is assumed to be valid. If an object is moving at velocity v' in frame Σ' itself moving at velocity u relative to frame Σ, as shown in Figure A.1, the position x and velocity v of the object in frame Σ are given by (for simplicity we use only one dimension):

$$\left.\begin{aligned} \mathrm{x} &= \mathrm{x}' + (\mathrm{u} + \mathrm{v}')\mathrm{t} \\ \mathrm{v} &= \mathrm{u} + \mathrm{v} \\ \mathrm{t}' &= \mathrm{t} \end{aligned}\right\} \tag{A.4}$$

and Newton's laws stay the same in both frames. This is also valid in a three-dimensional framework.

(b) *Gravity*

In Newton's approach the force of attraction between two objects of respective masses m and M is inversely proportional to the product of their mass as if they were concentrated at one point and inversely proportional to the square of their distance r:

$$\mathbf{F}_{gr} = -G\,\frac{mM}{r^2}\mathbf{r}, \tag{A.5}$$

where G is Newton's constant of gravitation equal to $6.67428(11) \times 10^{-11}[\mathrm{m^3\,kg^{-1}\,s^{-2}}]$. In the equation **r** (bold) is a unit vector giving the direction of the force, which is in a radial direction. In the case of homogeneous masses with spherical symmetries, the force is directed towards the centre of each mass. The expression can be used to calculate the gravitational field developed by mass M as:

$$\mathbf{E}_{gr} = -G\,\frac{M}{r^2}\mathbf{r}. \tag{A.6}$$

This makes the gravitational force on an object of mass m equal to:

$$\mathbf{F}_{\mathrm{gr}} = m\,\mathbf{E}. \tag{A.7}$$

It is very useful to define the gravitational potential as:

$$\Phi_{gr} = G\,\frac{M}{r}. \tag{A.8}$$

The field is then given by the gradient of the potential that is its rate of change with distance:

$$E_{\mathrm{gr}} = \nabla\Phi_{\mathrm{gr}}. \tag{A.9}$$

We define generally $\mathrm{GM/r^2}$ as the acceleration of gravity and call it g, which on planet Earth is equal to $9.8\,\mathrm{m/sec^2}$. The Earth has a radius of 6.4×10^6 metres. With the value of g and G given above it is possible to calculate the mass of the Earth. It is left as an exercise. The answer is 6×10^{24} kg, a large mass indeed.

(c) *Special relativity*

The speed of light has been found to be independent of the speed of the emitting source and observer. This observation, sometimes called a conspiracy of nature (Poincaré), has been made the main principle of Einstein's theory of relativity.[1]

In special relativity a four dimensional space is formed combining time and space into a single coordinate system called space-time. The length of a line element or interval 'ds' in space-time is given by the equation:

$$ds^2 = c^2dt^2 - dx_1^2 - dx_2^2 - dx_3^2, \tag{A.10}$$

where the x_i are coordinates in a three dimensional space, t is the time coordinate and c is the speed of light. In such a system of coordinates, time and space are entangled and motion in space implies a change in the time axis of such a nature and value that the interval ds is independent of the frame of reference. In other words, motion in space influences the rate of time flow in the frame of reference attached to the moving object as seen from a rest frame. This is shown as follows. To simplify we look at motion along one axis only and attach a frame of reference Σ' to an object that moves by dx in time dt in frame at rest Σ. In frame Σ

$$ds^2 = c^2dt^2 - dx^2, \tag{A.11}$$

In moving frame Σ' the time lapse is $d\tau$. Then, for an object at rest in that frame the invariant line element is

$$ds^2 = c^2d\tau^2, \tag{A.12}$$

The line element ds is invariant in frames in relative motion and simple arithmetic gives:

$$d\tau^2 = dt^2 - c^{-2}dx^2, \tag{A.13}$$

Just divide by dt^2 and manipulate to obtain:

$$c^2\left(\frac{d\tau}{dt}\right)^2 + \left(\frac{dx}{dt}\right)^2 = c^2. \tag{A.14}$$

If the object is at rest in Σ then, $(dx/dt)^2$ is zero, and $(d\tau/dt)^2$ is equal to 1. This means that $d\tau$ is equal to dt or that in the frame of reference Σ' attached to the object, time flows at the same rate than in the frame of reference Σ. If the object is moving at speed dx/dt, then $(dx/dt)^2$ is positive and the equation above says that the ratio $(d\tau/dt)^2$ must be less than one for satisfying the equation. This shows that time in frame Σ' attached to moving objects must flow at a slower rate than in frame Σ. In other words speed in space is compensated by a change in time to compensate. It is a trade off between space and time coordinates. And object at rest in a given frame may be said to move along the time axis at the speed of light. If an object is going nearly at the speed of light, that is to say $(dx/dt) \to c$, then we need $d\tau/dt \to 0$, meaning that time stops in frame Σ' as observed from Σ.

The mass of an object moving at speed v, as measured from a rest frame, is altered according to the formula:

$$m = \frac{m_o}{\sqrt{1 - (v/c)^2}}, \tag{A.15}$$

where m_o is the mass of the object measured at rest in a reference frame. As is seen readily, when speed v gets to be close to the speed of light mass appears to become very large. A very large force needs to be applied from the rest frame to increase its speed. Energy spent in attempting to accelerate it in that rest frame appears to be transformed into mass. It should be recalled that in the reference frame attached to the moving object, that object is at rest and its mass remains m_o. For example, the Earth moves at great speed around the Sun, which moves with the galaxy, which itself is found to move in the direction of the so-called great attractor at a speed of 600 km per second. For us, moving with the Earth, attached to it by gravity, its mass is m_o the rest mass in our reference frame.

In the case that v is much smaller than c Equation A.15 can be expanded in series and one obtains:

$$mc^2 = m_o c^2 + \frac{1}{2} m_o v^2 \dots, \tag{A.16}$$

where the first term is recognized as the rest mass energy and the second term as the kinetic energy of the rest mass now in motion at speed v.

An object moving at speed v in a frame of reference moving itself at speed w relative to a rest frame appears to move in that rest frame at a velocity given by:

$$V = \frac{v + w}{1 + v\,w/c^2}. \tag{A.17}$$

The result makes V always smaller than c even if v and w approach c themselves. For example, if $v = w = (3/4)\,c$, we obtain $V = (24/25)\,c$, smaller than c. This is called the velocity addition theorem.

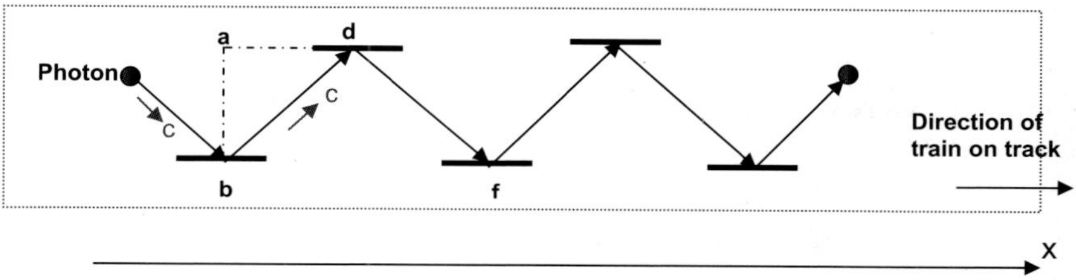

Figure A.2. The photon or light pulse clock.

(d) *Photon clock*

The difference in time lapse between systems in relative motion can also be easily demonstrated with the help of the photon or light pulse clock introduced in the main text and shown in Figure A.2. The derivation is based entirely on the constancy of the speed of light in both the rest frame Σ and the moving frame Σ'. In the Σ' frame attached to the moving mirrors, the photon travels a distance ab in half a cycle. However, as seen from the rest frame Σ the mirrors are moving and the photon travels a distance bd. Simple calculation gives:

$$(\mathrm{bd})^2 = (\mathrm{ab})^2 + (\mathrm{ad})^2. \tag{A.18}$$

In the moving frame as seen from the rest frame we have ab $= \mathrm{c}\Delta\tau$. The distance as measured from the rest frame is bd $= \mathrm{c}\Delta\mathrm{t}$ and ad $= \mathrm{v}\Delta\mathrm{t}$. Replacing in equation A.15 above we obtain:

$$\Delta t = \frac{\Delta\tau}{\sqrt{1 - v^2/c^2}}. \tag{A.19}$$

Consequently the time lapse $\Delta\mathrm{t}$ observed in the rest frame appears to depend on the relative speed of the moving frame and is longer than the time $\Delta\tau$ in the moving frame.

If the clock orientation is turned 90°, the photon or light pulse travels in the direction of the wagon motion. We will call it the horizontal clock. We thus have the situation shown in Figure A.3.

Assume it takes time t_1 for the photon to travel from mirror a to mirror b. While the photon travels, mirror b moves to position b′, a distance vt_1. The distance travelled by the photon is thus $\mathrm{L} + \mathrm{vt}_1$. The path travelled by the photon during that time at speed c, is

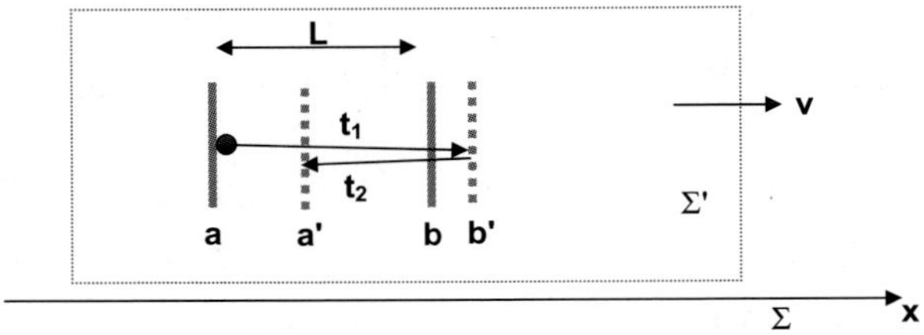

Figure A.3. The photon or light pulse horizontal clock moving at speed v. After one half of one period mirror b moves a certain distance and finds itself situated at position b′. After a full period, mirror a finds itself situated at a′.

obviously $\mathrm{ct_1}$. We have thus

$$\mathrm{ct_1 = L + vt_1.} \tag{A.20}$$

This can be written:

$$\mathrm{t_1 = L/(c - v).} \tag{A.21}$$

Similarly while the photon is moving back to mirror a in time $\mathrm{t_2}$, that mirror has also moved by a distance $\mathrm{vt_2}$. The path that must be travelled by the photon is shorter than if the mirrors were at rest. We can thus write:

$$\mathrm{ct_2 = L - vt_2} \tag{A.22}$$

and we have:

$$\mathrm{t_2 = L/(c + v).} \tag{A.23}$$

The total period of the clock as seen in rest frame Σ is $\mathrm{t_1}$+ $\mathrm{t_2}$ equal to:

$$\mathrm{\Delta t = t_1 + t_2 = 2Lc/(c^2 - v^2).} \tag{A.24}$$

This can also be written:

$$\Delta t = \frac{2L/c}{1 - v^2/c^2}. \tag{A.25}$$

This is apparently different from the period of the vertical clock given by Eq. A.18. However, we have forgotten something. The distance between mirrors a and b is altered by the motion of the clock at speed v. That distance L_o at rest, is altered under a Lorentz transformation (see below Equation A.40) to the value:

$$L = L_o \sqrt{1 - v^2/c^2}. \qquad \text{(A.26)}$$

Substituting in Eq. A.24 and calling $\Delta\tau = 2L_o/c$ the period in reference frame Σ' we obtain:

$$\Delta t = \frac{\Delta\tau}{\sqrt{1 - v^2/c^2}}, \qquad \text{(A.27)}$$

the same as Equation A.19.

(e) *Electrodynamics*

When objects contain electric charges, a force of attraction or repulsion exists between them, which is proportional to the product of the value of the charges q and q′ divided by the square of the distance r between the electrically charged objects. It is called Coulomb's law and is written as:

$$\mathbf{F} = \frac{q\,q'}{4\pi\varepsilon_o r^2}\mathbf{r}. \qquad \text{(A.28)}$$

The force is directed along the line joining the two charges whose direction is given by the unit vector **r**. It is similar to the law of gravitation, but in the present case nature has invented two types of charge which either can attract or repel each other. We have called those charges positive (+) or negative (−). Charges of opposite signs attract themselves while charges of the same sign repel each other. The constant ε_o is called the vacuum dielectric constant. It is equal to $8.854\ 187\ 817 \times 10^{-12}$ [F m^{-1}]. Since it enters in the expression of the speed of light, which is defined as exact, the value of ε_o is also exact. The factor 4π is introduced for questions of symmetry and simplification of notation in other equations resulting from the theory.

The size of that force can be compared to that of gravity in the case of two electrons separated say by 1 mm. The elementary charge e has a value of 8.85×10^{-19} coulomb while the mass of the electron is 9.11×10^{-31} kg. Using equations A.5 and A.28 we obtain a ratio of electric force to gravitational force equal to $\sim 10^{42}$. For two protons in the same circumstances we obtain $\sim 10^{36}$. Like in the case of gravity we define an electric field E given by a charge q′ as:

$$\mathbf{E} = \frac{q'}{4\pi\varepsilon_o r^2}\mathbf{r}. \tag{A.29}$$

A charge in motion produces a so-called magnetic induction **B.** It is a complex function of the movement of the charge. A charge q moving itself at speed **v** in a combination of electric field **E** and magnetic induction **B** is submitted to a force given by:

$$\mathbf{F} = q[\mathbf{E} + \mathbf{v} \times \mathbf{B}], \tag{A.30}$$

where in this case the multiply sign $\times$ means a vector product, a type a multiplication that takes into account the relative orientation of the speed **v** and the field **B** and produce an effect at right angle to both vectors **v** and **B.** This force accelerates charged particles according to Newton's law. Thus:

$$m\frac{d^2\boldsymbol{r}}{dt^2} = q[\mathbf{E} + \mathbf{v} \times \mathbf{B}] \quad \text{or} \quad \mathbf{a} = \frac{q}{m}[\mathbf{E} + \mathbf{v} \times \mathbf{B}]. \tag{A.31}$$

Here $\mathrm{d}^2\boldsymbol{r}/\mathrm{dt}^2$ is the second derivative of the position $\boldsymbol{r}$ in respect to time and is thus the acceleration **a** of the charged object. It is readily observed that the acceleration of the object is function of the ratio of its electric charge to its mass, (q/m). Thus, this ratio can be determined from the trajectory of the charge in a combined electric and magnetic field and the sign of its charge is determined by the sign of the curvature of its trajectory in the magnetic field. This is a basic technique used in detectors of large accelerators to determine the properties of electrically charged particles. The magnetic induction caused by a current I obeys Ampere's circuital law:

$$\oint \mathbf{B} \bullet \mathbf{dl} = \mu_o \mathbf{I}, \tag{A.32}$$

where the dot sign, $\bullet$, means a scalar product, which is the product of length dl by the projection of the induction **B** on the direction dl**.** The integral is on a closed circuit enveloping the current I and μ_o is the magnetic constant equal to $4\pi \times 10^{-7}$ henry per metre. It gives also the relation between the induction **B** and the magnetic field **H** as

$$\mathbf{B} = \mu_o \mathbf{H}. \tag{A.33}$$

It is practical to associate with the magnetic field a quantity called the flux Φ that measures also its intensity. A most important phenomenon discovered by Faraday concerns the effects of fluctuations of the fields with time: if the magnetic field varies with time, the flux also varies and an electric field E is induced in the surrounding space or vacuum. It is as if vacuum, stretched by the varying flux reacts orthogonally and creates an electric field to compensate. The size of the electric field E created is given by a complex formula:

$$\oint \mathbf{E} \bullet \mathbf{dl} = -\frac{\mathrm{d}\Phi}{\mathrm{d}t}, \tag{A.34}$$

where the integral is essentially a summation of all elements $\mathbf{E} \bullet \mathbf{dl}$ under its sign over a closed circuit. These are the basic laws of electricity obtained or derived from experimental observations. On their basis, Maxwell has developed an analysis that unifies completely electric and magnetic concepts into a single theory that is called Classical Electromagnetism. The final result consists essentially of four equations relating the vectors **E** and **B** to each other:

$$\left.\begin{array}{rr} \nabla \bullet \mathbf{E} = \dfrac{\rho_e}{\varepsilon_o}, & \nabla \bullet \mathbf{B} = 0, \\ \nabla \times \mathbf{B} = \dfrac{\mathbf{j}}{\varepsilon_o c^2} + \dfrac{1}{c^2}\dfrac{\partial \mathbf{E}}{\partial t}, & \nabla \times \mathbf{E} = -\dfrac{\partial \mathbf{B}}{\partial t}. \end{array}\right\} \tag{A.35}$$

Here the inverted delta symbol, called 'del' stands for the summation of second order derivatives relative to the three spatial coordinates and is a vector. The symbol ρ is the charge density in the space where the equations are applied. The symbol **j** stands for the density of current due to the motion of the charges in the same space. It is possible with these equations to describe essentially all possible experimental situations so long as we know the boundary conditions. They provide a complete description of the field **E** and **B** in the environment studied and consequently provide the basic equations for calculating the forces seen by charged particles submitted to those fields. Unfortunately it is possible to solve them only in simple symmetrical geometric situations. For more complex geometries it is necessary to use numerical solutions for which a digital computer is an ideal tool. Programs exist that solve those equations in the most complex situations, like resonant cavities, and provide a picture of the distribution of the fields **E** and **B.** That set of equations is normally completed with an equation that fulfils the condition of conservation of charges as discussed in the main text:

$$\frac{\partial}{\partial \mathrm{t}}\rho + \nabla \bullet \mathbf{j} = 0. \tag{A.36}$$

This equation says that the rate of change of the density of electric charge ρ in one region of space is exactly equal to the current $\mathbf{j}$ that leaves the region: no electric charges are created or destroyed in any region of space. Electric charge is conserved.

As mentioned in the main text, Equations A.34 are not symmetrical regarding E and B. In electrostatic we have the electric charge density ρ_{e}, but in magnetostatic we do not have magnetic charge density ρ_{m}. We have not found any monopoles yet in nature and have not been able to produce any in the laboratory. The theory is thus developed for div • B equal to zero, which means that there are no magnetic charges.

(f) *Electromagnetic wave equation*

A combination of Maxwell's equations and some algebra leads directly to the wave equation:

$$\left(\frac{\partial^2}{\partial x^2} + \frac{\partial^2}{\partial y^2} + \frac{\partial^2}{\partial z^2}\right)\boldsymbol{E} = \frac{1}{c^2}\frac{\partial^2}{\partial t^2}\boldsymbol{E}, \tag{A.37}$$

with c, the speed of light given by

$$c^2 = \frac{1}{\varepsilon_{o_o}\mu_o}. \tag{A.38}$$

A similar expression is obtained for the magnetic field **H.** A solution of Equation A.37 leads to waves propagating in free space at speed c, as given by the expressions:

$$\boldsymbol{E} = \boldsymbol{i}E_o cos(\omega_k t - \boldsymbol{k} \bullet \boldsymbol{r}), \tag{A.39}$$

$$\boldsymbol{H} = \boldsymbol{j}H_o cos(\omega_k t - \boldsymbol{k} \bullet \boldsymbol{r}), \tag{A.40}$$

where $\boldsymbol{i}$ and $\boldsymbol{j}$ are unit vectors at right angle to each other making the magnetic field $\boldsymbol{H}$ perpendicular to the electric field $\boldsymbol{E}$. $\omega_{\rm k}(= 2\pi\nu_k)$ is the angular frequency of the wave identified by the propagation vector $\boldsymbol{k}$ along direction $\boldsymbol{r}$ and equal to $2\pi/\lambda$. We also have the relation $\nu\lambda = $ c. The wave described by Equations A.39 and A.40 is said to be linearly polarized, the electric field $\boldsymbol{E}$ being along the unit vector $\boldsymbol{i}$, the x direction, while the magnetic field $\boldsymbol{H}$ is at right angle along the unit vector $\boldsymbol{j}$, the y direction, and oscillates in phase with it. In the case of a circularly polarized wave the electric field $\boldsymbol{E}$ and magnetic field $\boldsymbol{H}$ rotate around the direction of propagation.

We note that the speed of light c appears naturally in Maxwell's equations as a combination of the two constants $\varepsilon_{\rm o}$ and $\mu_{\rm o}$ characterizing vacuum, presumably a non substance. That speed appears to be independent of the reference frame that may be selected. It has a nature different from that of the speed of sound in air which is attached to the physical characteristics of air. If air moves, sound waves, which are pressure fluctuations, follow it and their speed relative to a fixed reference frame is altered.

(g) *Lorentz transformation*

If a change from one frame of reference to another in motion is made to Maxwell's equations by means of a Galilean transformation, they are altered. Consequently, they do not transform according to the Galilean rule of addition of speeds. Lorentz found that the appropriate coordinates transformation for those equations, from a rest frame Σ to a frame Σ' moving at speed u along the x direction is:

$$x' = \frac{x - ut}{\sqrt{1 - u^2/c^2}}, \quad y' = y, \quad z' = z, \quad t' = \frac{1 - ux/c^2}{\sqrt{1 - u^2/c^2}}. \tag{A.41}$$

When Maxwell's equations are transformed according to those rules they remain the same. Einstein and Poincaré proposed that mechanical and electromagnetic physical laws should be of such a nature that they stay the same under a Lorentz transformation. That meant that Newton's laws should be altered to obey that rule. Those equations will obey the rule if mass is interpreted as relativistic and is represented by Equation A.15. Galileo's transformation applies then as an approximation for speeds low compared to that of light making u^2/c^2 very small compared to 1.

(h) *Doppler effect*

When an emitter and an observer are at rest the wavelength and the frequency of the radiation emitted and observed are related by:

$$\nu_o \lambda_o = c. \tag{A.42}$$

If the emitter moves a distance x at speed v during one period T_o the wavelength emitted is reduced to:

$$\lambda = \lambda_o - x = \lambda_o - vT_o. \tag{A.43}$$

Since T_o is the inverse of frequency $(1/\nu_o)$, we obtain directly:

$$\nu = \frac{\nu_o}{1 - v/c} \cong \nu_o(1 + v/c) \quad (\text{for } v \ll c). \tag{A.44}$$

The frequency is raised by the motion of the emitter towards the observer. If the emitter is recessing from the observer, the velocity is negative and the frequency is lowered. Symmetrical statements apply for the observer's motion. It is all relative motion.

In the case of relativistic speeds, that is to say speeds that are a good fraction of the speed of light, we must take into account time dilation given by Equation A.19. The frequency becomes:

$$\nu = \sqrt{\frac{1 + v/c}{1 - v/c}}\, \nu_o, \tag{A.45}$$

which becomes A.44 for low velocities, or $v \ll c$.

B. Quantum Mechanics

(a) *Black body radiation*

The expression black body is meant to represent a specific object that absorbs all radiation that falls on it, thus looking black when it is cold. However, when hot, such an object emits radiation that can be in the visible. It is still called a black body because it is characterised by the same physical laws as when it is cold.

Quantum mechanics started with the assumption that the variation of the intensity of radiation as emitted by a blackbody is a discontinuous process. Before the introduction of that concept by Planck it was not possible to explain the shape of the spectrum of radiation emitted by a body at a given temperature. Planck solved the problem by assuming that variation of the intensity of electromagnetic radiation was not continuous, but was taking place by small steps. In that formalism, the density dU of the energy radiated by such a black body at temperature T, in the range of wavelength $d\lambda$, as obtained by Planck, is given by:

$$dU = \frac{8\pi hc}{\lambda^5} \frac{d\lambda}{(e^{hc/kT\lambda} - 1)}. \tag{B.1}$$

This is one of the most important and most significant equations in physics. At the moment of its presentation to the scientific community it introduced for the first time the concept of discontinuity or quantization in the variation of the intensity of electromagnetic radiation. It was a break with classical physics. Scientists used the result for introducing the concept of the photon of energy hν or hc/λ. It opened the door to the introduction of the field of quantum mechanics. A graph of the actual shape of the emitted radiation was given in the main text for the cosmic microwave background (see Figure 9.8). Objects at a given temperature T emit radiation with such a spectrum.

(b) *Wave mechanics*

The behaviour of the electron is guided by Schrödinger's equation known also as the wave equation of quantum mechanics:

$$\left(-\frac{\hbar^2}{2m}\nabla^2 + V\right)\psi = i\hbar\frac{\partial}{\partial t}\psi, \tag{B.2}$$

where m is the mass of the electron, and ψ is the wave function describing the behaviour of the electron in the potential V. The symbol "i" stands for $\sqrt{-1}$, the imaginary number. This shows effectively the strangeness of the quantum world when compared to Maxwell's world of electromagnetic waves (see Equation A.35). The quantum mechanical wave equation is totally different from the electromagnetic wave equation.

That equation may be accepted as such somewhat like the second principle for the response of a mass to an applied force. However, it can be obtained directly from operator algebra, a technique used at large in quantum mechanics. The idea is to interpret some classical observable as operators and give them a form according to some specific rules. The classical equation for the total energy E of a particle of mass m and momentum p is:

$$\frac{\mathrm{p}^2}{2\mathrm{m}} + \mathrm{V} = \mathrm{E}, \tag{B.3}$$

where V is the potential energy. This equation can be used as an operator applied to the wave function ψ of the particle. In quantum mechanics, the classical variables are assumed to be operators with the form

$$\mathrm{p_x} = -i\hbar\frac{\partial}{\partial \mathrm{x}}, \quad \text{similar expressions for } \mathrm{p_y} \text{ and } \mathrm{p_z} \tag{B.4}$$

and

$$E = i\hbar\frac{\partial}{\partial t}. \tag{B.5}$$

When these operators are replaced in Equation B.3 operating on ψ, one obtains Equation B.2. When we said that quantum mechanics was a strange theory, this derivation and Equation B.2 that dictates the behaviour of an atomic particle through the wave function ψ is a rather convincing demonstration.

In the early days after the development of that theory, questions were raised relative to the physical meaning of ψ. It has several specific properties. It is found to be a function that can be expressed in terms of a complete set of orthogonal unit vectors forming the basis of a multidimensional space called a Hilbert space. Those basis vectors are particular solutions of the Schrödinger equation. In this sense it is a function represented by a number at each point of space. A proposal made by Max Born in an effort to give a physical meaning to that wave function has survived up to these days. It is assumed that its squared absolute value, $|\psi|^2$, multiplied by an element of volume 'dv', represents the probability of finding the electron in that given element of volume. It is thus essentially a mathematical tool used to represent the probability of finding the position of the electron.

(c) *Hydrogen atom energy levels*

In the hydrogen atom the electron is assumed to be at a distance r from the proton. Its potential energy V is $-e^2/(4\pi\varepsilon_o r)$. When this value of potential is used Equation B.2 an exact solution of that equation can be obtained and it is found that the energy of the electron in that potential is quantized and equal to:

$$E_n = -\frac{me^4}{32\pi^2\varepsilon_o^2\hbar^2}\frac{1}{n^2}, \qquad \text{(B.6)}$$

where the various symbols have been defined earlier and n is the principal quantum number and can take integer values, $1, 2, 3, \ldots$ to infinity where the energy is zero. This is the same result as that obtained by Bohr in his technique of quantizing orbits of the electron circling the nucleus. However, the result is obtained on the basis that the electron behaves as a wave, and that wave is specific to every value of n. We do not have to make the assumption of the existence of so-called stable electron orbits around the nucleus, a situation that violates classical electrodynamics laws, since an electron in an acceleration path would emit radiation on a continuous basis and would fall into the nucleus. The number 'n' gives the general behaviour of the electron in the hydrogen atom. Light is emitted by the electron when it passes from one level E_n to another one with n smaller. Replacing in Equation B.6 the symbols by their appropriate values, the first energy levels occur for $n = 1$ and its energy is $E_1 = -13.6\,\text{eV}$, which simply means that an energy of 13.6 eV must be spent to knock an electron from its entrapment in the field of the nucleus. This is not much energy. As we said, it is the energy that an electron would acquire when accelerated under a potential of 13.6 volts, a small quantity by laboratory standards. Such a value is the kind of energy that we encounter in dealing with the binding in chemical compounds ($\sim$ eV). However, due to the enormous number of particles involved in reactions of chemical products with a mass of order of kg ($\sim 10^{25}$), the energy released can be enormous such as in the case of dynamite explosions.

(d) *Angular momentum of electron in orbit around the nucleus*

The solution of the wave equation, however, leads to more information on the behaviour of the electron. Its solution in three dimensional space leads directly to the presence of an angular momentum as if the electron were circling the proton in orbits, as in Bohr model. However, this angular momentum appears naturally and it may not take arbitrary values and orientations in that space. It is characterised by two quantum numbers that can take only very specific integral values as the principal quantum number does. These quantum numbers are identified as l, the so-called azimuthal quantum number and m_l the magnetic quantum number. The situation is illustrated in Figure B.1.

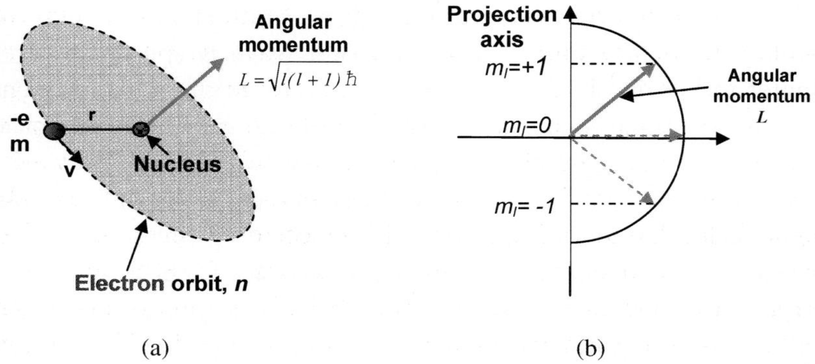

Figure B.1. In the planetary model of the atom, the electron moving around the nucleus at speed v describes an orbit and possesses an angular momentum. It also creates an electric current and thus has a magnetic moment like a current loop. In Schrödinger's model, the concept of a quantized angular momentum appears naturally. (b) Graphical representation showing the projection of the angular momentum on an axis. The case illustrated is that for which $l = 1$. Associated with the motion of the charge $-e$, a magnetic moment is created oriented in the same direction as the angular momentum.

In the theory, it is found that the number l can take values only from zero to a number equal to $(n - 1)$, otherwise a solution does not exist. Consequently, in the first Bohr orbit where $n = 1$, the quantum number l can only take the value 0. This means that the electron does not have angular momentum in that first orbit or in other words, it does not really circle the proton as illustrated.

It does not appear to be located precisely in space since we cannot define a precise orbit. According to Born's hypothesis the probability of finding the electron in a given region of space is given by the square of the wave function. In that case it is found that the wave function is spherically symmetric in the space around the proton: it is said to envelop completely the proton as a cloud and thus does not have angular momentum. A calculation shows, however, that the region where the probability of finding the electron is maximum is at a distance equal to Bohr radius, that is 0.05 nm from the proton. This is another comforting result.

For n larger than 1, l may have a value larger than zero and the electron may possess an angular momentum as illustrated in Figure B.1(a). The theory shows that the angular momentum is given by $\mathrm{L} = (\mathrm{l}(\mathrm{l}+1))^{1/2}\hbar$. Classically the electron's orbit could be oriented in any direction relative to an arbitrary equator around the nucleus. This means that the angular momentum could take any orientation. In the atomic world, it does not work that way: the orientation is quantized and the projection of the angular momentum on a vertical axis can take only discrete values dictated by the magnetic quantum number m_l, which can take values from $-l$ to $+l$ as illustrated in Figure B.1b. That quantum number is called magnetic for the following reason. The electron in orbit around the nucleus carries a charge $-\mathrm{e}$ and can be visualised as an electric current in a loop as shown in Figure B.1(a). This current-carrying loop possesses a magnetic moment the energy of which in an applied magnetic field is a function of its orientation. Consequently m_l, being a projection of the angular momentum on a given axis is a measure of the orientation of the atom magnetic moment relative to an applied magnetic field: it thus provides an indication of the energy of the electron in that magnetic field. For $l = 1$ we thus have three possibilities of orientation and if a magnetic field is present these three orientations have three different values of energy.

(e) *Quantum Mechanics and relativity*

When special relativity is introduced in Equation B.2, as was done by Dirac, it is found that the electron possesses an intrinsic angular momentum called 'spin' with its associated magnetic moment. The analysis incorporating relativity shows also the possibility of existence of particles identical to the electron but with a positive charge. They look like electrons running backward in time. They were given the name positron. The existence of such a particle was demonstrated in 1932 by Anderson. It is an antiparticle leading to the concept of antimatter.

(f) *Electromagnetic mass and classical radius of the electron*

The electron may be thought of as being a sphere of radius r_o with its charge distributed on the surface of that sphere. The charge produces an electric field in the space surrounding it and if it moves at a given velocity it produces a magnetic field. Those combined fields contain energy and the momentum of the resultant can be calculated. When this is done it is found that we can calculate an equivalent mass to the electron originating from that momentum. It is found to be inversely proportional to the electron radius assumed to start with. That mass is called the *electromagnetic mass.* Let us assume that it is the actual mass of the electron. It is then possible to calculate the electron radius that is consistent with that hypothesis. The result is:

$$r_o = \frac{e^2}{4\pi\varepsilon_o mc^2} = 2.82 \times 10^{-15}\,\text{metre.} \tag{B.7}$$

It is called the classical radius of the electron. We introduce above the concept of electron spin as if the electron was rotating on itself like a top. We may then calculate the magnetic moment of that spinning sphere with the charge distributed uniformly on its surface as above. The result can be compared to the magnetic moment obtained in the quantum mechanical calculation. The result obtained shows that points of the surface of that classical electron must travel at a speed greater than the speed of light showing that the model does not represent reality.

Classical pictures of the electron lead to difficulties. We believe that it is a point charge in space with a given mass. However, if we calculate the energy content of a point charge, we obtain an infinite result. It appears that in dealing with that sort of situations we need to do some kind of normalization by dealing only with differences of energy, rather than absolute values, since we do not deal anyway with infinities theoretically or experimentally. In doing this, infinities disappear like if we swept them under the rug. A similar problem arises in connection with the determination of the actual charge of the electron. We have seen that virtual particles are continuously created in vacuum. They appear as pairs, particle-antiparticle and live for a short time obeying the uncertainty principle. Thus, vacuum is filled with these virtual particles and in the presence of a real electron positive particles are attracted towards the electron and vacuum tends to polarize. The electron charge becomes shielded. What is then the meaning of the value of the electron charge that is measured experimentally? A mathematical treatment of the phenomenon leads as in the previous case to infinities. It is a problem that has haunted the theoretical physics community for a long period. A solution has been engineered in a so-called renormalization process. The charge of the electron is said to be the one that is measured experimentally and it appears that the problem is again swept under the rug.

(g) *Fermions and bosons*

Fermions and bosons are different primarily through their spin value. Bosons have integral spin like the photon that has a spin 1, while fermions have half integral spin like the electron that has a spin 1/2. This has consequence for the wave function representing those particles. Fermions are represented by antisymmetric wave functions. This means that when two interacting electrons having a common wave function are exchanged, that wave function changes sign. In the case of bosons the wave function is symmetrical and does not change sign upon exchange of particles.

This characteristic has important consequences for the distribution of a large number of those particles among energy states. In the case of fermions, the distribution of particles as a function of energy is given by the statistical expression:

$$f(E) = \frac{1}{e^{(E-E_r)/kT} + 1} \tag{B.8}$$

where E is the energy of the particle, E_r is a reference level called Fermi level in solid state physics, k is Boltzmann constant ant T is the temperature. This distribution has been given the name Fermi–Dirac statistical law. At high energy, $E \gg E_F$ it may be approximated by Boltzmann statistical law $f(E) = e^{-E/kT}$.

In the case of bosons the expression is:

$$f(E) = \frac{1}{e^{E-E_r/kT} - 1} \tag{B.9}$$

and is called Bose–Einstein statistical law. It follows that bosons can occupy the same energy state and pile up on each other like in a laser to form a beam of intense coherent radiation.

(h) *Uncertainty principle*

The quantum world is based on probability and uncertainty. This property is best expressed under the form of Heisenberg's uncertainty principle, which for position and speed takes the form:

$$\Delta p \times \Delta x \sim \hbar, \tag{B.10}$$

where Δp is the uncertainty of momentum mv, thus the uncertainty on speed v, Δx is the uncertainty on position and $\hbar$ is the reduced Planck's constant, or $h/2\pi$. This is an intrinsic property of the quantum world. An electron restricted to a space given by Δx acquires a momentum $\Delta p = \hbar/\Delta x$. This is also interpreted as the impossibility of determining both momentum and position of a particle to a precision better than $\hbar$. It is a minimum value.

Similarly, the same concept applies between the two observables, time t and energy E, expressed as:

$$\Delta E \times \Delta t \sim \hbar. \tag{B.11}$$

The energy of an elementary object that has a lifetime of Δt cannot be determined to better than $\hbar/\Delta t$. This property of the quantum world gives rise to the possibility of existence of virtual particles if their lifetime is very short. In a sense energy is borrowed from vacuum to create particles ($m = E/c^2$) and is reimbursed within a very short period. It is like borrowing money from the bank for a time so short that the accounting system does not have the time to detect it.

(i) *Analogy between optical polarization effects and Stern–Gerlach experiment with spin*

The Stern–Gerlach experiment deals with the wave–particle duality of the electron. It is possible to think of a similar experiment that is done with electromagnetic waves and a polarizing cube that splits an electromagnetic wave into two components with respective polarization along the two polarization axes of the cube. If the incident wave is polarized along one axis of the cube, it goes through the cube without alteration. In the Stern–Gerlach filter approach, this corresponds to the situation where the first filter providing an ensemble of spins all oriented in the same direction is followed by another filter oriented in the same direction. On the other hand, in the optical analogy, if the incident radiation polarization is oriented at 45 degrees between the polarizing angles of the cube, the incident radiation is split into two waves of equal amplitudes. If the incident wave amplitude is reduced to contain only one photon we then have a situation very similar to the case illustrated in Figure 6.14 where one spin goes through the Stern–Gerlach dipole magnet at right angle to the field. In the optical case we detect the photon with equal probability at the exit of either polarization channel as in the case of the spin. It is readily observed that the de Broglie wave associated to the electron plays a very similar role for the spin as the electromagnetic wave does for the photon. However, the de Broglie wave keeps its interpretation as a measure of the probability of finding the electron at a given position in space while in the case of electromagnetism the wave associated to the photon is a real physical entity characterised by electric and magnetic fields the amplitudes of which are determined by Maxwell's equations.

(j) *Quantum mechanics prediction for entangled spins*

In the main text we provided a simple proof of Bell's theorem using classical reasoning. We showed that if the spins are somehow programmed in pairs when they leave the source we would expect to detect them opposite at the detectors in more than half of the cases, or at least 55%. In quantum mechanics it is found that they are detected as correlated or opposite in exactly half of the cases. This is shown as follows. In a magnetic field an electron spin may be in either of two states, called for simplicity either up or down. Its wave function may be written:

$$\psi = \mathrm{a}_{\mathrm{up}}\, \mathrm{u}_{\mathrm{up}} + \mathrm{a}_{\mathrm{down}}\, \mathrm{u}_{\mathrm{down}} \tag{B.12}$$

where a_{up} is the probability coefficient associated with the spin up state, u_{up}, and similarly for $\mathrm{a}_{\mathrm{down}}$ and $\mathrm{u}_{\mathrm{down}}$. These states correspond to proper states for a spin oriented along an axis. When a measurement is made only those two directions may result. For a spin oriented at an angle α measured from the up direction of the axis, the coefficient is calculated as $\cos(\alpha/2)$. For $\alpha = 90°$, a spin at right angle to a given chosen axis, both coefficients are $\cos(45°)$ or $1/\sqrt{2}$. The probability of finding the spin in either state is the square of those coefficients and is thus $1/2$. Thus a large number of measurements on spins oriented at 90° to the axis of measurement would find the spins with equal probability in either states or, in other words, either directions.

This kind of reasoning may be applied directly to our entangled spin experiment. We may interpret the angle α as the difference of the angles that each Stern–Gerlach detector makes with the vertical. The measurements are then made for various directions taking into account the fact that the entangled spins have opposite directions.[2] When the detectors are opposite the angle is then taken as 0° and the probability of detection of a spin in both detectors is the square of $\cos^2$ (0°) or equal to 1. If the angle between the detectors is 120° as suggested in the text we then have as probability, $\cos^2(120°/2)$ or 1/4. We now assume that the detectors are set at random in any of the three directions, 0°, +120° and −120°, which says that the detectors are oriented along any of those directions 1/3 of the times measurements are made. The total probability of detecting a spin along those directions is thus $(1/3)(1) + (1/3)(1/4 + 1/4) = 1/2$. Consequently, quantum mechanics predicts detection of correlations between the spins for 50% of the measurements while the classical reasoning made in the main text with programmed spins predicts a correlation larger than 55%.

(k) *Interactions of electromagnetic waves and electrons*

It is standard practice to analyse the interaction of electrons with electromagnetic wave in a semi classical manner, which means that the electron is assumed to be quantized, that is to say, to exist in discrete energy levels while the electromagnetic wave is assumed to act in a classical way. The interaction then uses the solution of Maxwell's equation leading to a wave and the coupling takes place between the electron charge with the electric and magnetic fields of the wave written classically. It is then possible to calculate the transition probability of an electron from an energy level to another one under the influence of an incident electromagnetic wave. For example the transition probability (number of transitions per second) of the electron from the state n = 2 to state n = 1 in hydrogen is

$$P_{n=2\rightarrow n=1} = 0.416\,\pi r_o c I(\nu). \tag{B.13}$$

where I(n) is the intensity of the incident wave at the frequency of the transition, and r_o is the classical radius of the electron introduced above

The calculated transition probability, $P_{n=2\rightarrow n=1}$, agrees with experimental data and the approach is a real success. However, such an approach cannot be used to calculate the spontaneous emission rate for an electron in state n = 2 when no waves are present.

Experiment shows that the life time in such a state n = 2 is of the order of one nanosecond. Consequently something must interact with it, otherwise the electron would stay in that level forever and emission of light by atoms would be a rare event.

The solution is found in the use of a quantum approach for the electromagnetic wave as well. It consists in quantizing the electromagnetic field, an operation that is the foundation of the development of quantum electrodynamics, which studies the interaction of electrons with the electromagnetic radiation both in a quantum formalism.

The electromagnetic wave may be quantized by assuming that the wave behaves like an harmonic oscillator. The analysis leads directly to a representation of the oscillating electromagnetic field by a manifold of energy levels with spacing corresponding to the energy hν, ν being the frequency of the electromagnetic wave. This is shown in Figure B.2(a). We are thus in the presence of discrete excitation levels and the level of excitation, 'n', not to be confused with the quantum number 'n' of the atom, is a measure of the number of photons in the field. We may thus say that a wave in level n + 1/2 contains n photons. The number n for ordinary intensity of radiation is a very large number. For example, a small solid state laser emitting a radiation power of one milliwatt in the red (λ = 600 nm), emits a total of 3×10^{15} photons per second.

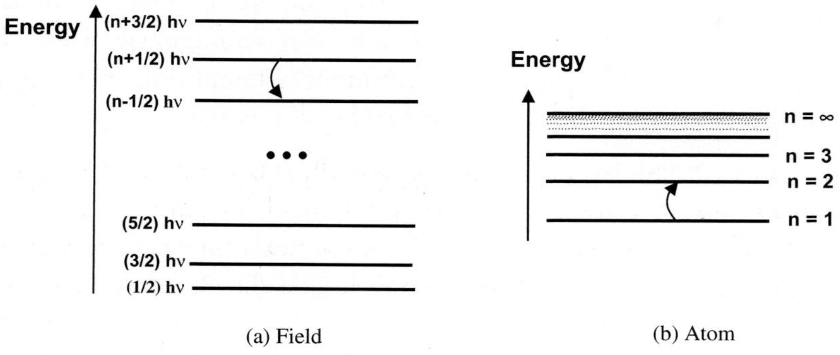

Figure B.2. (a) Representation of the electromagnetic field by a quantized manifold of levels representing the various levels of excitation of the field. (b) Representation of the energy level manifold of an atom. Two of the energy levels are shown in interaction with the field.

What is that 1/2 doing in the picture? It turns out that the lowest state of the electromagnetic field is not zero energy. It represents an excitation of the electromagnetic field in its lowest state, an excitation that cannot be extinguished and is not available for experimentation. Space is filled with those photons and is in permanent effervescence.

In figure B.2b, arrows show a transition within the atom and the field energy levels manifold. The atom is absorbing a photon and the field intensity decreases by passing from excitation level n+1/2 to n −1/2. Energy is conserved. The process is called absorption and its probability is the same as that calculated for stimulated emission in which the opposite effect takes place. Here the atom absorbs a photon passing from energy level n = 1 to level n = 2 and the field losing one passing from level n +1/2 to level n −1/2.

A very important effect results from the presence of the residual field excitation when no electromagnetic field is intentionally applied. If the atom is in an excited state such as n = 2 in the figure, it interacts with those photons filling space. The effect leads to an extra term in the final equation representing the interaction of the field and the atom, like an extra photon. Since there is a great number of possible electromagnetic field modes that can be present in space at various frequencies, the atom is induced to fall to its lower energy level emitting itself a photon: this is the origin of spontaneous emission. The effect is important. The probability that the atom decays to a lower state is given by the expression:

$$P_{21}^{\text{spont}} = \frac{\omega^3}{3\pi\varepsilon_o c^3 \hbar} |d_{12}|^2, \tag{B.14}$$

where ω is equal to $2\pi\nu$ and d_{12} is called the dipole moment of the atom and is a measure of the electron distribution around the nucleus. For example, when P_{21}^{spont} is evaluated in the case of the hydrogen atom for the transition giving Lyman α radiation with $\lambda = 121.6\,\text{nm}$, it gives rise to a lifetime of the electron in state n = 2 of 1.6 nanosecond in agreement with experimental measurement. We are thus rather confident in the approach consisting in quantizing the field.

(1) *Forces as exchange of particles*

This approach is pushed much further in a full development of the interaction between an electron and an electromagnetic waves represented by photons. This is the theory of quantum electrodynamics which is probably the most tested physical theory and which is found to be in agreement with experimental data to a very high precision. It leads, for example, to the full explanation of the shifts observed in atomic energy levels caused by the residual electromagnetic effervescence of vacuum, to the exact determination of the properties of the electron, properties that cannot be explained by classical electromagnetism. It leads also to a model of the interaction at distance of electrically charged particles by means of an exchange of virtual photons as represented in the main text by means of Feynman diagrams providing a visualization of the electromagnetic force. The photon is then interpreted as the mediator of the force. The approach is part of the broader one called Quantum field theory that includes the case of the strong and weak force working inside the nucleus, the mediators being gluons in the case of strong force and vector bosons, $W^{\pm}$, Z_o, in the case of the weak force.

The phenomenon can be visualised by means of the uncertainty principle relation B.11. The idea is that the process of emission of a virtual photon by an electron can violate the principle of conservation of energy if the process takes place over a short time that obeys the relation $\Delta E \times \Delta t \sim \hbar$. In the case of a photon, the wavelength can be as long as one wishes meaning that the energy $\Delta E = h\nu$ can be very small and for all purposes be zero. In that case the photon can exist for a very long time obeying the uncertainty relation and may travel to large distances or infinity at the speed of light.

In the case of nuclear forces, Yukawa proposed that the interaction takes place through a similar process by means of particles called mesons. We know that the range of interaction is limited to a very small volume leading to a radius or distance of interaction Δx_{max} of the order of 10^{-15} m. Assuming that the mesons travel at a maximum speed less than that of light, their lifetime in travelling that distance is given by

$$\Delta t = \Delta x_{max}/c. \tag{B.15}$$

Such a short lifetime can result if the mesons have mass, giving them energy $\Delta E = mc^2$. Combining the uncertainty relation and that hypothesis, one gets:

$$mc^2 = c\hbar/\Delta x. \tag{B.16}$$

This gives a mass of about 200 MeV. The meson responsible for the nuclear force has been found to have a mass of the order of 150 MeV.

The same argument can be used in the case of the W and Z bosons introduced as mediators of the weak force. In that case the range of interaction is assumed to be much shorter and the mass is evaluated to be in the range of 100 GeV, while the experimental value is of the order of 80 GeV.

(m) *International system of units (SI)*

The International System of units is based on a limited number of base units representing seven physical quantities. All physical measurements can be expressed as a combination of those units. They are, metre, for length, kilogram for mass, second for time, ampere for electric current, lumen for light intensity, Kelvin for temperature, mole for quantity of matter. The system has been the subject of fundamental changes in the last fifty years. Use has been made of fundamental properties of nature to redefine or represent the units and standards based on those fundamental changes have been implemented. In particular, the second is now defined n terms of the frequency of an atomic transition in Cs (ν = 9 192 631 770 Hz), and the metre is defined in terms of the speed of light (c = 299,792,458 ms^{-1}). Both numbers are now defined as exact numbers. The volt, a derived unit is implemented in practice in terms of Josephson effect with the constant equal to:

$$K_J = (h/2e) = 483\ 597.9\,\mathrm{GHz/V}, \tag{B.17}$$

where h and e are Planck's constant and the electronic charge respectively. The unit of electrical resistance, the ohm, is implemented in practice in terms of Quantum Hall effect with the constant given by:

$$R_K = (h/e^2) = 25\ 812.807\,\Omega. \tag{B.18}$$

These practical representations of the volt and ohm lead, by means of Ohm's law, to an implementation of the ampere in terms of fundamental constants. However, Equations B.17 and B.18 are not formal definition. They are used only for practical implementation of standards of physical measurements for the volt and the ohm.

(n) *Natural units*

The associations of fundamental constants that give rise to natural units as discussed in the main text are:

Length: $$l_P = \sqrt{\frac{\hbar G}{c^3}} = 1.62 \times 10^{-35}\,\text{m} \tag{B.19}$$

Time: $$t_P = \sqrt{\frac{\hbar G}{c^5}} = 5.39 \times 10^{-44}\,\text{s}. \tag{B.20}$$

Mass: $$m_P = \sqrt{\frac{\hbar c}{G}} = 2.18 \times 10^{-8}\,\text{kg}. \tag{B.21}$$

Electric charge: $$q_P = \sqrt{\hbar c 4\pi\varepsilon_o} = 1.87 \times 10^{-18}\,\text{C}. \tag{B.22}$$

Temperature: $$T_P = \sqrt{\frac{\hbar c^5}{Gk^2}} = 1.42 \times 10^{32}\,\text{K}. \tag{B.23}$$

C. Cosmology

(a) *Equivalence of inertial and gravitational mass*

Newton's second principle says that an object of mass m is accelerated by an applied force F. In a sense mass opposes to acceleration. It introduces inertia in the system. We can call it inertial mass m_{in} and write:

$$F = m_{in}\, a \tag{C.1}$$

On the other hand, an object of mass m on the surface of the Earth is exposed to a gravitational field and is submitted to a force given by:

$$F = m_{gr}(GM_{Earth}/r^2), \quad \text{(C.2)}$$

in which we have called the mass m_{gr}. Consequently, according to Equation C.1, that object in the Earth's field will be accelerated and we can equate the two forces to get the acceleration:

$$a = (m_{gr}/m_{in})(GM_{Earth}/r^2). \quad \text{(C.3)}$$

The two masses appear to originate from two totally independent phenomena. Are they equal? The question is at the basis of Galileo's famous experiment at the Leaning Tower in Pisa. He said they were. Einstein said so and gave the reason why it was so: gravity is simply equivalent to acceleration. We may think of it, for example, as the Earth being accelerated towards the object, which is at rest. According to this reasoning we thus have:

$$m_{gr} \equiv m_{in}. \quad \text{(C.4)}$$

Einstein made it a basic principle for his general theory of relativity: it is one representation of the so-called principle of equivalence. If 'r' is set to be the distance from the centre of the Earth to the surface, the quantity $GM_{Earth}/(\mathbf{r}_{surf})^2$ is the acceleration of gravity at the surface of the Earth and is called g. It is equal to 9.8 metres per squared second. Thus gravitational acceleration is independent of the mass of the object. The orbits of satellites are independent of their mass. This is the reason why an astronaut floats in the space station, the size of his orbit being independent of his mass as is the size of the orbit of the space station itself.

(b) *The concept of space curvature*

Space curvature can be visualised mathematically through the so-called n-sphere approach. For example we can imagine a 4-dimensional space, x, y, z, w, and a so-called 3-sphere in that space. Our three-dimensional space may be interpreted as the surface of that sphere at fixed distance R from the origin:

$$R^2 = x^2 + y^2 + z^2 + w^2. \quad \text{(C.5)}$$

The infinitesimal distance between two points or the line element in that four-dimensional space is:

$$dl^2 = dx^2 + dy^2 + dz^2 + dw^2. \tag{C.6}$$

Taking $r^2 = x^2+y^2+z^2$ within that sphere makes w the dependent variable given by $w^2 = R^2-r^2$, from which we obtain:

$$dw = \frac{-rdr}{(R^2 - r^2)^{1/2}}. \tag{C.7}$$

Using this relation and polar coordinates Equation C.6 may be written:

$$\mathrm{dl}^2 = \frac{dr^2}{(1 - r^2/R^2)} + r^2(d\theta^2 + \sin^2\theta\, d\varphi^2), \tag{C.8}$$

and for a space-time interval we have:

$$\mathrm{ds}^2 = \mathrm{c}^2 dt^2 - \frac{dr^2}{(1 - r^2/R^2)} + r^2(d\theta^2 + \sin^2\theta\, d\varphi^2), \tag{C.9}$$

in which R is interpreted as the radius of curvature in that space. It is usually normalized and written as 1/k. One sees readily that if R is very large, the denominator of the second term becomes 1 and we obtain a flat Minkowski space. If space is expanding, that equation can be written

$$\mathrm{ds}^2 = \mathrm{c}^2 dt^2 - a(t)^2 \left(\frac{dr^2}{(1 - r^2/R^2)} + r^2(d\theta^2 + \sin^2\theta\, d\varphi^2)\right), \tag{C.10}$$

where $a(t)$ is the expansion scale factor and should not be confused with the acceleration introduced above in Newton's second principle (equation C.1). This equation is known as the Friedman–Lemaître–Robertson–Walker metric. The expansion factor a(t) is obtained from a solution of Einstein's field equations.

(c) *Cosmos expansion*

The four dimensional space-time metric in an homogenous, isotropic, universe is written:

$$\mathrm{ds}^2 = c^2 dt^2 - a(t)^2 dx^2, \tag{C.11}$$

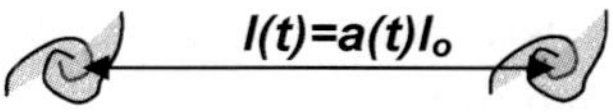

Figure C.1. The space between two galaxies at distance $l(t)$ from each other expands at a rate a(t).

where dx represents the three dimensional components of space in the proper coordinate system. In that expression, $a(t)$ is the scale factor of the Universe; it measures relative spatial dimensions. If dimensions remain constant with time, $a(t)$ would be a constant equal to unity. However, Hubble found that the Universe is expanding and in that context $a(t)$ increases with time. The distance $l(t)$ between two galaxies then increases with time as:

$$l(t) = a(t)l_o \tag{C.12}$$

where $l_{\rm o}$ is a constant for the two objects (galaxies). This is illustrated in Figure C.1. The rate of change of $l(t)$ is then the speed at which these objects recess from each other

$$v = \frac{d\,l(t)}{dt} = \dot{l}(t) = l_{\rm o}\dot{a}. \tag{C.13}$$

The dot on a symbol means a derivative with respect to time. This may be written as

$$\mathrm{v} = l(t)\frac{\dot{\mathrm{a}}}{\mathrm{a}} = \mathrm{H}l(t), \quad \text{with } \mathrm{H} = \frac{\dot{\mathrm{a}}}{\mathrm{a}} \tag{C.14}$$

H is called the Hubble constant and is a measure of the rate of recession of celestial objects from each other. Its present value is evaluated as ~72 km per second for two galaxies separated by 1 megaparsec or 3.26 million light years.

(d) *Einstein's field equations*

Einstein's field equations are a system of ten coupled equations connecting the geometrical properties of space to its content of mass and energy. They are given under the expression:

$$G_{\mu\nu} + \Lambda g_{\mu\nu} = \frac{8\pi\,G}{c^4}T_{\mu\nu}. \tag{C.15}$$

In that equation the left hand side represents the geometrical properties of space with its curvature while the right hand side represents the effect of mass on the geometry. $G_{\mu\nu}$ is a tensor representing those geometrical properties, Λ is the cosmological constant introduced by Einstein to make his solution agree with a static Universe; on the right hand side $T_{\mu\nu}$ is the stress-energy tensor representing the effect of mass and energy, and G is Newton's gravitational constant; $g_{\mu\nu}$ is the metric tensor characterizing space; indices μ and ν are the four coordinates for time and space; c is the speed of light. The set of equations that results from that equation represents essentially the effect of mass and energy on space geometry, as well as the effect of space geometry on mass behaviour. Although the equation is elegant and looks somewhat simple, its solution in particular cases is a major exercise.

(e) *Universe expansion: Hubble's constant*

A simple path may be used to make explicit the relativistic dynamical behaviour of the Universe considered as a homogeneous, isotropic fluid. This is done by means of Poisson's equation for the gravitational acceleration g in which the active matter density is obtained from the stress-energy tensor as $\rho + 3\mathrm{p}/\mathrm{c}^2$, where p is the pressure within the fluid.[3] That equation within a given region of space is:

$$\nabla \bullet g = -4\pi \mathrm{G}(\rho + 3\mathrm{p}/\mathrm{c}^2). \tag{C.16}$$

The solution of this equation is straightforward in a flat Minkowski space. The result is the same as that obtained by Friedmann (1924), from Einstein field equations for an homogeneous, isotropic Universe. It provides a relation between the expansion factor a(t) of Equation C.12 and the properties of the Universe, matter density and pressure. The result is:

$$\frac{\ddot{\mathrm{a}}}{\mathrm{a}} = -\frac{4\pi\,\mathrm{G}}{3}\left(\rho + \frac{3\mathrm{p}}{\mathrm{c}^2}\right), \tag{C.17}$$

It is possible to integrate this equation and using an equation of state coupling volume, density and pressure in the fluid, obeying conservation of energy, one obtains:

$$\left(\frac{\dot{a}}{a}\right)^2 = \frac{8\pi\, G}{3}\rho - \frac{k\, c^2}{a^2} \tag{C.18}$$

where the last term in Equation C.18 is a normalized constant of integration and is interpreted as the curvature of space. If the Universe is assumed static as it was believed in the early 1900s these two Equations C.17 and C.18 would be equal to zero since a would be constant. To satisfy this condition Einstein introduced the empirical cosmological constant Λ. He rejected it later when confronted with the evidence provided by Hubble's measurements that the Universe was not static but expanding. It is readily recognized that Equation C.18 is the square of Hubble's constant as defined in Equation C.14. It is standard practice to re-introduce a possible cosmological constant (Λ/3) to the right hand side of this last equation representing an additional component to the expansion, such as the one made evident recently in measurements on Supernovae Ia.

$$(H)^2 = \left(\frac{\dot{a}}{a}\right)^2 = \frac{8\pi\, G}{3}\rho - \frac{k\, c^2}{a^2} + \frac{\Lambda}{3} \tag{C.19}$$

These are the two equations (C.17 and C19) that describe the expansion of the Universe in the context of general relativity. The curvature of space given by k, using for simplicity a similitude to a two-dimensional space, can take the value 0 for flat space, +1 for the surface of a sphere, and −1, as for the surface of a saddle. It is directly related to the radius of curvature R introduced above in the calculation of the metric in a three-sphere. It is standard practice to define a critical density ρ_c as:

$$\rho_c = \frac{3H_o^2}{8\pi\, G}, \quad \Omega = \frac{\rho}{\rho_c} \tag{C.20}$$

where $\mathrm{H_o}$ is the value of Hubble's constant at the present moment. Neglecting the cosmological constant, Equation C.17 can then be written as:

$$\Omega(t) - 1 = \frac{k\,c^2}{\dot{a}^2}. \tag{C.21}$$

If the left hand side is equal to zero, Ω(t) is then equal to 1, and the curvature k is zero. In this case, the density is critical and the Universe is said to be flat. In the case of a flat space as is believed to be the present situation, the Universe will expand forever. If the density is larger than ρ_c, space is curved and closed on itself. The Universe's expansion will decrease with time and will result in a big crunch. The value of ρ_c is evaluated to be of the order of 10^{-29} grams per cubic centimetre, or about 6 protons or hydrogen atoms per cubic metre. This is very small, but we do not know its present value. Consequently we do not know the fate of the Universe, particularly in the light of the recent discovery of an acceleration of its expansion.

(f) *Schwarzschild radius*

A solution of Einstein field equations was obtained by Schwarzschild for a spherical non rotating mass (1916). In the case of large masses it leads to a possible singularity at the origin and an event horizon, inside which no objects whatever their speed, not even electromagnetic radiation, can escape. The distance from the singularity to the event horizon is known as Schwarzschild radius and is given by:

$$r_s = \frac{2GM}{c^2}, \tag{C.22}$$

where M is the mass and G is the gravitational constant. That radius is generally very small except for extremely massive objects. For example the Earth r_s is 9 mm. For the Sun it is 3 km. There is thus no problem, the mass of ordinary stars and planet being outside that radius. However, for extremely compact object resulting from a star collapse, a singularity may be created, and that radius defines a region of space from which no objects can escape from the gravitational attraction of that singularity. Such a singularity is known as a black hole.

(g) *The redshift factor*

(1) *Cosmological redshift*

The redshift factor z is defined as:

$$z = \left(\frac{\lambda_{\mathrm{obs}} - \lambda_{\mathrm{em}}}{\lambda_{\mathrm{em}}}\right) = \left(\frac{\lambda_{\mathrm{obs}}}{\lambda_{\mathrm{em}}} - 1\right) \tag{C.23}$$

or

$$1 + z = \frac{\lambda_{\mathrm{obs}}}{\lambda_{\mathrm{em}}} = \frac{\mathrm{a_o}}{\mathrm{a_e}}, \tag{C.24}$$

where λ_{obs} is the wavelength observed or measured and λ_{em} is the wavelength at emission or as it would be measured for a source at rest in the laboratory. The relation gives also the ratio of the expansion parameter at the time of observation ($\mathrm{a_o}$) relative to its value at the time emission ($\mathrm{a_e}$). The value of z for the cosmic microwave background is about 1,000, while the measured record for galaxies is about 8. In that last case it means that the radiation was emitted at a time when such a galaxy was at about 1/9 its present distance. For small recession velocities the effect is similar to Doppler effect and using C.14 we can relate z to Hubble's constant. We obtain:

$$\mathrm{z} = \frac{Hl}{c} \tag{C.25}$$

If the recession velocity of two galaxies is independent of time we can extrapolate back to a time when the galaxies were coinciding in space. This time is called Hubble time with its associated Hubble length $\mathrm{L_H}$ at which the recession expressed by Equation C.14 extrapolates to the velocity of light:

$$\mathrm{T_H} = \mathrm{H}^{-1} \quad \mathrm{L_H} = \mathrm{cH}^{-1} \tag{C.26}$$

Using a value of 72 km per second per Mparsec distance for the present value of H we obtain $\mathrm{T_H} \sim$ 14 billion years and $\mathrm{L_H} \sim$ 14 billion light years.

(2) *Gravitational redshift*

We have shown in Chapter III that a clock rate is found to be affected by acceleration. Consequently, according to the equivalence principle a clock rate is function of the surrounding gravitational potential or gravitational field. This is called the gravitational redshift.

In general for time keeping we are interested by the change in the rate of clocks at different height, h, or altitude above the surface of the earth. It is shown from argument on gravitational potential energy that the shift is given approximately by:

$$\Delta\nu = \frac{\mathrm{gh}}{\mathrm{c}^2}\,\nu_\mathrm{o} \tag{C.27}$$

where g is the acceleration of gravity, ν_o is the frequency of the clock on the surface of the Earth. The effect is small, of the order of 10^{-16} per metre of elevation. It is identified as a redshift because gravity slows down a clock, its frequency is lower and the wavelength of the radiation that it emits is longer than in the absence of gravity.

(h) *Universe temperature versus expansion*[4]

In the main text, we assume that the temperature was decreasing with the expansion of the universe. Can we make a simple calculation proving that conclusion? We use Planck's black body radiation law:

$$dU = \frac{8\pi hc}{\lambda^5}\frac{d\lambda}{(e^{hc/kT\lambda}-1)}. \tag{C.28}$$

During linear expansion by a factor "a", the wavelength λ increases by the same factor. Thus:

$$\lambda' = \mathrm{a}\lambda. \tag{C.29}$$

In that case, the Universe increases in volume as a^3 and the photon density decreases as a^{-3}. And since the energy of the photon decreases as "a" also, the energy density must decrease as a^4. Thus C.28 becomes

$$dU' = \frac{8\pi hc}{\lambda^5 a^4}\frac{d\lambda}{(e^{hc/kT\lambda}-1)}. \tag{C.30}$$

Using Equation C.29, this can be written as:

$$dU' = \frac{8\pi hc}{\lambda'^5}\frac{d\lambda'}{(e^{ahc/kT\lambda'}-1)}. \tag{C.31}$$

This equation is the same as Equation C.28 but with the interpretation that

$$T' = T/a. \tag{C.32}$$

We conclude that Planck's law still represents black body radiation in an expanding Universe, but with a temperature decreasing with expansion.

(i) *Energy equivalence*

In particle physics and cosmology the physics involved often introduces energy under various forms. The concept can be expressed as:

$$E = mc^2 = kT = h\nu, \tag{C.33}$$

where c is the speed of light, k is Boltzmann's constant, and h is Planck's constant. We see readily that we can express energy equivalently as mass, temperature or frequency. Often, we also represent that energy as "electron volt" (eV), that is to say, the energy acquired by a charge of 1 electron accelerated under a potential difference of one volt. For example, the mass of an electron, 9.11×10^{-31} kg, corresponds to an energy of 8.14×10^{-14} joule, a temperature of 5.90×10^{9} kelvin or still 0.51 million electronvolts. The equivalent photon frequency when the mass is transformed into electromagnetic radiation is 1.22×10^{20} Hz, a gamma ray. The equivalent temperature sets a point below which the electron is stable. The same exercise can be done with all particles.

Glossary

Absolute space: space viewed as immutable not affected by its content. Space acts as a permanent grid in which physical phenomena occur.

Absolute time: concept under which time flows naturally and continuously as a quantity independent of space and other influences such as forces and motion of objects.

Absolute zero: the lowest theoretical temperature. It is 0 kelvin or −273.15 degrees Celsius.

Acceleration: change of velocity per unit time achieved by means of an applied force; units are metre per second squared.

Accelerator: see particle accelerator.

Aether: also called ether; substance that was assumed to fill all space to support electromagnetic waves in a similar way as air supports sound waves; the Michelson-Morley experiment disproved its existence.

Anthropic principle: controversial principle under which a version says that the laws of nature are such as to favour the advent of intelligent life.

Antimatter: type of matter that, upon interaction with its matter counterpart, reacts with it, the result being the destruction of both, producing electromagnetic radiation. It is made of antiprotons, anti neutrons and antielectrons.

Atom: the smallest fractional part of an element, below which the element loses its identity.

Atomic clock: clock whose principle of operation is based on transitions between atomic energy levels. In such an instrument the pendulum in a classical clock is replaced by an oscillating atom.

***Atomic mass unit (amu)*:** normalized mass of nucleons by which C^{12} is said to have 12 atomic mass units.

***Atomic number*:** number of protons in the nucleus of a given element.

***Baryons*:** nuclear particles such as neutrons and protons made of three quarks that satisfy the exclusion principle and interact through the strong force.

***Beta decay*:** process in which a nucleus emits an electron and an antineutrino; in the process the atom changes its nature and becomes another element.

***Big Bang*:** in the standard model of cosmology, a scenario in which the evolution of the Universe started by a very energetic and fast expansion looking like an explosion.

***Big Crunch*:** possible end of a scenario of the evolution of the Universe in which the expansion will reverse under the force of gravity and the Universe will collapse to a singularity similar to the one that existed at the origin of the Big Bang scenario. The big crunch scenario requires that the density within the present Universe be larger than the critical density.

***Black hole*:** entity assumed to be so massive that all objects closer than a certain distance, called the event horizon, are trapped whatever their speed; even light within that distance cannot escape, making the black hole invisible.

***Boson*:** class of particles with an integral value of spin; typically it is a carrier of force; the photon is a boson.

***Broken symmetry*:** concept by which a system, under a change of environmental conditions, loses rotational or translational symmetry. For example, rotational symmetry is broken in cooling liquid water that becomes ice crystals with hexagonal symmetry.

***Caesium clock*:** atomic clock using Caesium as the controlling element.

***Casimir Force*:** quantum force that develops between two plates when they are placed very closed to each other (micrometres apart). Theory concludes that the force is created by 'zero field fluctuations'.

***Classical physics*:** in a broad sense that part of physics that describes the dynamical behaviour of nature without making appeal to its quantum nature.

***Collapse*:** see wave function collapse.

***Colour charge*:** name given to a property of quarks, which is assumed to be the origin of the strong force; three colours are assumed to exist; the word colour has nothing to do with its meaning relative to sight in our macroscopic world.

***Cosmic horizon*:** in cosmology, a boundary that delimits events that can be observed; light from more distant objects did not have time to reach our location since the beginning of the Universe expansion.

***Cosmic microwave background radiation (CMB)*:** radiation produced during the early stage of the Big Bang scenario, which became free some 380,000 years after the Big Bang; the CMB has been propagating through space in the expanding Universe and its wavelength has increased through the expansion of space.

***Cosmological constant*:** a constant introduced by Einstein in his General Relativity field equations to compensate for gravitational attraction; it looks like negative pressure and make the whole Universe static; it has been discredited by Einstein himself when he was confronted with Hubble's discovery that the Universe is expanding.

***Cosmological principle*:** hypothesis under which the Universe is homogeneous and isotropic.

***Cosmology*:** study of the matter in the world around us and of the dynamic evolution of the Universe.

***Critical density*:** density of matter in the Universe, for which the University will expand to an infinite size (eternally!). Above that density the Universe will contract on itself causing what is called a Big Crunch.

***Curvature*:** property of space-time introduced by the theory of General Relativity and caused by the presence of matter.

***Cyclotron*:** particle accelerator formed of two electromagnets in form of Ds, which force charged particles to effectuate a circular motion while accelerated by an electric field between the Ds; particles reach high energies after several revolutions.

***Dark energy*:** energy, whose nature is not yet identified and which is assumed to exist throughout space exerting negative pressure and causing an accelerated expansion of space; it has an effect similar to Einstein's cosmological constant.

***Dark matter*:** a type of matter not yet identified that is assumed to exist in the Universe and that would explain the peculiar speed of outlying stars in their revolution around galaxies' core.

***Dimension*:** a property of various models of space; we see our world as a three-dimensional world, but according to certain theories, such as String theory, it is possible that our world could be multidimensional, up to ten dimensions.

***Doppler effect*:** effect in which the wavelength of radiation, as measured by an observer, is altered by the relative motion of the observer and the source.

***Electric charge*:** internal property of particles that respond to electromagnetic forces; the charge can be positive or negative.

***Electric field (see Field)*:** field assumed to emanate from an electrically charged particle, which transmits the electric force.

***Electromagnetic field (see Field)*:** electric and magnetic fields normally associated with the transmission of electromagnetic radiation.

***Electromagnetic radiation*:** energy carried by an electromagnetic wave.

***Electromagnetic wave*:** a wavelike fluctuation of an electromagnetic field, travelling at the speed of light.

***Electron*:** elementary particle, negatively charged and assumed to be extremely small, like a point in space (no dimension): it possesses one unit of quantized charge, $-1e$, has mass and a spin angular momentum, $1/2\hbar$ and a magnetic moment.

***Electron-volt*:** energy acquired by an electron under the effect of a potential difference of 1 volt. It is a very small quantity of energy, corresponding to 1.6×10^{-19} joule (see joule).

Electroweak theory: quantum field theory unifying electromagnetic interaction and weak interaction responsible for beta decay.

Entropy: a measure of disorder in the Universe.

Equivalence principle: see principle of equivalence.

Event horizon: boundary around a black hole where the escape velocity is larger than that of light.

Exclusion principle: see Pauli exclusion principle.

Families of particles: grouping of particles that have common properties, for examples neutrons, protons and mesons under the name hadrons, electrons and neutrinos under the name leptons.

Fermions: particles that have half integer spin; fermions obey the exclusion principle and cannot be at the same place at the same time; within an atom they cannot have the same quantum numbers or be in the same state; electrons and quarks are fermions.

Feynman diagram: graphical figure representing the interaction of particles.

Field: immaterial substance assumed invisible, which impregnates space or a region of space and which may influence the state of particles that may be present; it is like a 'fog' spreading through space.

Fine structure constant: a combination of fundamental constants (e, h, c, ε_o), which plays an important role in quantum electrodynamics.

Flat space: geometry of a space that obeys Euclidian geometry; it may be represented by a smooth table top; in such a space the sum of the internal angles of a triangle is 180 degrees.

Flavour (particles): name given to grouping of particles with common properties; similar to family; flavour has nothing to do with our macroscopic sense of tasting.

Force: Physical principle of action. Cause of modification of the properties (shape, movement, etc) of physical objects. (See ***interaction***)

Force carrier: see force mediator.

***Force mediator*:** a virtual particle that transmits force between two particles; for example the photon is the force mediator of the repulsive electromagnetic force between two electrons.

***Four-dimensional space-time*:** grouping of the three dimensions of space with time as an added dimension, forming a framework in which all objects of our Universe evolves.

***Frame of reference*:** a particular geometrical arrangement of axes to which position and motion of an object can be referred. A framework with axes x, y and z at right angle to each other forms such a frame of reference in three-dimensional space.

***Frequency*:** measure of the repetition rate of a cyclic phenomenon.

***Gauge symmetry*:** symmetry that exists in some theories, like electromagnetism, in which specific physical entities and laws are invariant under various local transformations.

***General theory of relativity*:** theory of gravitation developed by Einstein based on the equivalence principle; in that theory space-time is curved by the presence of matter.

***Geosynchronous orbit*:** specific orbit of a satellite around a planet, whose orbital period is the same as the rotation period of the planet.

***Gluon*:** massless mediator of the strong force between quarks.

***Gravitational field*:** assumed field that transmits the force of attraction between objects with mass.

***Gravitational force*:** force of attraction developed by a mass on other masses; pressure and energy are also a cause of gravitational attraction.

***Gravitational waves*:** excitation of the gravitational field, which propagates through space at the speed of light; none have been detected yet (2010).

***Graviton*:** an hypothetical particle associated to a gravitational wave as the photon is associated to an electromagnetic wave; it is introduced in the framework of quantum gravity, as the mediator of the gravitational force, is massless and must have a spin of 2 because gravitational waves originate from quadrupole moments; it has not been observed yet (2010).

***Heisenberg uncertainty principle*:** principle of quantum mechanics developed by Heisenberg under which the value of two complementary variables can be measured only with a certain precision; position and speed, energy and time are such complementary variables; due to the existence of such a principle, at very short times, fields fluctuate widely and vacuum becomes effervescent with so-called virtual photons and virtual particles.

***Higgs field*:** assumed field existing in all of space and giving mass to particles such as electrons, quarks, $W^{\pm}$ and Z bosons.

***Higgs particle, Higgs boson*:** particle associated with the Higgs field; it is expected to have a large mass (~200 GeV); it is sometimes called God's particle; it has not been observed yet (2010).

***Homogeneity*:** state of objects, fields or other entities that are characterized by properties having identical values over space; the scale at which homogeneity is determined must be made explicit for physical meaning; for example, if temperature is the same over a given system volume, the system may be said to have homogeneity relative to temperature.

***Horizon (cosmic)*:** distance over which signals from distant objects did not have time to reach us.

***Horizon problem*:** problem raised by the realization that very distant objects in our Universe have nearly identical properties; it is claimed that those objects could not have communicated in the past due to the age of the Universe and the limited speed of light; consequently it is not obvious why the Universe should be homogeneous, without making an added assumption such as inflation in the early stage of expansion.

***Hubble constant*:** rate of expansion of space. It is not truly a constant since it varies with time.

***Hubble's law*:** law discovered by Hubble that says that the speed of recession of a galaxy from an observer is proportional to its distance from that observer.

***Hubble length*:** in cosmology, distance from an observer on Earth where objects are recessing at the speed of light.

***Hubble sphere*:** sphere centred on an observer inside which galaxies recess at a speed less than that of light. Observers on different galaxies have different Hubble spheres.

***Hydrogen maser*:** very stable atomic clock using the principle of stimulated emission of radiation between the two lowest hyperfine energy levels of the hydrogen atom.

***Inertia*:** quality of objects with mass, which opposes to acceleration.

***Infinity*:** a situation without limit. In physics a situation where our physical laws lose their applicability: for example, the field of a point charge becomes infinite when the distance at which the calculation is made reduces to zero.

***Inflation*:** exponential increase of a quantity with time; in a state of inflation a quantity doubles in size at fixed time intervals; it is assumed that inflation was the Bang that took place at the first moment of expansion, smoothing out our Universe, making it flat and homogenous and solving the horizon problem.

***Inflaton*:** intense field that is assumed to have been present at the beginning of the Universe expansion and that caused inflation.

***Ion*:** an atom that has lost one or more electrons and that consequently becomes positively charged.

***Interference*:** phenomenon by which waves add constructively or destructively depending on their relative phase.

Interaction*:** reciprocal action of physical entities. In physics it leads to four forces. It can be *gravitational* as in the attraction between two objects with mass, *electromagnetic* as in the repulsion between two electrons, *weak* as in beta decay, *strong* as in the binding of nuclear particles. (See ***force)

***Isotropic*:** qualification of space that looks identical in all directions.

***Isotopes*:** elements whose nuclei have the same number of protons but have a different number of neutrons; these elements have the same chemical properties.

***Joule*:** SI units of energy; a 100 watt light bulb radiates an energy of 100 joules every second.

Kelvin scale: absolute temperature scale resulting from basic considerations on heat exchanges; its zero is fixed at $-273.15°$ on the Celsius scale and is the lowest achievable temperature.

Laser: acronym for Light Amplification by Stimulated Emission of Radiation; source of coherent electromagnetic radiation at wavelengths shorter than in the case of the maser (see Maser); the photons emitted are all in phase and form a narrow beam of intense radiation.

Leptons: family of elementary particles that have a spin 1/2 (fermions) and that are sensitive to electromagnetic and weak interactions. Electrons and neutrinos are leptons.

Light year: distance travelled by light in vacuum during one terrestrial year; abbreviated 'ly'.

Log scale: representation often used in graphics to cover a very large range of data. On that scale a factor of ten corresponds to an increase by one unit. For example 10 corresponds to 1, 100 to 2, 1000 to 3 and so on. Numbers smaller than unity are similarly represented but with negative numbers.

Lorentz contraction: reduction of the length of objects in uniform motion due to relativistic effects as seen by an observer at rest.

Mach's principle: imprecise hypothesis, which in one version says that inertia is created by the presence of all objects with mass distributed throughout the Universe.

Macroscopic: in the present text, large scale situation in which classical physics normally applies.

Microscopic: in the present text, small scale situation to which quantum mechanics applies.

Magnetic field: field developed around a magnet or a current carrier wire, which creates a force on moving electrically charged particles or other magnetic objects.

Maser: acronym for Microwave Amplification by Stimulated Emission of Radiation; as an example see ***Hydrogen Maser***.

Maxwell's theory of electromagnetism: theory developed by Maxwell in which electric and magnetic interactions are united under a single formalism.

Mediator of force: in quantum field theory, generally a boson that transmits force; the photon is a mediator of electromagnetic force.

Meson: particle formed of two quarks member of the hadrons family; mesons are not stable particles and disintegrate in a short time.

Microwave background: see ***Cosmic Microwave Background***.

Molecule: an assembly of atoms held together by quantum mechanical and electromagnetic interactions.

MRI instrument: imaging instrument using Nuclear Magnetic Resonance of protons in water molecules as indicator; it is of great use in medical diagnostics.

Neutrino: elementary particle introduced by Pauli and whose existence was inferred from the behaviour of emitted particles in beta decay; it was believed to be massless, but recent observations on neutrinos emitted by the Sun have led to the possibility that it has a mass, although very small.

Neutron: neutral particle made of three quarks, that respond to the strong force and sticks to protons to form complex and heavy nuclei; when free it has a lifetime of about 15 minutes and disintegrates into a proton, an electron and an antineutrino (beta decay).

Newton's principles: physics principles introduced by Newton in the 18th century forming the basis of classical mechanics.

Nuclear force: force that holds the constituents of the nucleus, protons and neutrons, together; this force appears to originate from a leak of the strong force between the quarks forming the protons and neutrons; it is mediated by the π meson.

Nucleons: constituents of the nucleus, protons and neutrons.

Nucleus: core of atoms made of nucleons, that is to say neutrons and protons.

Observable universe: ensemble of objects detectable with instruments within the sphere formed by the cosmic horizon.

Observer: instrument or ensemble of instruments (may include a person) used to measure the properties of a physical system; in quantum mechanics the action of the observer results in the collapse of the wave function.

***Parallax*:** phenomenon by which the position of distant objects appear to move relative to closer objects; the phenomenon is used to measure the distance of relatively close objects like nearby stars in the Milky Way.

***Parity symmetry*:** property of a system or a physical law, which do not change when observed under mirror reflection.

***Parsec*:** unit of measurement of the distance of a celestial object from earth measured by the parallax technique; it corresponds to a distance of 3.26 light-year.

***Particle accelerator*:** instrument used to accelerate particles to high energy.

***Pauli exclusion principle*:** principle introduced by Pauli, which says that Fermions, such as electrons, cannot occupy the same space or the same energy state.

***Phase*:** 1) state of evolution of a wave in space and time, which dictates its amplitude at given coordinates of time and space; 2) also used for qualifying the physical state of matter, such as gaseous, liquid and solid phases.

***Phase transition*:** change of the physical property of a physical system such as the passage from the gas phase to the liquid phase upon cooling.

***Photoelectric effect*:** effect by which a photon with a minimum energy extracts an electron from a solid.

***Photon*:** particle of electromagnetic radiation; it is interpreted as the level of excitation of an electromagnetic wave; it is also interpreted as the mediator of the electromagnetic force.

***Photon clock*:** fictitious clock made of two perfectly reflecting mirrors in which a photon oscillates between the two mirrors and whose travelling time between the mirrors provides the period of the clock.

***Planck charge*:** charge deduced from a combination of vacuum fundamental constants through dimensional analysis; it is equal to 1.87×10^{-18} coulomb.

***Planck length*:** length deduced from a combination of vacuum fundamental constants through dimensional analysis; it is equal to 1.62×10^{-35} metre.

***Planck mass*:** mass deduced from a combination of vacuum fundamental constants through dimensional analysis; it is equal to 2.18×10^{-8} kilogram.

***Planck temperature*:** temperature deduced from a combination of vacuum fundamental constants through dimensional analysis; it is equal to 1.42×10^{32} kelvin.

***Planck time*:** time deduced from a combination of vacuum fundamental constants through dimensional analysis; it is equal to 5.39×10^{-44} second.

***Planck's constant*:** constant introduced by Planck to explain the shape of the spectrum of black body radiation, making explicit that electromagnetic radiation can raise in energy only $h\nu$ at a time. It is the fundamental constant of quantum mechanics.

***Plasma*:** an ensemble of ionized gaseous atoms; it is sometime called the fourth state of matter, in addition to gaseous, liquid and solid.

***Positron*:** antiparticle of the electron.

***Potential energy*:** energy that is stored in a field or an object; example: an object raised to a certain height above the surface of the earth acquires potential energy that is released when dropped to the ground.

***Primeval atom*:** concept introduced by Lemaître describing the original entity present at the creation of the Universe; it could be infinitely small with infinite energy.

***Principle of equivalence*:** principle of general relativity under which the presence of a gravitational field is indistinguishable from accelerated motion; it follows that inertial mass is equivalent to gravitational mass.

***Proton*:** primary constituent of atomic nuclei; it is electrically charged, +1 e.

***Quantum*:** term introduced in quantum mechanics representing a small discrete quantity of a physical entity such as energy and electric charge.

***Quantum agitation*:** turbulent state of vacuum or of a system reflecting the presence of fluctuating fields or virtual particles created from nothing, a situation made possible through the uncertainty principle.

***Quantum chromodynamics*:** quantum field theory of the strong force; the theory includes the effects of special relativity.

Quantum effervescence*:** see ***quantum agitation.

***Quantum electrodynamics*:** quantum field theory of the interaction of electrons with quantized electromagnetic fields including the effect of special relativity.

***Quantum field theory*:** theory in which fields are quantized; in that theory particles such as electrons are assumed to be quantized point-like excitation of a supporting field; the theory includes the effects of special relativity.

Quantum fluctuations*:** see ***quantum agitation.

***Quantum gravity*:** theory under development in which the general theory of relativity is joined to quantum mechanics; the gravitational field is quantized and leads to the possibility of the existence of gravitons; no such particles have been observed yet (2010).

***Quantum mechanics*:** generic terminology representing a theory of the microscopic world, which describes the behaviour of the world of atoms, nuclei, elementary particles and fields in terms of probability, wave-particle duality and the quantum concept.

***Quantum numbers*:** integer or half integer numbers characterizing the state and property of an atomic and subatomic entity.

***Quantum physics*:** generic terminology that includes all theories that uses the concept of quantization of the microscopic world.

***Quarks*:** elementary particles, constituents of hadrons; they respond to the strong force; there are six flavours, up, down, charm, strange, top and bottom.

***Radiation*:** broad meaning term for electromagnetic waves in general as well as corpuscles emitted by atoms or radioactive nuclei.

***Relativity, principles of*:** 1) in special relativity, first principle: the laws of mechanics and electromagnetism are the same in any reference frame in relative motion; second principle: the speed of light is independent of the motion of the observer or of the source; 2) in general relativity, the core principle is the equivalence of acceleration and gravitation.

Reference frame*:** see ***frame of reference.

***Rotation symmetry*:** property of an object or a physical law that is unaltered when observed from a different perspective; for example a sphere has rotational symmetry.

***Scalar field*:** a field whose properties are independent of direction and that can be represented simply by a number or scalar value at each point of space. Temperature within a room may be interpreted as a scalar field.

***Schrödinger equation*:** core equation that governs the behaviour of the probability wave (wave function) in quantum theory (wave mechanics).

***Schwarzschild radius*:** event horizon of a non rotating black hole.

***Second law (in thermodynamics)*:** entropy either increases or stays constant in all physical processes that take place in an isolated system; another version says that it is impossible to build a machine that performs work by doing nothing else than extracting heat form a reservoir.

***SI*:** International System of units in which the base units are: metre (m), kilogram (kg), second (s), ampere (A), kelvin (K), candela (cd), mole (mol).

***Space curvature*:** deviation of space from a flat shape; Euclidean geometry does not apply in a curved space. Such a space was first studied by J. Gauss and G. Riemann.

***Space-time*:** framework in which space and time are joined to form a single entity or grid in which objects evolve; it is a four dimensional world.

***Spectral line*:** radiation concentrated around a narrow range of frequencies normally emitted by an atom.

***Speed*:** rate of change of the position of an object per unit of time; units are metre per second.

Speed of light: fundamental constant called c and equal to 299 792 458 m/s. Particles with zero mass travel at the speed of light.

Sphere: outer surface of a globe; in general the dimension of the sphere must be made explicit; a one-sphere is a circle; a two-sphere is the surface of a globe; a three-sphere is the surface of a four dimensional globe.

Spherical symmetry: property of objects or physical laws that remain unchanged when viewed from different directions.

Spin (in quantum mechanics): property by which particles have an intrinsic angular momentum.

Standard candle: a standard of luminosity that can be used in astronomy as the basis for determining the distance of far celestial bodies.

Standard model of cosmology: broad assembly of plausible scenarios, models and concepts describing the evolution of the Universe from the Big Bang to the present epoch.

Standard model of particle physics: union within quantum field theory of the electromagnetic, weak and strong interactions describing the dynamics of atomic and subatomic particles.

String theory: mathematical theory based on the hypothesis that particles are constructed of elementary vibrating strings; the theory has not yet received experimental confirmation (2010).

Strong force: strong short range force (10^{-15} m) that keeps quarks locked within hadrons.

Supersymmetry: a hypothetical symmetry in which all particles with half integer spins have a partner particle with an integer spin or vice-versa.

Supernova: explosion of a star that may outshine an entire galaxy; this may happen when the core of the star has consumed its nuclear fuel or when the star has accreted sufficient material from a companion.

Symmetry: property of systems or physical laws that are not altered when submitted to a specific transformation.

Symmetry breaking: see ***broken symmetry***.

Système International (SI): system of units of physical measurements agreed upon by the international community. Its basic units are metre, kilogram, second, ampere, kelvin, candela and mole. All others units are derived from those.

Theory of Everything (TOE): a theory that would join together all interactions, strong, weak, electromagnetic and gravitational into a single framework; not yet achieved as of 2010.

Theory of gravity: theory elaborated by Newton relative to the weakest of the four forces of nature, gravitational attraction, and further developed by Einstein in his General Theory of Relativity.

Thermodynamics: branch of physics concerned broadly with the macroscopic properties of matter; thermodynamics is largely concerned with the exchange of energy under its various forms and provides basics laws for these exchanges.

Three-sphere: see ***sphere***.

Time arrow: direction of the flow of time, claimed to be the direction in which entropy increases.

Time dilation: phenomenon deduced from the principle of special relativity in which time flows more slowly for a moving observer when referenced to an observer at rest.

Time inversion symmetry: property that characterizes a physical law that remains unaltered by an inversion of time flow.

Translation symmetry: property that characterizes a physical law that remains unaltered by a displacement in space or a change of coordinates.

Two-sphere: see ***sphere***.

Uncertainty principle: see Heisenberg uncertainty principle.

Unification, Grand (GUT): theory that joins weak, electromagnetic and strong forces under the same framework.

Unit of measurement: concept providing a framework in which the value of a quantity is determined; the results of a measurement can be represented by a number that is a multiple or a submultiple of a defined quantity or unit (see ***Système International SI***).

***Vacuum energy*:** energy originating form quantum vacuum fluctuations.

Vacuum fluctuations*:** see ***quantum agitation.

***Vector*:** entity characterized by a length and direction.

***Vector field*:** a field whose properties depend on direction; for example a magnetic field has direction and is a vector field.

***Velocity*:** vector specifying speed and direction of an object in motion.

***Virtual particles*:** particles created in vacuum for a short time; these particles live on borrowed energy and have a lifetime that is dictated by the uncertainty principle.

***Wave function*:** mathematical entity attached to a particle, which is interpreted as a probability amplitude, foundation of the theory of quantum mechanics; its evolution in time and space is governed by Schrödinger's equation.

***Wave function collapse*:** process by which the wave function probability takes a definite value through detection or a direct interaction with the particle being observed.

***Wave mechanics*:** alternative name of quantum theory dealing specifically with the wave nature of particles; Schrödinger's equation is the core of that approach.

***Wavelength*:** characteristic distance between two maxima or two minima of a wave.

***Weak force, weak interaction*:** one of the four interactions of nature; the force is mediated by $W^{\pm}$ and Z_o bosons. Beta decay is one effect of a weak interaction.

***$W^{\pm}$, Z bosons*:** mediators of the weak force; they have a large mass, $\sim 90\,\text{GeV}$.

***Zero-sphere*:** a sphere that is reduced to a point in space, see sphere.

Notes and References

Preface

1. Hugh Everett. See, for example: 'The Many Worlds of Hugh Everett', P. Byrne. *Scientific American*, December, 2007.

Chapter I

1. See, for example: *The Scientific Study of Unidentified Flying Objects*, E.U. Condon, Bantam Books, New York, 1969.
2. For a discussion of that matter see: *The Trouble with Physics*, L. Smolin, Houghton Mifflin Company, Boston, 2006.

Chapter II

1. We will address in detail that relativistic effect on time including the twin paradox question in the following chapter. We will see later that relativity has consequences that overwhelm our imagination and provides means for rather exotic space travel.
2. Several hundred planets have been detected around stars generally through indirect measurements. For example the wobbling of a star or its change of luminosity is a sign of the presence of an object orbiting the star in question. Those planets must of course be enormous to cause such a massive object as a star to alter its position and as well to be occulted even in a minor way. Recently planets have been seen directly by means of telescopes as reported by C. Marois, B. Macintosh, T. Barman, B. Zuckerman, I. Song, J. Patience, D. Lafrenière, R.Doyon, 'Direct Imaging of Multiple Planets Orbiting the Star HR 8799', *Science*, **322** (5906), 1348, 2008. See also: 'Direct Imaging Reveals Exoplanets in Orbital Motion' B. Schwarzschild, *Physics Today*, 11, Jan. 2009. A new satellite, NASA's Kepler, dedicated entirely to the discovery of planets around stars of the Milky Way has been launched in March 2009. Its aim is to identify planets of the size of Earth and the presence of water on them.
3. For a discussion on the question of the advent and nature of life the reader is invited to consult the book: *Time, Space, Stars & Man*, M.M. Woolfson, Imperial College Press, London, 2009.

4. See for example Robert A. Millikan, 'The Electron and the Light Quant from the Experimental Point of View', Nobel Lecture, May 23, 1924.
5. Present-day technology partly solves that problem by making available a dynamic encyclopaedia on Internet (Wikipedia). It is continuously updated by contributors at the frontier of knowledge.
6. See for example: *The Cosmic Landscape,* L. Susskind, Little Brown Book Group, New York, 2006; *The Trouble with Physics*, L. Smolin, Houghton Mifflin Company, Boston, 2007; *The Charm of Physics*, S.L. Glashow, The American Institute of Physics, 1991.
7. DNA. This is an abbreviation for the chemical Deoxyribo Nucleic Acid, a molecule formed as a helix, containing the code of each living being.
8. S.L. Miller, 'Production of Amino Acids under Possible Primitive Earth Conditions', *Science*, **117**, 528, 1953.
9. See for example: M.M. Woolfson *Time, Space, Stars & Man*, Imperial College Press, London, 2009.

Chapter III

1. For the mathematically inclined reader: Acceleration or deceleration a is the rate of change of speed v during period Δt evaluated at a given point on the object trajectory. It is written $a = \Delta v/\Delta t$ or the derivative dv/dt, where Δv is the change in speed. It can be positive or negative making a positive (acceleration) or negative (deceleration).
2. The reader interested in the underlying mathematics is invited to consult Annex A, section *a*.
3. For the mathematically oriented reader please see Annex A, section *e*.
4. On the condition, however, that we stay loyal to them for many years with a monthly payment.
5. A.A. Michelson and E.W. Morley, 'On the relative Motion of the Earth and the Luminiferous Ether', *American Journal of Science*, **34**, 333, 1887.
6. R. P. Feynman, R.R. Leighton, M. Sands, *The Feynman Lectures on Physics*, Addison Wesley, USA, 1965.
7. See for example Georg Joos on the use of an interferometer with 21 metres arms, *Theoretical physics*, Blackie and Son, London and Glasgow, 1954.
8. For the mathematically inclined reader: the mathematical expression is thus, energy equals frequency multiplied by Planck's constant, or $\mathrm{E} = \mathrm{h}\nu$. However, since frequency is inversely proportional to wavelength through the relation, frequency multiplied by wavelength equals the speed of light or $\nu\lambda = \mathrm{c}$, the energy can also be written as $\mathrm{E} = \mathrm{hc}/\lambda$.
9. A periodic motion is characterised first by its frequency, or cycle per second, which has been given the name 'Hertz', from the name of a physicist who has been the first to demonstrate experimentally the existence of electromagnetic waves.
10. We will see later that space is filled with microwave radiation originating from the Universe early moments of existence, the so-called cosmic microwave background (CMB). Interestingly enough, that radiation background appears

to provide a kind of reference frame in which the motion of our Milky Way can be determined by means of Doppler effect. It should be mentioned that this is not in contradiction with the Michelson-Morley results, these being concerned with the speed of the Earth relative to the speed of light, while, in the case of the CMB, the motion of the Earth or Milky Way is determined by means of the effect of their speed on the observed CMB wavelength by means of Doppler effect. It is realized that the measured motion of the Milky Way is done relative to the CMB and is still a motion relative to something else.

11. The reader interested in the underlying mathematics is invited to consult Annex A, section *a*.
12. The Lorentz transformations are given in Annex A, section *g*.
13. A. Einstein, '*Zur Elektrodynamik bewegter Körper*', ('On the Electrodynamics of Moving Bodies'), *Annalen der Physik*, **17**, 1905. See also: *The Principles of Relativity, Einstein and Others*, Dover Publications Inc., S81.
14. H. Minkowski, *Space and Time*, address delivered at the *80th Assembly of German Natural Scientists and Physicians*, Cologne, 21 September, 1908.
15. Adapted from R.P. Feynman, R.B. Leighton, M. Sands, *The Feynman Lectures on Physics*, p15, Addison Wesley, USA, 1965.
16. It should be mentioned that this is a truly fictitious experiment impossible to execute in practice. It is not possible to photograph the travel path of the photon or pulse of light unless it is reflected on some particles like dust or a mist. In such a case the path and average speed of the photon or light pulse would be altered and the argumentation would become obsolete. This is the kind of experiment that is often proposed to explain physical effects difficult to observe and interpret or is used to demonstrate principles. It is not done in practice. It is called a *Gedankenexperiment* or 'Thought Experiment'. Such fictitious experiments were used extensively by Albert Einstein and Niels Bohr in their discussions. It should also be mentioned that the kind of speeds that are achieved with trains have no measurable effects on clocks that could be attributed to relativity. The speed of a train such as a TGV may be as high as 300 km per hour while the speed of light is 300,000 km per second. This makes the ratio v^2/c^2 equal to 7.7×10^{-14}. The resulting change in the length of the second in the train as seen from an observer at rest would be 3.4×10^{-14} second. This is thus totally negligible and the measurements made in the text by Mary and Joe are only for illustration. Real experiments with measurable effects obtained with airplanes on long flights are described in the main text.
17. The reader interested in the underlying mathematics is invited to consult Annex A, section *d*.
18. A. Einstein, '*Zur Elektrodynamik bewegter Körper*', *Annalen der Physik*, **17**, 1905. See translation: 'On the Electrodynamics of Moving Bodies', in *The Principles of Relativity, Einstein and Others*, Dover Publications Inc., S81.
19. C.O Alley, 'Relativity and Clock', *Proc. 33rd Frequency Control Symposium*, p4, 1979.
20. We may also imagine that Joe makes two trips of differing lengths. Assume that in the two trips he goes at the same speed, but for lengths differing by a factor of two, and does his turning around at the same rate and in the same

space. Suppose Joe goes on his first trip at a speed of 0.87 c. After 1/2 hour of his time he turns around at a certain rate and comes back for 1/2 hour of his time at the same speed. When he is back after 1 hour he finds that his clock has retarded by 1/2 hour relative to that of Mary, which of course was thus running faster than his own. Now, suppose he goes back on a similar trip at the same speed but for two hours and to turn around he does exactly the same thing as in the previous trip. He finds that his clock now has retarded by 1 hour in agreement with the time dilation theory developed in Annex A. Although his acceleration for turning around was exactly the same in both cases his clock has now retarded by twice as much. Consequently, it could not be due to the acceleration, the same exact cause producing two different results. The time dilation came from the high speed he was travelling at.

21. J.C. Hafele and R.E. Keating, 'Around the World Atomic Clocks: Predicted Relativistic Time Gains; Observed Relativistic Time Gains'. *Science*, **177**, 166–168, 1972.
22. The reader interested in the underlying mathematics is invited to consult Annex A, section *d*.
23. H. Minkowski, 'Space and Time', address delivered at the *80th Assembly of German Natural Scientists and Physicians*, Cologne, 21 September, 1908.
24. The reader interested in the underlying mathematics is invited to consult Annex A, section *c*.
25. The interested reader will find a calculation making explicit the origin of Doppler effect in Annex A, section *h*.
26. Military contracts sometimes make things easy!
27. The relativistic law of the addition of velocities is given in Annex A, section *c*.
28. We will see later that in fact such a photon will change but for another reason than aging. It will change wavelength due to the expansion of space, which has nothing to do with the principles we are examining here.
29. There is sometime confusion in the literature on the writing of the equivalence of mass and energy. We will not elaborate on that question, which on occasions looks like semantic. For more insight into the question, the reader is referred to the following article: L. Okun, 'The Concept of Mass', *Physics Today*, 31, June 1989.
30. Some elementary mathematics showing this expansion is developed in Annex A, section *c*.
31. An electron-volt (eV) is a very useful unit of energy when dealing with atomic particles. It is the energy acquired by an electron accelerated under a potential difference of 1 volt.
32. The reader interested in the exact mathematical expression is invited to consult Annex A, section *b*.
33. A question may be raised concerning the distribution of the filings appearing to follow discrete lines while the field is obviously uniformly distributed although decreasing with distance from the magnet. The reason is that the filings become little magnets. They have size and tend to stick to each other creating gaps. In principle if the fillings were extremely fine and form a very fine powder there would be a uniform distribution around the magnet and it would be difficult to observe the phenomenon.

34. The reader interested in the underlying mathematics is invited to consult Annex C, section *a*.
35. There is sometimes confusion raised in the use of the word 'redshift'. This is because it is used in different contexts. In general it means an increase in wavelength. However, the effect may be caused by various physical effects. A radiation source recessing from an observer causes Doppler effect seen as a redshift of the radiation detected by the observer. In cosmology, the expansion of space causes lengthening of the wavelength of the radiation travelling in free space and thus causes a redshift. Finally, gravity has the property of altering the frequency of a clock and the wavelength of the corresponding radiation. The effect is also called a redshift. The mathematically oriented reader will find mathematical expressions for those three effects in Annex A section *h* and Annex C, section *g*.
36. R. Pound and G.A. Rebka Jr., 'Apparent Weight of Photons', *Phys. Rev. Lett.*, **4**, 337, 1960.
37. C.O. Alley, 'Relativity and Clocks', *Proc. Of the 33rd Annual Symposium on Frequency Control*, **4**, 1979.
38. R.F.C. Vessot *et al.*, 'Tests of Relativistic Gravitation with a Space-Borne Hydrogen Maser', *Phys. Rev. Lett.*, **45**, 2081, 1980.
39. A. Einstein, '*Die Grundlage der allgemeinen Relativistatstheorie*', ('The Foundation of the General Theory of Relativity'), *Annalen der Physik*, **49**, 1916. See also *The Principles of Relativity, Einstein and Others*, Dover Publications Inc, S81 for an English translation.
40. R.P. Feynman, R.R. Leighton, M. Sands, *The Feynman Lectures on Physics*, Addison Wesley, USA, 1965.
41. For a more complete description of such a situation, the reader is referred to B. Greene, *The Elegant Universe*, W.W. Norton, New York, 2003.
42. E.F. Taylor and J.A. Wheeler, *SpaceTime Physics*, W.H. Freeman and Co. New York, 1995.
43. Attempts have been made in the early 1960s by J. Weber at detecting those gravitational waves by means of a large resonant metallic cylinder suspended in such a way as to be sensitive to those waves. The experiment was not successful. (It may be mentioned that Weber was one of the first scientists to propose the possibility of masers). At the present time, large interferometers are being designed by Caltech and MIT groups in joint efforts to detect those waves. The instrument is called 'Laser Interferometer Gravitational wave Interferometer', (LIGO). Another venture, a collaboration between NASA and ESA, the European Space Agency, is to use a giant Michelson interferometer in space, called Laser Interferometer Space Antenna (LISA) with arms millions of km long to increase sensitivity.
44. The first observation of a pulsar occurred in 1967 by Jocelyn Bell-Burnell who was a graduate student at Cambridge under the direction of A. Hewish. Hewish was awarded the Nobel prize in 1974 with his collaborator A. Ryle for pioneering research in astrophysics. The nature of an object in the sky pulsating at a rate shorter than a second remained for sometime a challenging puzzle. Aside from the far-fetched idea that it could be a signal from another civilization, a scenario that was soon rejected, it raised interest in the deep

physics involved in such a phenomenon. The rate was very stable and fast. This observation required at the same time that it originated from a massive object and that the object was small. There was a lot of speculation and finally the only logical answer to the origin of such a pulsating signal was a rotating neutron star, an object made of closely packed neutrons about 10 to 30 km in diameter, the residue of a supernova. See for example: G. McNamara, *Clocks in the Sky, the Story of Pulsars*, Praxis Publishing Ltd, Chichester, UK, 2008.

45. The pulsar binary system described in the text is known as PSR B1913–16. It was discovered by R.A. Hulse and J. Taylor. They share the 1993 Nobel Prize for their work on the subject.

Chapter IV

1. G. Shortley and D. Williams, *Elements of Physics*, Prentice Hall, USA, 1971.
2. See for example N.D. Mermin, *It's About Time*, Princeton University Press, Princeton, N.J., 2005.
3. LHC stands for Large Hadron Collider and CERN stands for *Centre Européen de Recherche Nucléaire*. The LHC is a large accelerator situated near Geneva. Hadron is a technical word representing the class of particles such as protons, neutrons, mesons and other heavy particles.
4. We are more familiar in our diets with the unit calorie. The calorie is just, due to peculiar definition of units, 4.18 joules. We normally use the much larger unit, 1,000 calories, that is also called Calorie, which may be confusing.
5. R.P. Feynman, R.R. Leighton, M. Sands, *The Feynman Lectures on Physics*, Addison Wesley, USA, 1965.
6. See for example: S. Hawking, *Une brève histoire du temps* and *Commencement du temps et la physique*, Flammarion, France, 1991 and 1992.
7. B. Decaux and B. Guinot, *La Mesure du temps*, Que sais-je, Presse Universitaires de France 1969.
8. G. Shortley and D. Williams, *Elements of Physics*, Prentice Hall, USA, 1971.
9. Note that the situation is different from that when we consider the Universe as a whole and examine the concept of time including the irreversible process of increase in entropy, which we examine in an earlier paragraph. In the present discussion we are examining the law of physics not the evolution of the Universe.

Chapter V

1. The reader interested in the underlying mathematics is invited to consult Annex A, section *e*.
2. As will be seen later there is presently an active field of research, which interprets gravity in the extreme cases as a quantum phenomenon. In that interpretation, gravitational particles, called gravitons, would exist and be the mediators of the gravitational force.

3. The unit of magnetic field in the International System of Units (SI), is the 'tesla' defined as a combination of units such as the metre, the second and other electrical units. The tesla is worth 10,000 gauss and in the present description it is sufficient to know that a field of a few hundred gauss is sufficient to lift small magnetic objects such as paper clips surpassing for example the gravitational attraction of the whole Earth on the same object.
4. In a metal, due to the strong interactions between the atoms, the electrons are dislodged from their mother atoms and are essentially free like atoms in a gas.
5. The reader interested in the mathematical form of Maxwell's equation is invited to consult Annex A, section *e*.
6. See article by B. Schwarzschild, 'Search for magnetic monopoles at the Tevatron sets new upper limit on their production', *Physics Today*, p16, July 2006.
7. It is recalled that the symbol 'tera' means 1 followed by twelve zeros. Note that to obtain frequency from wavelength, one just needs to divide the speed of light by the wavelength.
8. A. Einstein, *Investigation on the Theory of the Brownian Movement*, Dover Edition, 1956, ed. R. Fürth, translated by A.D. Cowper from *Annalen der Physik* **17**, 549, 1905.
9. A fine description of the experiment is provided in Millikan's Nobel lecture, 1924, 'The Electron and the Light Quant from the Experimental Point of View'. The lecture provides also information on Millikan's contribution to the experimental verification of the photo-electric theory developed by Einstein.

Chapter VI

1. The reader interested in the underlying mathematics is invited to consult Annex B, section *b*.
2. The reader interested in more details is invited to consults Annex B, section *d*, that provides more visual information on the properties of those quantum numbers.
3. Those names have been chosen in recognition of two physicists, Satyendra Nath Bose and Enrico Fermi who have played an important role in the development of the theories leading to the phenomenon described.
4. For those interested, the form of the distributions is given in Annex B, section *g*.
5. It is common practice in engineering and physics to call a 'black box' a device which has a specific well determined function, without the need to know what is inside. What is important is then only what it does. For a similar description of the experiment done with electrons, see D.M. Harrison, *Physics Courses*, Department of Physics, University of Toronto, Canada. Some illustrations are inspired from those Courses. See also R.P. Feynman, R.R. Leighton, M. Sands, *The Feynman Lectures on Physics*, Addison Wesley, USA, 1965.

6. The reader familiar with optics will find in Annex B, section *i*, the description of an experiment with photons, which transpose into optics the essential of the Stern-Gerlach experiment.
7. R.P. Feynman, R.R. Leighton, M. Sands, *The Feynman Lectures on Physics*, Addison Wesley, USA, 1965.
8. This may seem in contradiction with the statement made above where the probability of finding the electron was said to be maximum at distance a_o. However, there is no contradiction because a_o is the distance where the electron is most likely to be found through an electronic density calculation requiring the introduction of a volume element that grows with distance from the nucleus centre. Although the wave function is a maximum at the centre of the nucleus the element of volume becomes negligible there and the probability of finding the electron there is negligible.
9. This type of radiation with a 21 cm wavelength is also observed originating from the sky and is very important for astronomical research. It is the reason why the use of a range of frequency around that particular value is protected in order to avoid contaminating the spectrum that would make astronomical observations difficult.
10. The reader interested in a more elaborate discussion of the effect of zero-field fluctuations is invited to consult Annex B, section *k*.

11. W.H. Louisell, *Radiation and Noise in Quantum Electronics*, McGraw-Hill Book Co. USA, 1964.
12. H. Casimir, 'On the attraction between two perfectly conducting plates', *Proc. Royal Netherlands Academy of Arts and Science*, **51**, 793, 1948.
13. T.H. Boyer, 'Derivation of the Blackbody Radiation Spectrum Without Quantum Assumptions', *Phys. Rev.*, **132**, 182, 1969.
14. H.E. Puthoff, 'Ground State of Hydrogen as a Zero-point Fluctuation-determined State', *Phys. Rev. D*, **35**, 3266, 1987.
15. L. McTaggart, *The Field*, HarperCollins, UK, 2002.
16. C. Cohen-Tannoudji, J. Dupont-Roc, G. Grynberg, *Photons et atomes, Introduction à l'électrodynamique quantique*, InterEditions, CNRS, France, 1987.
17. See for example the article by F. Wilczek, 'Quantum Field Theory', in *Reviews of Modern Physics*, **71**, S85, 1999. Available also at arXiv:hep-th/9803075v2, 19 May 1998.
18. D. Bohm and Y. Aharonov, 'Discussion of Experimental Proof for the Paradox of Einstein, Rosen and Podolsky', *Phys. Rev.*, **108**, 1070, 1957. See also D. Bohm, *Quantum Theory*, Prentice-Hall Inc, Englewood Cliffs, N.J., 1951.
19. A. Einstein, B. Podolsky, N. Rosen, 'Can Quantum-mechanical Description of Physical Reality be Considered Complete?', *Phys. Rev.*, **47**, 777, 1935.
20. N. Bohr, 'Can Quantum-mechanical Description of Physical Reality be Considered Complete?', *Phys. Rev.*, **48**, 696, 1935.
21. J.S. Bell, 'On the Einstein Podolsky Rosen Paradox', *Physics I*, 195, 1964.
22. N.D. Mermin, 'Bringing Home the Atomic World: Quantum Mysteries for Anybody', *Am. J. Phys.*, **49**, 940, 1981. See also: D. Harrison, 'Bell's Inequality and Quantum Correlations', *Am. J. Phys.*, **50**, 811, 1982.

23. The calculation is outlined in Annex B, section (j).
24. A. Aspect, P. Grangier, G. Roger, 'Experimental Test of Realistic Local Theories via Bell's Theorem', *Phys. Rev. Lett.*, **47**, 460, 1981. A. Aspect, P. Grangier, G. Roger, 'Experimental Realization of Einstein–Podolsky–Rosen–Bohm *Gedankenexperiment*: A New Violation of Bell's Inequalities', *Phys. Rev. Lett.*, **49**, 91, 1982.
25. A. Aspect, J. Dalibard, G. Roger, 'Experimental Test of Bell's Inequalities Using Time-Varying Analysers', *Phys. Rev. Lett.*, **49**, 1804, 1982.
26. W. Tittel, J. Brendel, H. Zbiden, N. Gisen, 'Violation of Bell Inequalities by Photons More than 10 km Apart', *Phys. Rev. Lett.*, **81**, 3563, 1998.
27. C.H. Bennett and G. Brassard, "Quantum Cryptography: Public key distribution and coin tossing", *Proceedings of the IEEE International Conference on Computers, Systems, and Signal Processing, Bangalore*, p. 175, 1984.
28. Many scientists have contributed to the theoretical understanding of superconductivity. However, the complete theory is due to J. Bardeen, L.N. Cooper, and J.R. Schrieffer and is described in their paper: 'Theory of Superconductivity', *Phys. Rev.*, **108**, 5, 1175, 1957. Bardeen is the scientist who, with W. Shockley and W. Brattain invented the transistor. He received the Nobel Prize twice, for the invention of the transistor and for his contribution to the theory of superconductivity.
29. The book on the subject by D. Sobel and W.J.H. Andrewes is strongly recommended: *The Illustrated Longitude*, Walker and Company New York, 1998.
30. For the truly mathematically inclined reader the book, J. Vanier and C. Audoin, *The Quantum Physics of Atomic Frequency Standards*, Adam Hilger, Bristol, UK, 1989, is recommended.
31. Norman Ramsey received the Nobel Prize in Physics in 1989 for the invention of that cavity and his contribution to the field of atomic clocks.
32. It should be emphasized that the 'atomic' clock uses electronic properties. It has nothing to do with radioactivity or other nuclear processes such as fission or fusion, as is sometimes referred to in the popular literature.
33. The term Maser is an acronym for Microwave Amplification by Stimulated Emission of Radiation. The maser was invented in 1955 by C.H. Townes in USA and Bazov and Prokhorov in the Soviet Union using ammonia. They were recognised for that work by means of the Nobel Prize.
34. For the mathematically inclined reader the book, J. Vanier, *Basic Theory of Lasers and Masers*, Gordon and Breach, New York, USA, 1971, is recommended.
35. W. Phillips, C. Cohen-Tannoudji and S. Chu were awarded the Noble prize in physics in 1996 for their realisation.
36. E.A. Cornell and C. Wiemann and Keterle were awarded the Noble prize in physics in 2001 for their realisation.
37. For a description of such an activity the reader may consult the article: J. Vanier, 'Standards of Physical Measurement: The Role of the National Research Council', *Physics in Canada*, **44**, 3, 1988.
38. The exact definition of the units of voltage and electrical resistance in terms of Josephson and quantum Hall effect are given in Annex B, section m.

Chapter VII

1. Its magnetic properties have been studied in detail in such a trap: R.S. Van Dyck, Jr, P.B Schwimberg, H.G. Dehmelt, 'Precise Measurement of Axial, Magnetron, Cyclotron, and Spin-Cyclotron Beat Frequencies on an Isolated 1-MeV Electron', *Phys. Rev. Lett.*, **38**, 310, 1977.
2. For the mathematically oriented reader: The conventional unit of measurement of the electron magnetic moment is the so-called Bohr Magneton, μ_B. It is obtained from elementary theory as a combination of constants: $\mu_B = e\hbar/2m_e$. In the case of the nucleus a similar entity is defined with the mass of the electron replaced by that of the proton and is called the Nuclear Magneton.
3. The mathematically inclined reader will find an elementary calculation of the meson mass in Annex B, section *l*.
4. For a description of the technique see for example F. Close, *The New Cosmic Onion*, CRC Press, USA, 2007.
5. For more precision on the definition of strangeness, see S.L. Glashow, *The Charm of Physics*, p145, *Am. Inst. of Phys. USA*, 1991.
6. It should be mentioned that the phenomenon of positron or antimatter emission is used in medicine as a diagnostic tool. Traces of positron emitting isotopes, such as ^{11}C are incorporated into a chemical substance called a tracer and injected in the body. The tracer gets fixed to problematic cells. The positrons emitted by the attached tracer recombine within a few mm with electrons and a flash or radiation is emitted. An analysis of the resulting radiation provides information on the localization of the problematic cells.
7. The concept of left or right handed for an electron is connected to the sense of spin rotation or sign relative to the direction of motion as viewed from an observer behind or preceding the electron on its trajectory.
8. Sheldon Glashow, Abdus Salam and Steven Weinberg were awarded the Nobel prize in Physics in 1979 for the development of the electroweak theory combining electromagnetism and weak interactions.

Chapter VIII

1. It may be mentioned that in the development of his theory of relativity Einstein was influenced by Mach's reasoning as made evident in his 1916 paper, '*Die Grundlage der allgemeinen Realtivitätstheorie*' *Annalen der Physik*, **49**, 1916. He even raised the idea to the level of a principle although later upon further reflection he appeared to have abandoned it. See translation: A. Einstein, 'The Foundation of the General Theory of Relativity' in *The Principles of Relativity, Einstein and Others*, Dover Publications Inc., S81.
2. For an interesting discussion of the concept of mass see for example: F. Wilczek, *The Origin of Mass*, MIT Physics Annual, 2003.

Chapter IX

1. The solution to the paradox appears to be the finite age of the Universe. For a rather complete discussion of the paradox the reader is referred to J.M. Overduin and P.S. Wesson, *The Light Dark Universe*, World Scientific Publishing Co., Singapore, 2008.
2. The reader interested in the underlying mathematics is invited to consult Annex C, section *c*.
3. The mathematical definition of the red shift is given in Annex C, section *g*.
4. For the reader interested in the mathematical representation of space curvature, a calculation for a three-sphere is given in Annex C, section *b*.
5. S. Weinberg, *The First Three Minutes*, Basic Books, New York, USA, 1977.
6. I would believe that most physicists would have the same feeling of awe and admiration in visiting that place. See for example the description of a similar experience by George Smoot in *Wrinkles in Time*, HarperCollins, New York, 2007.
7. The metric is given in mathematical form in Annex A. section *c*.
8. R.F.C. Vessot *et al.*, 'Tests of Relativistic Gravitation with a Space-borne Hydrogen Maser', *Phys. Rev. Lett.*, **45**, 2081, 1980.
9. S. Weinberg, 'Einstein's Mistakes', *Physics Today*, Nov. 2005, p31. See also P.J.E. Peebles, *Principles of Physical Cosmology*, Princeton University Press, 1993, p67.
10. The reader interested in the underlying mathematics is invited to consult Annex C, section *e*.
11. The metric is derived in Annex C, section *b*.
12. G. Gamow, 'The Origin of Elements and the Separation of Galaxies', *Phys. Rev.* **70**, 572, 1946; R.A. Alpher, H. Bethe, G. Gamow, 'The Origins of Chemical Elements', *Phys. Rev.*, **73**, 803, 1948; R.A Alpher, R. Herman, G. Gamow, 'On the Relative Abundance of the Elements', *Phys. Rev.*, **74**, 1737, 1948.
13. Annex C, section *h*, provides a mathematical demonstration of that property, a decrease of temperature with expansion.
14. A.A Penzias and R.W. Wilson, 'A Measurement of Excess Antenna Temperature at 4,080 MHz', *Astrophysical Journal*, **142**, 419, 1965.
15. R.H. Dicke, P.J.E Peebles, P.G. Roll, D.Y Wilkinson, 'Cosmic Black-Body Radiation', *Astrophysical Journal*, **142**, 414, 1965. It is said that Dicke learned from Penzias and Wilson the results of their findings through a phone call during a meeting with his colleagues. Having worked for sometime on a project for detecting the CMB radiation, he turned to his colleagues and said something like: 'We have been scooped'. The physics field is a competitive activity and one learns rapidly that some situations may be quite frustrating.
16. See for example: G. Smoot, *Wrinkles in Time,* HarperCollins, New York. 1993.
17. See for example: C. Bonneau, S. Brunier, V. Greffoz, *Qu'y avait'il avant le Big Bang?, Science et Vie*, **1054**, 54, 2005.

18. For a detailed description of questions connected to the horizon concept in cosmology the reader is referred to E. Harrison, *Cosmology, the Science of the Universe*, Cambridge University Press, UK, 2000.
19. See for example: A.H. Guth, 'Inflationary Universe: A Possible Solution to the Horizon and Flatness Problem', *Phys. Rev. D*, **23**, 347, 1981. See also http://nedwww.ipac.caltech.edu/level5/Guth/Guth3.html.
20. A.D. Linde, 'A New Inflationary Universe Scenario: A Possible Solution of the Horizon, Flatness, Homogeneity, Isotropy and Primordial Monopole Problems', *Phys Rev. Lett.*, **108B**, 389, 1982. A. Albrecht, P. Steinhardt, 'Cosmology for Grand Unified Theories with Radiatively Induced Symmetry Breaking', *Phys. Rev. Lett.*, **48**, 1220, 1982. See also A.D. Linde, *Scientific American*, Nov. 1994, p48.
21. There are alternatives, but it appears that the Inflation proposal, although not perfect, provides the best scenario. See R. Brandenberger, 'Alternative to Cosmological Inflation', *Physics Today*, March 2008, p44.
22. A.H. Guth, *Inflation and the New Eras of High Precision Cosmology*, MIT Physics Annual, 2002.
23. Doubling of size repeatedly has dramatic consequences. It does not take many doublings to ensure an out-of-control growth. For example, you can fold a piece of paper by half to double its thickness and repeat it many times. The paper is thin (0.08 mm), but after a while the thickness of the stack increases rapidly. You may wonder how many times you need to fold it to cover the distance between the Earth and the moon? Amazingly, although the exercise may be rather acrobatic, it takes only 42 times of folding.
24. G. Gamow, *My World Line*, The Viking Press, New York, 1970. The idea of a resultant zero total energy Universe, appearing out of nothing has been proposed by several authors over the years. See for example an interesting popular description of the inflation scenario by John Gribbin, 'Inflation for the beginner', at www.lifesci.sussex.ac.uk/home/John_Gribbin/cosmo.htm. See also: 'Was Cosmic Inflation the Bang of the Big Bang', A.H. Guth at www.ipac.caltech.edu/level5/Guth/Guth1.html.
25. In some scenarios such as one proposed by Linde, the process of inflation could have taken place on a patch of the starting primeval entity and our Universe could be one of a large number of them that have evolved in parallel. Conditions in those universes could be very different from those in our Universe, with possibly totally different physical laws. It is possible that those Universes would not be appropriate for the kind of evolution that has given rise to beings like us. Such proposals are the results of calculations based on particular assumptions that are speculative.
26. See for example: R. Lehoucq, *L'univers a-t-il une forme*, Flammarion, Paris, 2002.
27. P.A.M. Dirac, 'Quantized Singularities in the Electromagnetic Field', *Proc. Roy. Soc. A*, **133**, 60, 1931.
28. The reader interested in the underlying mathematics providing a relation between temperature and expansion is invited to consult Annex C, section *h*.

29. For a more elaborate discussion of the scenarios proposed relative to inflation and expansion the reader is referred to articles assembled in a special issue of *Scientific American* titled: 'The Cosmic Life Cycle, Origins of the Universe', #SCA50607. See also, M.S. Turner, 'The Universe', *Scientific American*, Sept. 2009, p36.
30. For a description of several scenarios of stars and planets formation the reader is referred to M.M. Woolfson, *Time, Space, Stars & Man*, Imperial College Press, London, 2009.
31. For a detailed description of the application of string theory to the origin of our Universe the reader is referred to the interesting book: M. Gasperini, *The Universe Before the Big Bang, Cosmology and String Theory*, Springer-Verlag, Berlin, 2008.
32. G. Gamow *et al.,* 1946 and following years.
33. For an extensive description of the work done on pulsars the reader is referred to: G. McNamara, *Clocks in the Sky, The Story of Pulsars*, Springer-Parxis Books, Berlin, 2008.
34. The actual value of the radius of the event horizon is given in Annex C, section f.
35. S.B. Giddings, M.L. Mangano, 'Astrophysical Implications of Hypothetical Stable TeV-scale Black Holes', *Phys. Rev. D*, **78**, 035009, 2008.
36. See also for example: D.J. Dicxen *et al.*, 'The Cosmic Microwave Background Spectrum from the Full Cobe Firas Data Set' arXiv:astro-ph/9605054v1, 10 May, 1996.
37. G. Smoot, *Wrinkles in Time*, HarperCollins, New York, 1993. G. Smoot and J. Mather received the Nobel prize in 2006 for their work on the CMB.
38. The satellite was named in honour of D.T. Wilkinson, a scientist who played an important role in the elaboration of experiments in connection with the CMB.
39. See also G. Hinshaw *et al.*, 'Five-Year Wilkinson Microwave Anisotropy Probe (WMAP), Data Processing', Sky Map & Basic Results, arXiv 0803, 0732v2 [astro-ph], 17 Oct. 2008.
40. F. Zwicky, 'On the Masses of Nebulae and Clusters of Nebulae' *The Astrophysical Journal*, **86**, 217, 1937.
41. There are several types of galaxies and they are classified according to their shape. For example there are some galaxies that are called elliptical because of their ellipse-shape profile. The spiral galaxy looks like a disc with arms, made of stars and dust, which appear to extend from a central bulge. They look somewhat like whirlpools with the central bulge acting as a centre of attraction. The stars in the arms are held by gravitational forces from the central bulge and revolve around it. The spiral shape appears to originate from the fact that the stars orbit around the central core in elliptical trajectories but with the direction of axes varying slowly and coherently from orbit to orbit. It looks very much like a wave propagating through the ensemble of stars, which makes them look like grouping in spiral arms.
42. V.C. Rubin, N. Thonnard, W.K. Ford, Jr. 'Rotational Properties of 21 SC galaxies', *Astrophysical Journal*, **238**, 471, 1980. See also T.S. van Albada and

R. Sancisi, 'Dark Matter in Spiral Galaxies', *Phil. Trans R. Soc. London A*, **320**, 447, 1986.

43. D.Clowe, M. Bradac, A.H. Gonzalez, M. Markevitch, S.W. Randall, C. Jones, D. Zaritsky, 'A Direct Empirical Proof of the Existence of Dark Matter'. *Astrophys. J. Lett.*, **648**, L109, 2006. See also report by B. Schwarzschild, 'Collision between Galaxy Clusters Unveils Striking Evidence of Dark Matter', *Physics Today*, Nov 2006, p21.
44. It may be of interest to mention that a new technique using a *frequency comb* implemented initially for the measurement of laser frequencies could in principle be used to measure more accurately the redshift of the radiation coming from those galaxies and make the measurements more reliable. This again shows the importance of realizations in the field of atomic frequency standards to precise measurements in various fields. (Walsworth, 7th Symposium on Frequency Standards and Metrology, Asilomar, Pacific Grove CA October 2008). The Nobel prize was given to T. Hansch and J. Hall in 2006 for the realization of that instrument. Please see: *Proc. of the 7th Symposium on Frequency Standards and Metrology*, ed. L. Maleki, World Scientific, Singapore, 2009.
45. S. Perlmutter, 'Supernovae, Dark Energy, and the Accelerating Universe', *Physics Today*, April 2003, p53.
46. The reader interested in the underlying mathematics is invited to consult Annex C, section *e*.
47. An Ethiopian tale tells us that the Phoenix was a bird that died on a stake and recovered from its own ashes.

Chapter X

1. For a further discussion of this subject the reader is referred to the monograph: H. Fritzsch, *The Fundamental Constants*, World Scientific Publishing Co., Singapore, 2009.
2. L. Smolin, *The Trouble with Physics*, Houghton Mifflin Co, Boston, 2006. In that book the author raises systematically several of the questions discussed in the present text.
3. S. Weinberg, *The First Three Minutes*, Basic Books, New York, USA, 1977.
4. T.X. Thuan, *Le chaos et l'harmonie, la fabrication du réel*, Folio Essais, Gallimard, 2007.

Annex

1. *The speed of light is a constant c. It has been made exactly equal to 299 792 458 [ms^{-1}] by definition. In doing so we have defined the unit 'metre' as the distance travelled by light in (1/299 792 458) second or in about 3.3 nanoseconds.
2. We note that we are saying 'opposite' for the spins because this is the way they are entangled to obtain zero angular momentum. The demonstration was done, however, in connection with the coincidence of colours for the light bulb

on the devices. This is immaterial since one of the Stern–Gerlach detectors could very well be identified with a reversed arrow without any effect on the physics involved and the whole proof remains the same.

3. See for example: P.J.E. Peebles, *Principle of Physical Cosmology*, Princeton University Press, 1993.
4. The derivation is inspired from P.J.E. Peebles, *Principles of Physical Cosmology*, Princeton University Press, USA, 1993, and S. Weinberg, *The First Three Minutes*, Basic Books, USA, 1977.

Suggested Reading

General and Popular

S. Hawking, *A Brief History of Time*, Bantam Press, London, 1988.

S. Hawking, *Black Hole and Baby Universes*, Bantam Press, London, 1993.

G. McNamara, *Clocks in the Sky*, Springer-Praxis, Chichester UK, 2008.

S. Hawking, *Commencement du temps et fin de la physique*, translation of *Is the End in Sight for Theoretical Physics?* and *The Edge of Space Time*, both Cambridge University Press, Cambridge, UK, 1980 and 1989 respectively.

S. Weinberg, *Dreams of a Final Theory*, Pantheon Books, New York, 1992.

A. Jacquard, *La science à l'usage des non-scientifiques*, Calmann-Levy, France, 2001.

J. VanCleave, *Physics for Every Kid*, John Wiley & Sons, Inc., New York, 1991.

D. Bohm, D.F. Peat, *Science, Order, and Creativity*, Routledge, 2000.

R.P. Feynman, *The Character of Physical Law*, MIT Press, 1967.

L. Susskind, *The Cosmic Landscape*, Little Brown Book Group, New York, 2006.

B. Greene, *The Elegant Universe*, W.W. Norton & Company, New York, 2003.

B. Greene, *The Fabric of the Cosmos*, A.A. Knopf, New York, 2004.

L. McTaggart, *The Field*, HarperCollins Publishers Inc., New York, 2002.

H. Fritzsch, *The Fundamental Constants: A Mystery of Physics*, World Scientific Publishing Co., Singapore, 2009.

R.P. Feynman, *QED, The Strange Theory of Light and Matter*, Princeton University Press, Princeton, N.J., 1985.

L. Barnett, *The Universe of Dr Einstein*, Bantam Books, New York, 1968.

J. Gribbin, *The Universe. A biography*, Penguin Group, London, 2006.

M. Gasperini, *The Universe Before the Big Bang*, Springer-Verlag, Berlin, 2008.

M.M. Woolfson, *Time, Space, Stars & Man*, Imperial College Press, London, 2009.

Y. Terzian, E. Bilson, *Carl Sagan's Universe*, Cambridge University Press, Cambridge, UK, 1997.

G. Smoot, K. Davidson, *Wrinkles in Time*, Harper Perennial, New York, 2007.

Elementary Level

G. Shortley, D. Williams, *Elements of Physics*, Prentice Hall, Inc. Englewoods Cliffs, N.J., 1971.

A. Sadoff, *Questioning the Universe*, CRC Press, 2009.

S. Weinberg, *The First Three Minutes: a Modern View of the Origin of the Universe*, Basic Books, 1977.

L. Smolin, *The Trouble with Physics*, A Mariner book, Houghton Mifflin Co, 2007.

Intermediate Level

C.W. Sherwin, *Introduction to Quantum Mechanics*, Holt, Rinehart and Winston, New York, 1960.

N. David Mermin, *It's About Time, Understanding Einstein's Relativity*, Princeton University Press, Princeton, N.J., 2005.

A. Einstein, *La relativité*, translated from German by M. Solovine, Petite Bibliothèque Payot, Paris, 2001.

S.L. Glashow, *The Charm of Physics*, American Institute of Physics, New York, 1991.

J. Baggott, *The Meaning of Quantum Mechanics*, Oxford University Press, Oxford, UK, 1992.

C. Emiliani, *The Scientific Companion*, John Wiley and Sons, New York, 1995.

R. Penrose, *The Road to Reality*, Vintage Books, London, 2004.

Advanced Level

J. Vanier, *Basic Theory of Lasers and Masers*, Gordon and Breach, 1971.

S. Weinberg, *Gravitation and Cosmology*, John Wiley & Sons, New York, 1972.

J.D. Walecka, *Introduction to General Relativity*, World Scientific Publishing Co, Singapore, 2007.

C. Cohen-Tannoudji, B. Diu, F. Laloë, *Mécanique quantique*, Hermann, Paris, 1973.

P.J.E. Peebles, *Principles of Physical Cosmology*, Princeton University Press, Princeton, N.J., 1993.

C. Cohen-Tannoudji, J. Dupont-Roc, G. Grynberg, *Processus d'interaction entre photons et atomes*, InterEditions/Editions du CNRS, 1988.

A. Zee, *Quantum Field Theory in a Nutshell*, Princeton University Press, Princeton, N.J., 2003.

D. Bohm, *Quantum Theory*, Prentice-Hall Inc, Englewood Cliffs, N.J., 1951.

J.S. Bell, *Speakable and Unspeakable in Quantum Mechanics*, Cambridge University Press, Cambridge, UK, 1987.

J.M. Overduin, P.S. Wesson, *The Light/Dark Universe*, World Scientific Publishing Co., Singapore, 2008.

B.W. Petley, *The Fundamental Physical Constants and the Frontier of Measurement*, Adam Hilger, Bristol, 1985.

Jim Bagott, *The Meaning of Quantum Mechanics*, Oxford University Press, Oxford, UK, 1992.

F. Close, *The New Cosmic Onion*, CRC Press, Taylor and Francis, 2007.

J. Vanier, C. Audoin, *The Quantum Physics of Atomic Frequency Standards*, Adam Hilger, 1989, Bristol, UK.

B.G. Sidharth, *The Thermodynamic Universe, Exploring the Limit of Physics*, World Scientific Publishing Co., Singapore, 2008.

Index

About the Author

Jacques Vanier completed his undergraduate studies in physics at the University of Montreal before moving to McGill University to undertake his graduate studies. During his career he has worked in industry (Varian, Hewlett-Packard), taught physics and carried out research at Laval University, Québec, and also been an active member of the National Research Council of Canada, in Ottawa. His research work was oriented towards the understanding and application of quantum electronics phenomena and he has been a consultant for several companies engaged in the development of atomic clocks. Jacques has also been very active on the academic circuit, giving lectures and presenting at numerous conferences in Universities, National Institutes and Summer Schools around the world. He has written more than 120 publications and is the author of review articles and books on masers, lasers and atomic clocks. His book

The Quantum Physics of Atomic Frequency Standards, written with C. Audoin, is recognized as a main reference in the field. Jacques is a Fellow of the Royal Society of Canada, the American Physical Society and the Institute of Electrical and Electronic Engineers. He has been the fortunate recipient of several awards for his contributions to the field of measurement science. He is currently Adjunct Professor at the Physics Department, University of Montreal.